S0-AFP-241

The Curious Map Book

Longhead point
Solway Firth
Cumberland
Westmorland
Guzzleguts Bay
Wolney
YORK-SHIRE
Scarborough
LANCASHIRE
Pontefract
R. Dee
LINCOLN SHIRE
Humber
DERBY SHIRE
CHESHIRE
The Wash
SHROP-SHIRE
STAFFORD SHIRE
NORFOLK
CAMBRIDGE SHIRE
RAD-NOR
HERE-FORD-S
WAR-WICK S
SUFFOLK
HART FORD
ESSEX
Middles
GLOCESTER SHIRE
SURRY
HAMP SHIRE
River Thames

The Curious Map Book

ASHLEY BAYNTON-WILLIAMS

THE UNIVERSITY OF CHICAGO PRESS

CHICAGO

Ashley Baynton-Williams is an antiquarian map dealer and researcher based in London and the author of several books.

The University of Chicago Press, Chicago 60637

Printed in Hong Kong

24 23 22 21 20 19 18 17 16 2 3 4 5

ISBN-13: 978-0-226-23715-2 (cloth)

First published in 2015 by The British Library, 96 Euston Road, London, UK, NW1 2DB.

Library of Congress Cataloging-in-Publication Data

Baynton-Williams, Ashley, author.
The curious map book / Ashley Baynton-Williams.
pages : maps ; cm
This collection is a curation of cartographic curiosities from the British Library collections.
Includes bibliographical references and index.
ISBN 978-0-226-23715-2 (cloth : alk. paper) — ISBN 978-0-226-23729-9 (e-book) 1. Maps—Miscellanea. 2. Cartography—Miscellanea. 3. British Library—Map collections. I. Title.
GA108.7.B39 2014
912.074'41—dc23
2014035710

∞ This paper meets the requirements of ANSI/NISO Z39.48-1992 (Permanence of Paper).

Designed by Bobby Birchall

Note to the Reader
Square brackets around dates indicate information that can be inferred, but is not printed on the original items.

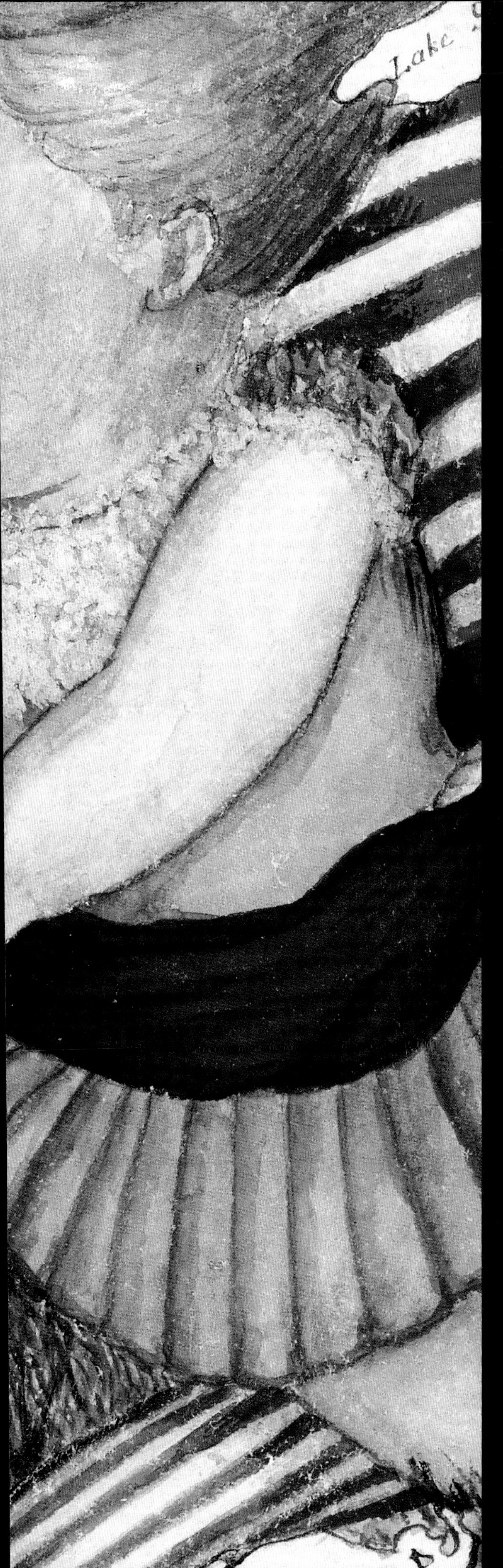
Lake

Introduction

MAP-USERS are accustomed to looking at and treating maps as geographical tools – for position-finding, way-finding and similar purposes – but maps have long been used in other ways. Many is the map that, by word or design, promoted the importance of a region, a country or a ruler through their pictorial representation in the map. This has long been part and parcel of mapmaking. Until at least the late eighteenth century most mapmaking was conducted by private individuals, reliant on generating income and perhaps on sponsorship; it is really only from the nineteenth century that state-financed mapping agencies supersede private endeavour.

If we regard the quest for geographical accuracy as the cartographer at work, then this second vein of mapmaking might well be called the cartographer at play: able to give full rein to his or her imagination to produce images that at one and the same time entertain the viewer and convey the message of the mapmaker, while frequently having little or no real useful geographical function.

There is no widely accepted umbrella term for such maps: colloquially they are often termed 'cartographic curiosities', as being both cartographic and curious, but perhaps this casual term underplays the significance of many of the maps, particularly those with an overt political or religious agenda. Yet no better term has emerged for what is quite a disparate grouping – as will be seen in the selection of items in this volume.

The selection of maps can be broken down into five broad bands, not necessarily mutually exclusive:

- game maps
- maps in animal form
- maps in human form
- maps on objects
- allegorical maps – moral, political or religious.

GAME MAPS

The first published map-game seems to have emerged in England, which is somewhat of a surprise, as England was (relatively speaking) a publishing backwater during the period in question – the 1590s. A set of geographical playing cards was published in London in 1590; these were much-reduced copies of the maps of the counties of England and Wales taken from Christopher Saxton's atlas of 1579. The cards were engraved by Augustine Ryther, one of the principal engravers for Saxton; the pack is of considerable rarity but the British Library was able to add a set to its collections as recently as 2013. A second set of map cards, attributed to the same man, appeared in 1605.

Beyond cards, the earliest recorded board-games featuring maps appeared in Paris, published by Pierre Duval from 1645 onwards, based on the Game of the Goose. This game was immensely popular and a familiar sight throughout Europe from at least the sixteenth century and probably much earlier. The players raced round a spiral course, their moves determined by the throw of dice, with particular hazards or advantages assigned to some of the individual squares or circles on the boards. In the Duval version the individual circles contained sections of a map. Such games were part of a general stock of maps aimed at a mature market. Duval's game map of France (1659) is overtly a gambling game: the players decide upon the financial stake and play for money.

Race games of this kind were particularly suited to map themes, and cartographic variants of the Game of the Goose predominate, although by the mid-eighteenth century British publishers were discarding the more structured format of the Goose game in favour of games based on travelling around actual maps, reflecting a change in emphasis. By the 1750s, and certainly by the 1780s, English publishers were deliberately recasting their map-games as educational, increasingly aimed at a junior audience by emphasising learning through pleasure.

Although earlier continental precursors are known, the first commercially produced jigsaw maps seem to have appeared in England, made and sold by John Spilsbury. Their actual inventor may have been French. A London trade directory of 1763 contains an entry for someone called simply Leprince, described as the 'Inventor of the dissection of maps on wood'; this may refer to Jeanne-Marie le Prince de Beaumont (1711–1780), a French tutor and educationalist who was certainly using dissected wooden maps in her London school in the 1750s but who returned to France in 1762, or one of her half-brothers:

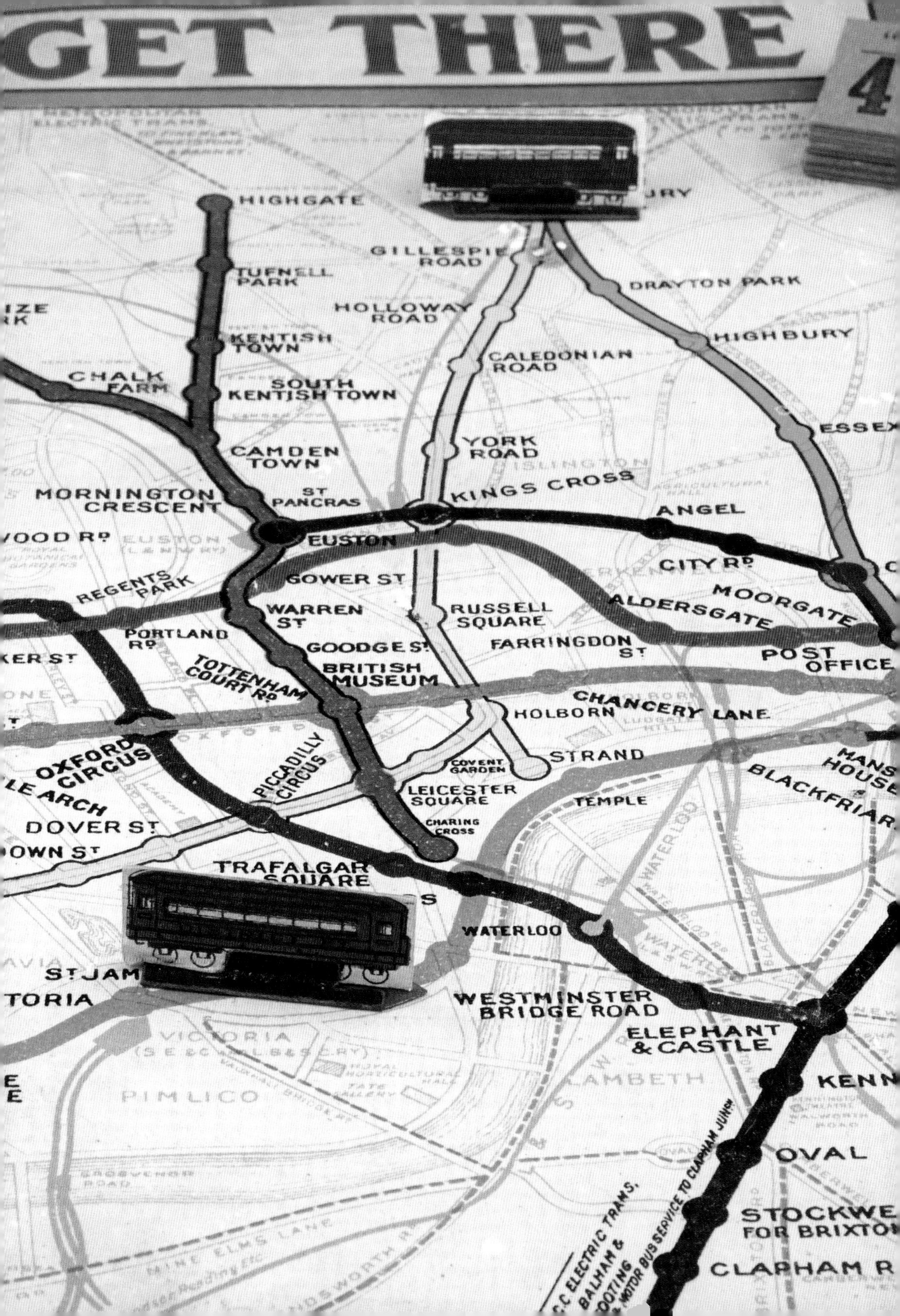

GET THERE
4
HIGHGATE
TUFNELL PARK
KENTISH TOWN
SOUTH KENTISH TOWN
CHALK FARM
CAMDEN TOWN
MORNINGTON CRESCENT
ST PANCRAS
EUSTON
GILLESPIE ROAD
HOLLOWAY ROAD
CALEDONIAN ROAD
YORK ROAD
KINGS CROSS
DRAYTON PARK
HIGHBURY
ESSEX
ANGEL
CITY RD
MOORGATE
ALDERSGATE
FARRINGDON ST
POST OFFICE
GOWER ST
WARREN ST
RUSSELL SQUARE
REGENTS PARK
PORTLAND RD
GOODGE ST
BRITISH MUSEUM
TOTTENHAM COURT RD
CHANCERY LANE
HOLBORN
STRAND
COVENT GARDEN
LEICESTER SQUARE
CHARING CROSS
OXFORD CIRCUS
PICCADILLY CIRCUS
DOVER ST
TRAFALGAR SQUARE
TEMPLE
BLACKFRIARS
WATERLOO
ST JAM
TORIA
WESTMINSTER BRIDGE ROAD
ELEPHANT & CASTLE
VICTORIA
PIMLICO
LAMBETH
OVAL
STOCKWE
FOR BRIXTO
CLAPHAM R
MOTOR BUS SERVICE TO CLAPHAM JUNCn
C.C. ELECTRIC TRAMS, BALHAM & TOOTING

the engraver Jean-Baptiste le Prince (1734–1781) or Jean-Robert le Prince, reportedly a geographer who died in London in or about 1762.

Maps certainly lent themselves to being used as jigsaw puzzles, even if the early makers struggled with the intricacies of faithfully following the boundaries of counties or countries in the dissection process. With mechanisation these processes became easier, but this was coupled with a decline in the overall quality of the jigsaws, away from the heavier wooden backing to the lighter and flimsier cardboard of modern games.

MAPS IN ANIMAL FORM

The most frequently encountered of these maps is the depiction of countries in animal – zoomorphic – form; the most famous of these form a tight grouping: the Leos. During the Dutch Revolt against Spanish rule, Amsterdam publishers produced a series of propaganda maps – 'Leo Belgicus' – representing the Low Countries (Holland and Belgium) as an invincible lion, standing firm against the oppressor. The lion, a common symbol in the arms of the towns and cities of the Low Countries, encapsulated the armed struggle of the (small) Netherlands against the might of the Spanish Empire. As the struggle evolved a subsidiary group emerged; it became clear that the southern provinces – modern Belgium – would not win their independence, while the northern provinces – modern Holland – might. As Holland's cause advanced, 'Leo Hollandicus' emerged, depicting the seven United Provinces as a triumphant lion.

Later mapmakers popularised the octopus as a not-too-subtle means of portraying a country as evil and grasping, with Russia a frequent choice, but Falmouth Town Council and the landowners of London were similarly portrayed. Satirical maps from the First World War often portrayed countries as animals reflecting national characteristics: the British bulldog, the French poodle and Germany as either a dachshund or an eagle. The first map in this book exemplifies regions of the world through imagery created from legendary medieval accounts, Bible history, travellers' tales and the like.

MAPS IN HUMAN FORM

Maps in human (anthropomorphic) form are more readily understood; Britain was often portrayed as John Bull, possibly with Churchillian features, while the United States could appear as Uncle Sam. Earlier maps portrayed Europe as a queen, while a Filipino cartographer sought to portray the Spanish Empire in similar form.

Other mapmakers took their inspiration from shapes of continents, countries or regions: Olof Rudbeck when he depicted the Baltic Sea as Charon, the boatman from the classical tale, or Robert Dighton in his depiction of England and Wales, Scotland and Ireland in human form, but Lilian Lancaster is perhaps the most famous exponent of caricature maps. Several of her maps are represented here, including two manuscript maps finding humour in the 1880 United States Presidential election campaign.

MAPS ON OBJECTS

The invention of transfer-printing, a process whereby an image would be created on a flat-printing plate, printed to paper and while the ink was still wet transferred to a curved surface, is credited to the engraver Robert Hancock (1731–1817), who worked at the Worcester porcelain factory in the early 1750s. Many of the maps of this type were commemorative, for example a simple jug celebrating Nelson's great triumph over the French at the Battle of Trafalgar, but elaborate dinner services were also made. Sadly, because of the fragility of porcelain, very few of the oldest examples survive.

Another genre which ill survives is that of map screens: room-dividers of canvas stretched over wooden frames and decorated with maps. These were particularly popular in the eighteenth century as draught-excluders and personal displays of wealth and culture, in which the most important element was the impression created, rather than the maps themselves. Today four screens with printed maps are known; the British Library has two, with the most dramatic described here.

ALLEGORICAL MAPS

Perhaps the smallest category represented here are those maps made for overt political purposes. Many propaganda maps – a number sufficient to form a study of their own – were prepared at the time of the Second World War, by both Axis and Allied Powers; but those with more crude and unpleasant religious and racist overtones are not represented here.

One exception to this policy of excluding the more extreme religious manifestations is also one of the most spectacular maps included – a very rare sixteen-sheet wall map of the world of 1566. This is a vicious Protestant attack on the Papacy, portraying the many 'crimes' of the Catholic Church as the central part of the map, within the jaws of the Devil, while the (righteous) Protestant rulers lay siege to the Catholic world.

While one thinks of the Victorian age as being one of an increasingly moralistic religious tone in both public and private life, the roots of this are often said to lie in the late Georgian period of the eighteenth century. Authors and mapmakers of that time exhorted the reader to follow a narrow path through life, where diversion from the path was to risk one's soul. Yet these roots go back even further; Sir Thomas More, in 1516, used a map to illustrate his description of Utopia. John Bunyan's *Pilgrim's Progress* ... (1678) was suited to illustration by a map, but it was only in the eighteenth century that such depictions were inserted, and inspired subsequent generations of moralising authors.

Perhaps surprisingly, one subject that seems ideally suited to being mapped is the course of true love. Madame de Scudéry, in 1655, created a map of the 'Land of Tenderness' as a moral guide for a young lady, although the emphasis was not on the reality of love as an emotion but on its spirituality: on the propriety of doing the right thing, as dictated by the head not the heart. Later mapmakers, among them Robert Sayer in *c.*1772, saw love as a journey by sea, through treacherous and uncharted waters. In contrast, Georg Matthaüs Seutter, in the 1730s, portrayed love as war: the heart as a castle besieged, to be defended at all costs.

Maps could also be used to portray love gone wrong, as in Joseph Onwhyn's 'Map of Green Bag Land', a satire on the lengthy and increasingly bitter divorce battle between King George IV and his consort Queen Caroline, during which she was barred from his coronation (physically locked out on the day), and which ended only with her sad demise three weeks later.

If it is hard to see divorce as a subject of political humour, it is even more perplexing that the outbreak of the First World War spawned a number of satirical maps and games, as if the rival powers were participating in an elaborate game of fancy dress. The harsh reality of an increasingly brutal war of attrition saw these games disappear, but the war was depicted as a battle between different breeds of dog, John Bull against the German eagle, a throwing game with Germany as the target, and a game of dexterity.

The boundaries of the uses that maps could be put to seem to have been limited only by the imagination of their makers, as will be seen in the following pages.

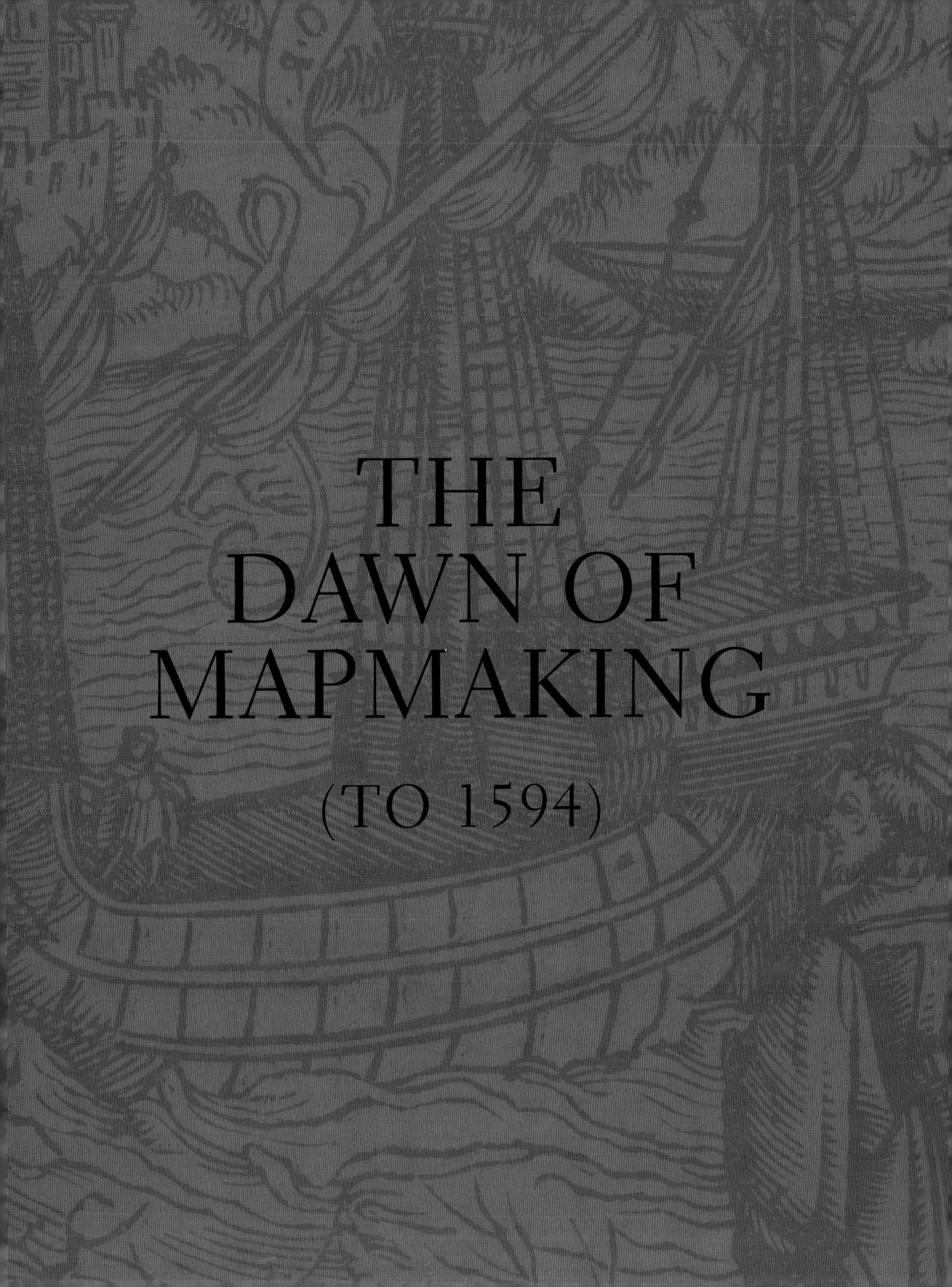

THE DAWN OF MAPMAKING

(TO 1594)

Ptolemaic Map of the World, 1493

Hartmann Schedel, Nuremberg

Woodcut; 308 × 428 mm (12 1/8 × 16 7/8 in)

HARTMANN SCHEDEL'S (1440–1514) *Liber cronicarum*, known in English as the 'Nuremberg Chronicle', was the most ambitious and substantial book printed and published before 1500, containing more than 1,800 woodcut illustrations cut by Michael Wohlgemuth and Willem Pleydenwurff, perhaps with the assistance of their even more famous pupil, Albrecht Dürer. The 'register' of the book describes it as a chronicle with figures and images from the beginning of the world. The multiple images include portraits, genealogical trees, historical and biblical scenes, and the accrued myths and legends of the preceding centuries. There are many fine town views, and two maps: one shows Germany and the adjacent countries of northern Europe, and the other is this map of the known world.

The map is characteristic of the Ptolemaic style current at the time, derived from the work of the Greek astronomer and cartographer Claudius Ptolemy (or Ptolemaeus) who was active at Alexandria in Egypt around AD 150–160. The European rediscovery of Ptolemy's work in about 1400 was one of the earliest and most significant developments of the Renaissance, that period of the rebirth and re-assimilation of classical learning and the gathering together of manuscript texts 'lost' for centuries. Unknown in the west for more than 1,000 years, Ptolemy's explanation of the scientific basis of mapmaking and the primacy of tabulating accurate co-ordinates was first put into printed form at Bologna in 1477. During the intervening centuries there had been little advance in the mapping of the countries of the known world, which, at the time, wholly excluded the Americas, southern Africa, the East Indies and the farthermost parts of Asia. The great voyages of exploration of the European navigators such as Christopher Columbus and Vasco da Gama were yet to take place.

Schedel's map of the world was in the press while Columbus was at sea on the first of his four voyages of discovery to the Americas (1492–1502). Although initially Columbus thought he had landed on the western shores of Asia, the world was never to look the same again. Columbus knew the world was round: that fact had been proved by the Greek Erastosthenes, who calculated the circumference of the Earth with startling accuracy in about 250 BC. The error Columbus made, perhaps following Ptolemy, was to underestimate its circumference. Schedel would also have known the world was round. His Ptolemaic projection clearly represents a window on to only a portion of a globe. In 1492 a fellow citizen of Nuremberg, Martin Behaim, drew a terrestrial globe, and Schedel claimed to have assisted him in its construction.

Schedel depicts the world as divided among the sons of Noah, from the biblical story, and enlivens the left-hand margin of the map with illustrations from classical and medieval legend. The figures include a six-armed man, a centaur (half-man, half-horse) from Greek mythology, a four-eyed man, and a man with the neck and head of an ostrich-like bird. In the text on the reverse are similarly fantastic figures. Although a thoroughly modern production – a map produced on a better scientific understanding than had been available since classical times, and produced as part of the most complex piece of printing yet attempted with the brand-new technology of the printing press – this map, perhaps more than any other, marks the final flowering of the medieval world view, at the dawn of the modern understanding of Europe and the wider world.

Secunda etas mundi — Secunda etas mūdi — Foliū XIII

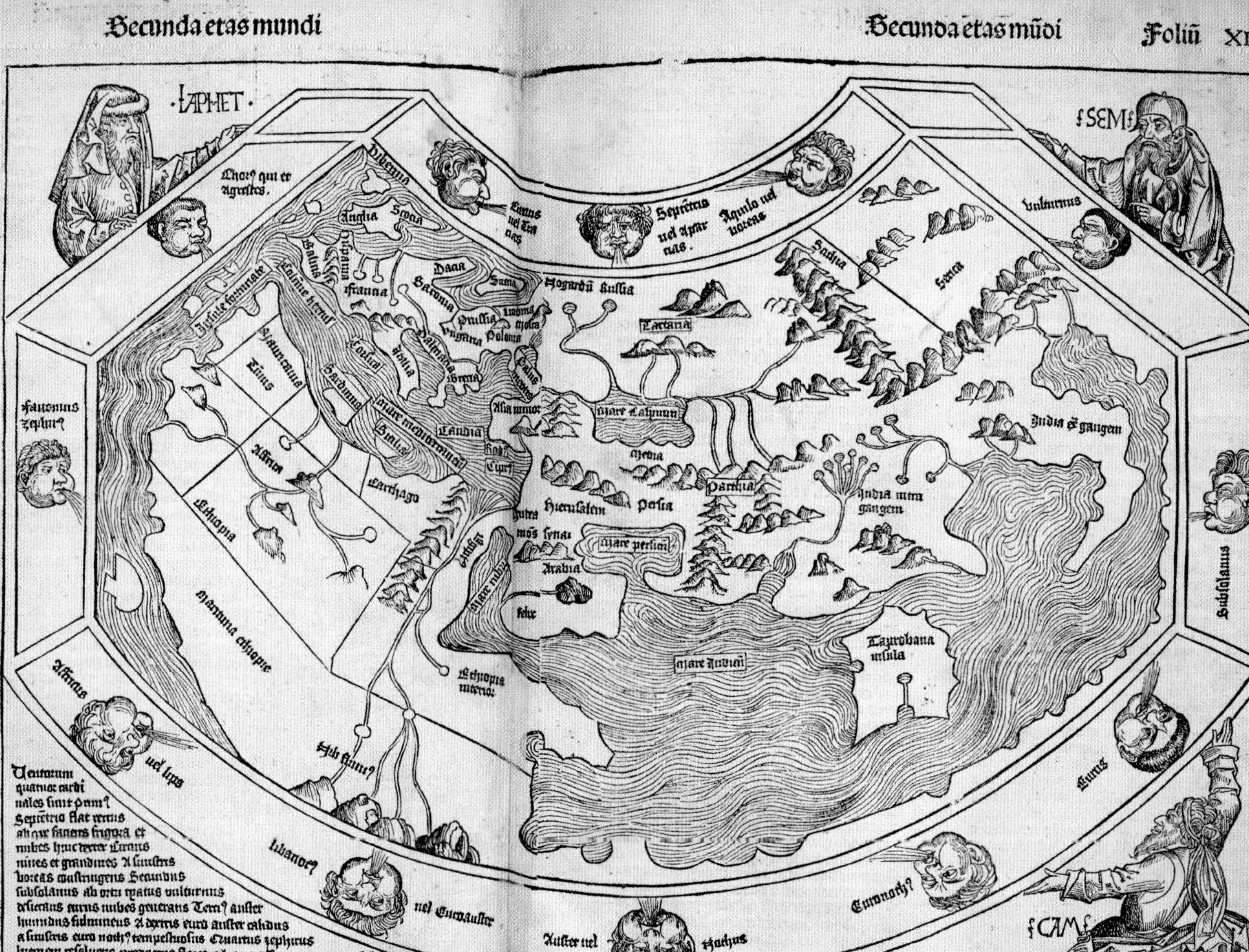

Orbis dicitur a rota ꝛ est q̄libet figura sperica ꝛ rotunda. Et ideo mūd⁹ orbis dr̄. qꝛ rotūd⁹ ē: ꝛ dr̄ orb terre vl' orbisterrarū. Dicūt āt bm vincē. filij sem obtinuisse asiā. filij chā affricā ꝛ filij iaphet europā. Isid. in li. Ethy. asserit ꝙ orbis diuisus ē in tres partes ꝫ nō eq̄liter. Nā asia a meridie ꝑ orientem vsqꝫ ad septētrionem ꝑuenit. Europa ṽo a septētrione vsqꝫ ad occidentē ꝑtingit. Sed affrica ad occidentem ꝑ meridiez se extendit. Sola quoqꝫ Asia continet vnam partem nostre habitabilis. s. medietatem: alie ṽo ptes. s. affrica ꝛ europa aliam medietatez sunt sortite. Inter has autem partes ab occeano mare magnū progreditur. easqꝫ intersecat: quapropter si in duas partes orientis ꝛ occidentis orbem diuidas in vna erit asia in alia ṽo affrica ꝛ europa. Sic autem diuiserunt post diluuiū filij Noe: inter quos Sem cum posteritate sua asiam. Japhet europam: cham affricam possederunt. vt dicit glo. super Gen. x. ꝛ super libro Paralippo. primo. Idem dicit Crisostomus Isidorus ꝛ Plinius.

Illustration of the Island of Utopia, 1518

Thomas More, Basel

'Utopiae insulae tabula'

Woodcut; border: 181 × 120 mm (7 ⅛ × 4 ¾ in)

SIR THOMAS MORE (1478–1535) was an English statesman, lawyer, social philosopher and Catholic saint and martyr. As Archbishop of Canterbury, he earned Henry VIII's grave displeasure when he opposed the king's divorce from his wife Catherine of Aragon. The divorce set in motion the English Reformation, the break with Roman Catholicism, the dissolution of the monasteries, and the establishment of English Protestantism and the Church of England. It also led to More's execution on trumped-up charges.

More's 1516 book *Utopia* describes, not without satire, an idealised society – one in which, ironically, divorce was permissible – located on an imaginary island. It is one of the great political tracts, and to this day Utopia (Greek, 'no-place') gives its name to an earthly paradise. Although the book was written in England, for various reasons the different editions were published on the continent, first in Antwerp and then (this edition) in Basle. Johannes Froben's 1518 printing is the third edition of the text, but the second with a map. This is a more elaborate version of the map than that of the first edition; it is generally attributed to the woodcutter Ambrosius Holbein, but may equally be the work of Hans Holbein the younger, his brother, who signed the woodcut used on the title page.

The geography of the island of Utopia is described by More in a section labelled 'Sit & forma Vtopiae novae insulae' ('the location and shape of the new island of Utopia'). He writes that the island was described to him by the traveller Hythlodaeus, who had voyaged there. Hythlodaeus and More are depicted in the bottom left corner of the map.

At first glance this appears to be simply a nicely performed map of the imaginary island, with a large ship in the foreground. There is more to it: concealed in the image, and best seen with one eye half-closed, is a death's head skull, a *memento mori*, possibly suggested by a wordplay on More's surname. While the map in the first (Antwerp) edition seems to have been intended to resemble a skull, the execution was not so successful, and it can be more easily seen in this second map. The skull should be imagined as facing slightly to the right of the viewer. The figures of Hythlodaeus and More, with the left-hand ship, form the neck and right ear. The segmented hull of the foreground ship form the teeth, with the prow and stern the jaw bone, the bay below the ship the chin, and the central mast the nose. The island itself gives outline to the skull proper, with 'Fons Anydri' and 'Ostium anydri' – the source and mouth of the Anydrus (Greek, 'no-water') River – and their surrounding hills creating the effect of eyes and sockets; 'Amarotũ urbs', ('Mist Town') with its label, is the forehead.

This image has been studied for many years, but it was in the end a dentist, with his highly technical understanding of the bones around the jaw area, who first identified the skull in 2005. The same element is found in other works by the Holbeins, notably Hans Holbein's painting *The Ambassadors*.

 VTOPIAE INSVLAE TABVLA.

New Map of the Papist World, 1566

Pierre Eskrich (?), Geneva

'Mappe-Monde Nouvelle Papistique'

Woodcut with handcut and type-lettering, on fourteen sheets (of sixteen); each sheet, outer border approximately: 335 × 430 mm (13 ¼ × 16 ⅞ in), with twelve sections of letterpress text.

THIS SIXTEEN-SHEET allegorical map of the papist world is a virulent attack on the Roman Catholic Church from a staunch Protestant viewpoint. It is a cartographic treasure, among the largest and most elaborate maps of its day, and is set apart by the skill of the woodcutter.

The map is anonymous – the mapmaker, the printer and the publisher are not identified. It accompanied a book, which was issued with the participants' identities concealed by pseudonyms: *Histoire delà mappemonde papistique, composée par M. Frangidelphe Eschorche-messes ... Imprimé en la ville de Luce nouvelle, par Brifaud Chasse-diables* ... Recent research has suggested that the name *M. Frangidelphe Eschorche-messes* is a punning amalgamation of the names of two men: Francesco Negri and Pierre Eskrich. Thus *Frangidelphe* combines the 'Fran' of Francesco with 'delphe' (i.e. 'Guelph'), a reference to the Black [or Negri] Guelphs of Florence, with 'Eschorche' a corruption of Eskrich. Eskrich was a famous designer and draughtsman working both in France and Geneva, while the text and map were inspired by the writings of Jean-Baptiste Trente (Giovanni Battista Trento) an Italian also living in the Protestant stronghold of Geneva. The letterpress type is identified as belonging to the printer François Perrin of Geneva, while the woodcutter was Christoffel Schweytzer; a figure (perhaps a self-portrait) is shown with a shield labelled '*Christoph Schwytz Proplates Tigurinus*' ('Christoffel Schweytzer woodcutter from Zurich').

The Protestant Reformation – a religious reaction against the corruption and venality of the medieval Catholic Church – had much earlier origins but is generally dated from 1517 when Martin Luther pinned his famous ninety-five theses to the door of the Castle Church of Wittenberg. The Catholic response was aggressively spearheaded by the Society of Jesus, or Jesuits; the Protestant leaders, thinkers and scholars (pacifist at first) in turn became militant in reply. It was a religious dispute that led to well over a hundred years of bigotry, bloodshed and appalling warfare. It was in that milieu that this fine map was created.

The *monde papistique* – an allegorical map with elements of Hell, Rome and the Catholic world – is depicted confined within the jaws of the Devil, dribbling with saliva. This world is under siege; outside the walls, the Protestant forces are lined up, with their cannons labelled 'Parole de Dieu' ('Word of God'). The leading Protestant kingdoms are aligned on the upper border, as noted in text at the top left:

> These are the faithful of all the nations who are here around the Papistique world, & are come to attack it some with cannon and artillery shots of the word of God, and others with other arms, right up until the time they see it totally destroyed. And these faithful are all under the command of their Kings, Princes, Seigneurs and Republics, as firstly under the command of Elizabeth [I] Queen of England and an infinite numbers of Seigneurs and Princes, English evangelicals, and under the direction of the very-powerful King of Poland ...

The broad theme is the corruption, immorality and venality of the Catholic Church, and the need for Protestants to be forceful in their response. Sebastian Münster, Konrad Pellikan, Wolfgang Musculus and others are all described by contemporaries as men of peace, but all are shown here resorting to violence – Münster threatening an opponent with his clenched fist.

CITE 30.
Courtisans du Roy F. Arb.
Raison & Volontez meres.
45. Le ROY FRANC ARBITRE.
Acte Eli cite.
43 Sophistes.
Mõtagne de Perdition.
Discours Humain
HERMI. de S. Appollonien, c.27.
dragores.
HERMI. de S. Theõ, c.26
44. Dame Bõne-Intention.
D. de CONGRVO féme.
D. de CONDIGNO fille.
CITE 31.
42. L.N.
osiques.
92. ADVENT, cité 58.
90. Libraire.
Page du Roy.
MER du Ieusne.
59.
88. La CARESME, cité 57.
Liures Canonistes & Scholastiques.
Marché de la cité de Caresme. de 46. iours.
86. IEVSNE, Prou. 15.
89. Artisans.
SECONDE partie du Royaume des Bonnes-œuures.
96. ABSTINENCE par deuotion, c.62.
91. Le Seigneur Caresme-Prenant auec des Inquisiteurs ses compagnons.
usne.
Montagne de Plaisirs & Dissolutions.

MAPPE-MONDE NOUV
L'ORIGINE ET COMMENCEMENT
de ceste Mappe-Monde nouuelle Papistique, & comment
elle a esté trouuee.

LLE PAPISTIQVE

'The Whole World in a Cloverleaf' [1581]

Heinrich Bünting, Helmstadt

'Die gantze Welt in ein Kleberblat, Welches ist der Stadt Hannover, meines lieben Vaterlandes Wapen'

Woodcut with letterpress; border 270 × 360 mm (10 5/8 × 14 1/8 in)

HEINRICH BÜNTING (1545–1606) of Hanover was a Professor of Theology who composed a book on the Bible entitled the *Itinerarium sacrae scripturae*, a narrative of 'The Travels of the Holy Patriarchs, Prophets, Judges, Kings, our Saviour Christ and his Apostles, as they are related in the Old and New Testament; with a Description of the Places to which they travelled, and how many English miles they stood from Jerusalem' (as an English edition rendered the title). It was illustrated with woodcut maps. They are an unusual mix of 'formal' geographical maps, including those of the world, the Holy Land, the Exodus and the eastern Mediterranean, and three cartographical curiosities, all discussed in this volume: the world in the shape of a three-leafed clover, Europe depicted as a queen (p. 22) and Asia depicted as Pegasus, the winged horse (p. 24).

This cloverleaf map is perhaps the most unusual. Each leaf is one of the 'Old World' continents: Europe, Asia and Africa. The fourth continent, the Americas, is depicted as an isolated landmass in the lower corner. The idea for the cloverleaf design came from the arms of Bünting's home city of Hanover, which included the plant, as alluded to in the title: 'a cloverleaf, which is the arms of the City of Hanover, my dear home' (literally 'my Fatherland'). In an age obsessed with emblem, symbol and allegory, the three-leaf clover also commonly represented the Holy Trinity, adding an additional layer of religious meaning to a devoutly Christian depiction of the world.

The map depicts Jerusalem at the focal point of the three continents, at the centre of the world – an idea which would have resonated with biblical scholars and is only a small step removed from medieval 'T-O' maps, in which the world is depicted as circular, surrounded by ocean (the 'O'). The three known continents are shown divided by three tracts of water (forming the 'T'): the Mediterranean, the Don River and the Red Sea. Indeed, the first printed European map, from Isidore of Seville's *Etymologiae* (Augsburg, 1472) was of the 'T-O' design, and it proved a popular and well-known illustration for printings of early classical texts.

Bünting's map is a woodcut. One of the problems associated with wood as a printing platform was the difficulty of cutting curved letters, such as 'c', and 'e'; an early and common solution, particularly in Germanic countries where the woodcut medium was preferred to copperplates, was to insert letterpress plugs (moulded metal type, forming letters or whole words) into recesses in the wooden block. This accounts for the differing appearance of the lettering in various places and the lines of the map itself. Another benefit of this approach was that the lettering could be reset between editions, particularly if editions in different languages were produced; this map can also be found with a Latin running title.

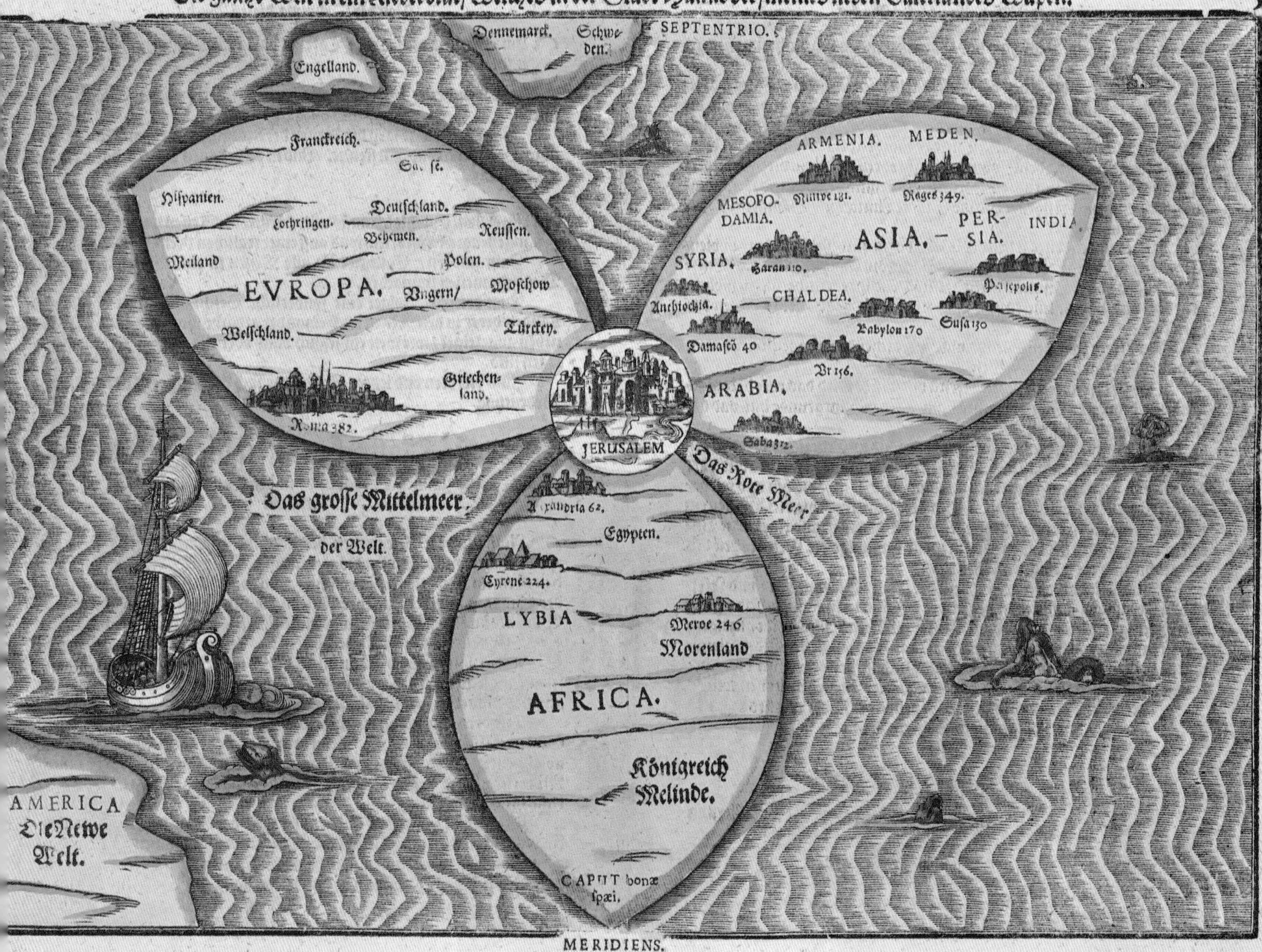
Die gantze Welt in ein Kleberblat/Welches in der Stadt Hannover/meines lieben Vaterlandes Wapen.
SEPTENTRIO.
Dennemarck.
Schweden.
Engelland.
Franckreich.
Hispanien.
Deutschland.
Lothringen.
Behemen.
Reussen.
Meiland
Polen.
EVROPA.
Ungern/
Moschow
Welschland.
Türckey.
Griechenland.
Roma 382.
ARMENIA.
MEDEN.
MESOPODAMIA.
Ninive 181.
Rages 349.
ASIA.
PERSIA.
INDIA.
SYRIA.
Haran 110.
CHALDEA.
Persepolis.
Anthiochia.
Babylon 170
Susa 150
Damasco 40
Ur 156.
ARABIA.
JERUSALEM
Saba 312.
Das Rote Meer
Das grosse Mittelmeer,
der Welt.
Egypten.
Cyrene 224.
LYBIA
Meroe 246
Morenland
AFRICA.
Königreich
Melinde.
AMERICA
Die Newe
Welt.
CAPUT bonæ
spæi.
MERIDIENS.

‘Europe, the First Part of the World, in Female Form’ [1581]

Heinrich Bünting, Magdeburg

‘Europa prima pars terrae forma virginis’

Woodcut with letterpress; border: 235 × 347 mm (9 ¼ × 13 ⅝ in); widest, with text: 268 × 360 mm (10 ½ × 14 ⅛ in)

HEINRICH BÜNTING’S cloverleaf map of the world (p. 20) and his map of Asia in the form of Pegasus, the winged horse (p. 24) were completely original designs, but this map depicting Europe in the form of a queen was a recycling of an earlier motif, apparently introduced by Johannes Bucius (a Latinised form of Johann Putsch) in 1537 and popular throughout the sixteenth century. In allegorical depictions of the four continents used on the elaborate frontispieces of atlases by Abraham Ortelius, Johannes Blaeu and others, ‘Europa’ was frequently portrayed as a female figure, with crown and sceptre, symbolising temporal and spiritual domination over the other continents. The motif of ‘Europa’ as a queen had considerable longevity.

If the queen stands upright, west is at the top of the map. Spain and Portugal are the crown and head, France and Germany are the upper torso, Italy and the Jutland peninsula the arms, and southern and eastern Europe the lower body and feet (demurely covered by her flowing dress), with the Peloponnese as an appendage to the dress. The British Isles, Scandinavia, the Danish islands, Africa and Asia are all to be seen in the waters around the queen. The neckline of the dress is formed by the Alps, which – with a certain amount of artistic licence – extend from the Mediterranean coast to the English Channel; the queen’s necklace is the River Rhine, with the mountainous ring round Bohemia as the bauble suspended from the necklace, but also representing the heart.

The queen’s crown, that of the Holy Roman Empire, is labelled ‘Hispania’; her orb is formed by ‘Sicilia’, with the cross representing Christian supremacy, while her left hand, formed by Denmark, holds her sceptre, signifying her authority to rule.

When Bucius was working in 1537, Charles V was not only the head of the House of Habsburg, but also heir to the House of Valois-Burgundy of the Burgundian Netherlands, and the House of Trastámara of the Crowns of Castile and Aragon: he was King of Spain, Holy Roman Emperor, Archduke of Austria (and therefore ruler of the Low Countries) and King of Italy, thus holding sway over large parts of Europe as well as the vast Spanish Empire overseas. This is why the crown is labelled ‘Hispania’. Spain was the source of his unprecedented wealth and power. It is said that Bucius modelled his queen on Charles’ wife Isabella, but the depiction may well be more generic than specific and, by the time of Bünting (and rival delineations by Sebastian Münster and Matthaüs Quad) it is likely that the personal significance of the portrait may have been forgotten.

The map illustrated here is from a later (1598) edition of Bünting’s text, published in Magdeburg; a new woodcut was prepared for the Magdeburg printings, although it is very similar to the 1581 original.

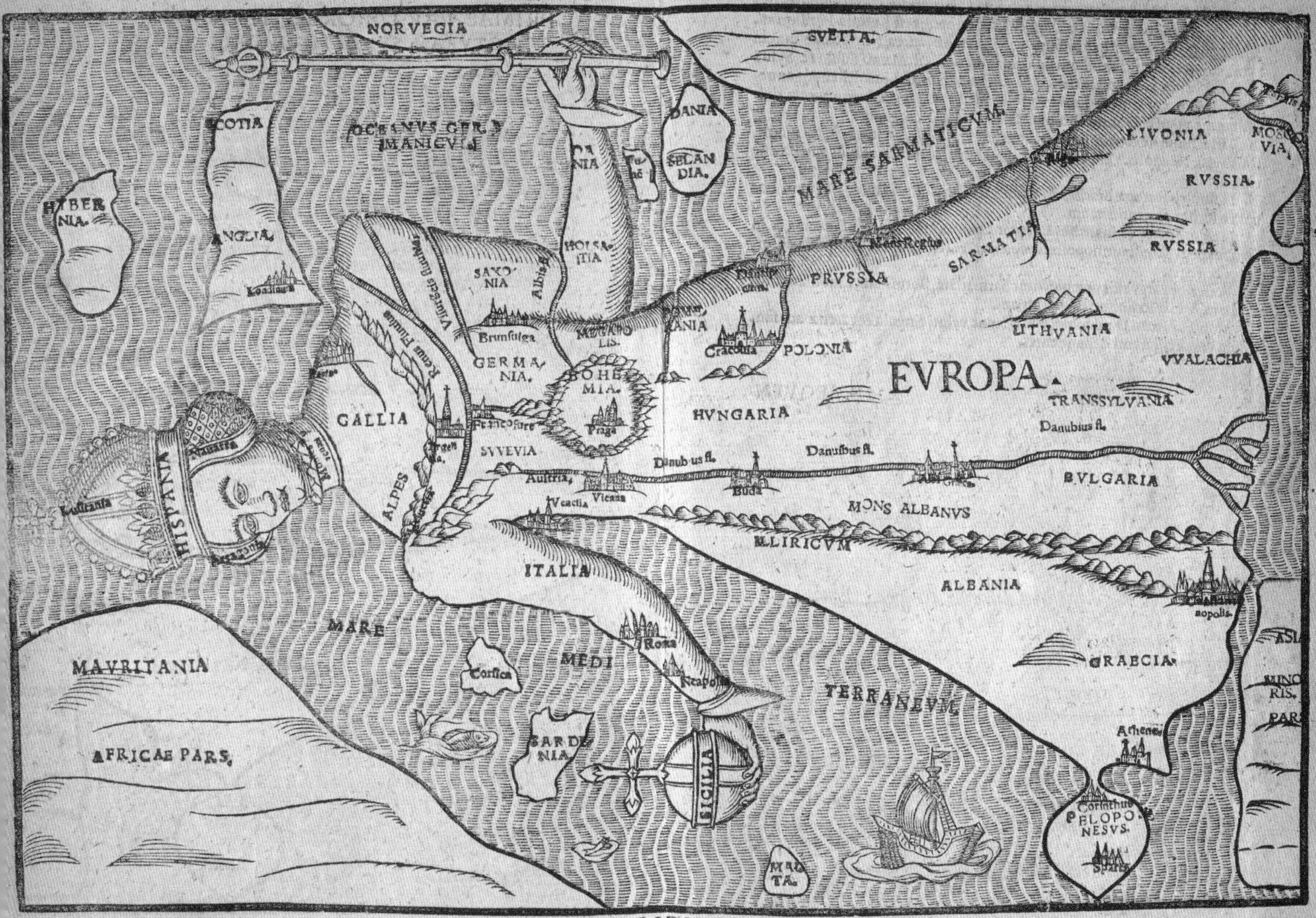

30
EVROPA PRIMA PARS TERRAE FORMA VIRGINIS.
31
SEPTENTRIO.
NORVEGIA
SVETIA
SCOTIA
HYBERNIA.
ANGLIA.
DANIA
SELAN DIA.
MARE SARMATICVM
LIVONIA
RVSSIA.
RVSSIA
PRVSSIA
SARMATIA
SAXONIA
HOLSATIA
GERMANIA.
BOHEMIA
Praga
POLONIA
EVROPA.
VVALACHIA
TRANSSYLVANIA
HVNGARIA
GALLIA
SVEVIA
ALPES
HISPANIA
Danubius fl.
BVLGARIA
MONS ALBANVS
ITALIA
ALBANIA
MARE
MEDI
TERRANEVM.
MAVRITANIA
AFRICAE PARS.
Corsica
Roma
SICILIA
GRAECIA.
PELOPONESVS.
MERIDIES.
En tibi, formosæ sub forma Europa Puellæ
Viuida fœcundos pandit vt illa sinus.
Ridens Italiam dextrâ Cimbrosq; sinistrâ
Obtinet, Hispaniam fronte geritq; solum.
Pectore habet Gallos, Germanos corpore gestat,
Ac pedibus Graios, Sauromatasq; fouet.

'Asia, the Second Part of the World, in the Form of Pegasus' [1581]

Heinrich Bünting, Magdeburg

'Asia secunda pars terræ in forma Pegasir'

Woodcut with letterpress; border of map: 245 × 350 mm (9 ⅝ × 13 ¾ in); at widest, with text: 280 × 367 mm (11 × 14 ½ in)

LIKE HIS cloverleaf map of the world (p. 20), Heinrich Bünting's hippomorphic map of Asia is an original creation of the author. Here he depicts Asia, the 'second part of the world' as he terms it in the title, as Pegasus. In classical mythology, Pegasus was a winged horse, sired by the Greek god of the sea and of horses, Poseidon. Poseidon raped Medusa on the floor of a temple dedicated to the goddess Athena. A late version of the story is that Athena was so enraged that she turned Medusa into a Gorgon, her head covered by snakes instead of hair, and so hideous that anyone who looked on her face would be instantly turned to stone. The hero Perseus was sent to kill her. Perseus achieved his task by looking into a highly polished shield, given to him by Athena, and was able to cut off Medusa's head by looking at her reflection. In that very moment, Pegasus and his brother Chrysaor sprang from her body. The winged horse was tamed by the hero Bellerophon with the assistance of the goddess Athena, and featured in a number of Bellerophon's exploits, including his fight with the Amazons, and his defeat of the Chimera, a terrible monster.

Bellerophon subsequently provoked the anger of the gods when, carried away by his success in killing the Chimera, he tried to ride Pegasus up Mount Olympus to the home of the gods. Zeus caused Bellerophon to be dismounted, and he ended his life as an outcast, blinded and crippled by the fall back to earth. Pegasus completed the flight, and was kept by Zeus, before being turned into the constellation of that name.

Pegasus was used as a symbol both of fame and of wisdom in the Middle Ages, and in later periods of poetic inspiration, but Bünting was probably influenced in his choice by the myth of the birthplace of the winged horse which, according to Hesiod, took place at the 'springs of Oceanus, which encircles the inhabited Earth, where Perseus found Medusa'. In the map, Pegasus is facing left. His muzzle and head are Asia Minor (modern Turkey). Arabia forms the front legs, the Indian subcontinent and South-East Asia the rear, the Levantine countries, Persia and Inner Asia the main barrel of the body and the Far East the rump, with Scythia and Tartary the wings, and southern Persia part of the saddle blanket.

Bünting seems to have been the only author to use the Pegasus motif for Asia, but there were a large number of editions of the book, utilising several different woodblocks and even a copperplate version. This example is from a later, replacement woodblock prepared for Magdeburg printings of the book, although it is a very close copy of the original.

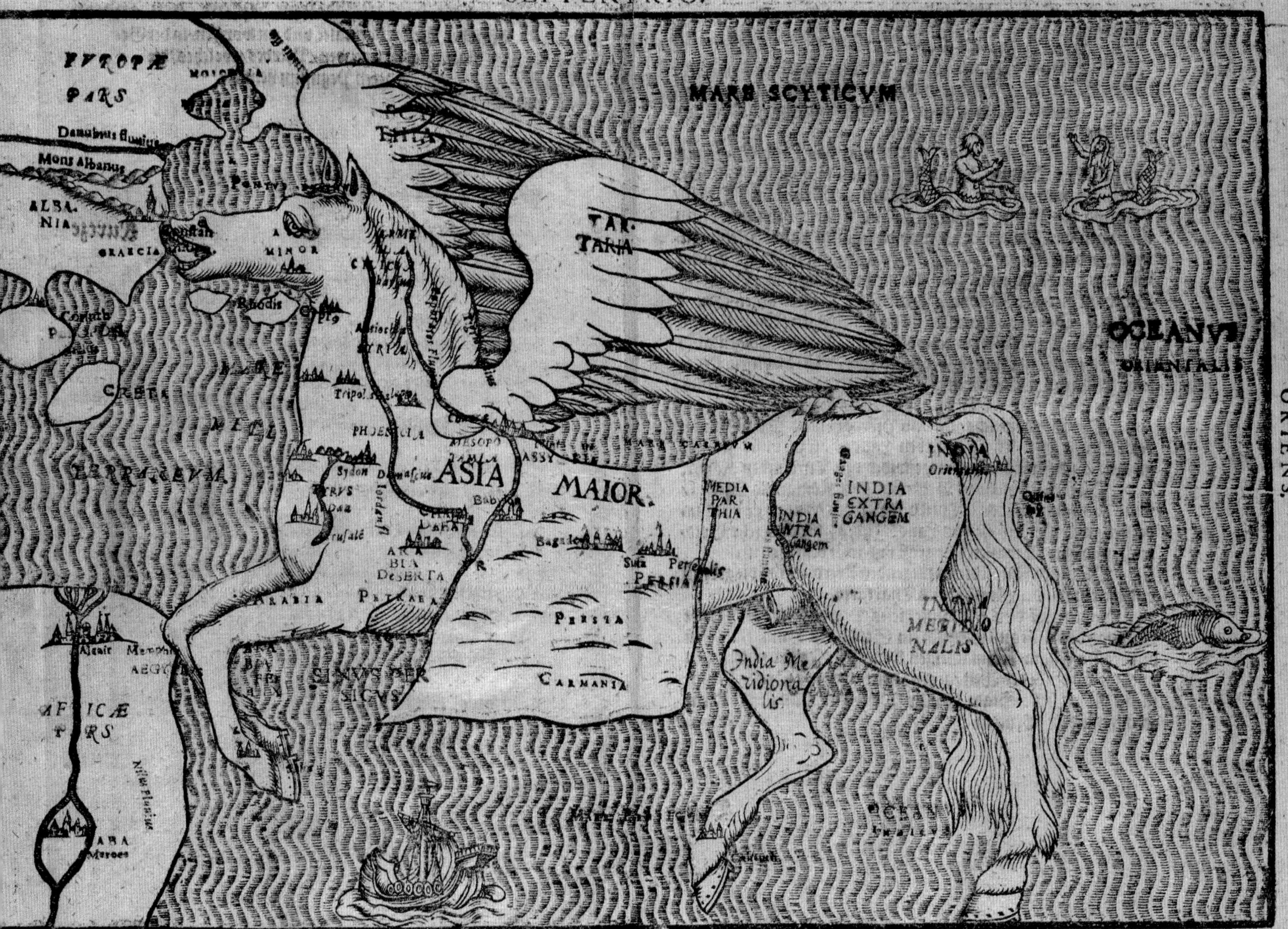
ASIA SECUNDA PARS TERRÆ IN FORMA PEGASIR.
17
SEPTENTRIO.
EVROPÆ PARS
Danubius fluuius
Mons Albanus
ALBA NIA
GRAECIA
MINOR
Rhodis
CRETA
MARE SCYTICVM
TAR TARIA
OCEANVS ORIENTALIS
ORIENS.
ASIA
MAIOR.
Damascus
Sydon
TYRVS
MEDIA PAR THIA
INDIA INTRA Gangem
INDIA EXTRA GANGEM
INDIA Orientalis
Susa
Persepolis
PERSIA
CARMANIA
ARA BIA DeSERTA
ARABIA
AEGY
AFRICÆ PARS
Nilus fluuius
Meroes
India Me ridiona lis
INDIA MERIDIO NALIS
MERIDIES.
sus Christus magnus ille Beller opontes, omnium malorum occisor ascendens Pegasum, hoc est, in Asia fontem doctrinæ aperies Solimos vicit, & chimæram interfecit
bile monstrum quod flammas evomens caput & pectus Leonis habuit, ventrem autem Capræ, & caudam Draconis, hoc est, Superavit ac interfecit filius antiquum
raconem Diabolum, sublato peccato more ac inferno.

'The Shape and Position of New Guinea' [1593]

Cornelis de Jode, Antwerp

'Novæ Guineae Formæ, & Situs'

Copper engraving; map: 285 × 192 mm (11 ¼ × 7 ½ in);
at widest: 340 × 215 mm (13 ⅜ × 8 ½ in)

GERARD DE JODE (1509–1591) and his son Cornelis (*c.*1568–1600) were leading mapmakers, mapsellers and publishers working in Antwerp. The family published a large number of important maps, ranging from single sheets to eight-sheet wall-maps, and seems in the main to have had a highly successful business.

Gerard embarked on the compilation of a world atlas, the *Speculum orbis terrarum*, which was completed by about 1571. In one of those accidents of history, however, Abraham Ortelius produced a rival atlas at just the same time. Ortelius was better known and had more influence, obtaining the necessary encouragement and privileges to publish his atlas in 1570; de Jode was denied these privileges until 1577 or 1578, by which time Ortelius' *Theatrum orbis terrarum* had achieved a stranglehold on the market. De Jode's atlas was a commercial failure.

Undeterred, Gerard embarked on preparations for a second edition, completed after his death by Cornelis in 1593. For this edition, a number of new maps were engraved, several of great importance, including this one, the first printed map of New Guinea. In the short text on the verso, Cornelis admitted that he knew little about New Guinea, and we have to be careful about interpreting the geographical depiction given. For some commentators, the large landmass to the south is the earliest printed depiction of Australia, evidently relying on considerably earlier reports than that of the first undisputed sighting of Australia by a European, made by Willem Janszoon, the Dutch captain of the *Duyfken*, in 1606. For others, it is simply an imaginative attempt at depicting the 'Great Southern Continent', which geographical theorists believed had to exist, a dense landmass necessary to counterbalance the weight of the landmasses in the northern hemisphere and keep the world rotating smoothly on its axis.

While the map is a geographic conundrum, it also exemplifies the old expression 'Here be dragons'. (There is actually only one recorded appearance of this famous phrase on an early map: it is rendered as 'Hic sunt Dracones' on the Lenox globe from about 1505, in the New York Public Library.) In the map shown here, it has been suggested that, despite the barbed tongue, batwings and eagle's claws of the conventional European heraldic dragon, the mapmaker has tried to show an Indonesian Komodo dragon – recent research has indicated that the Komodo dragon originated in Australia and spread north, before the water-levels of the region rose sufficiently to turn the continental shelf into a series of islands.

Komodo dragons, actually lizards, are fearsome creatures, averaging about 2.6 metres (8½ feet) in length, and capable of hunting pigs, deer and even humans. Rather like boa constrictors, they can swallow their prey whole and then regurgitate those bits they are unable to digest; or, alternatively, they use their strong teeth to bite chunks out of their prey. All in all, the dragon is an intimidating creature, akin to the monster depicted in de Jode's map.

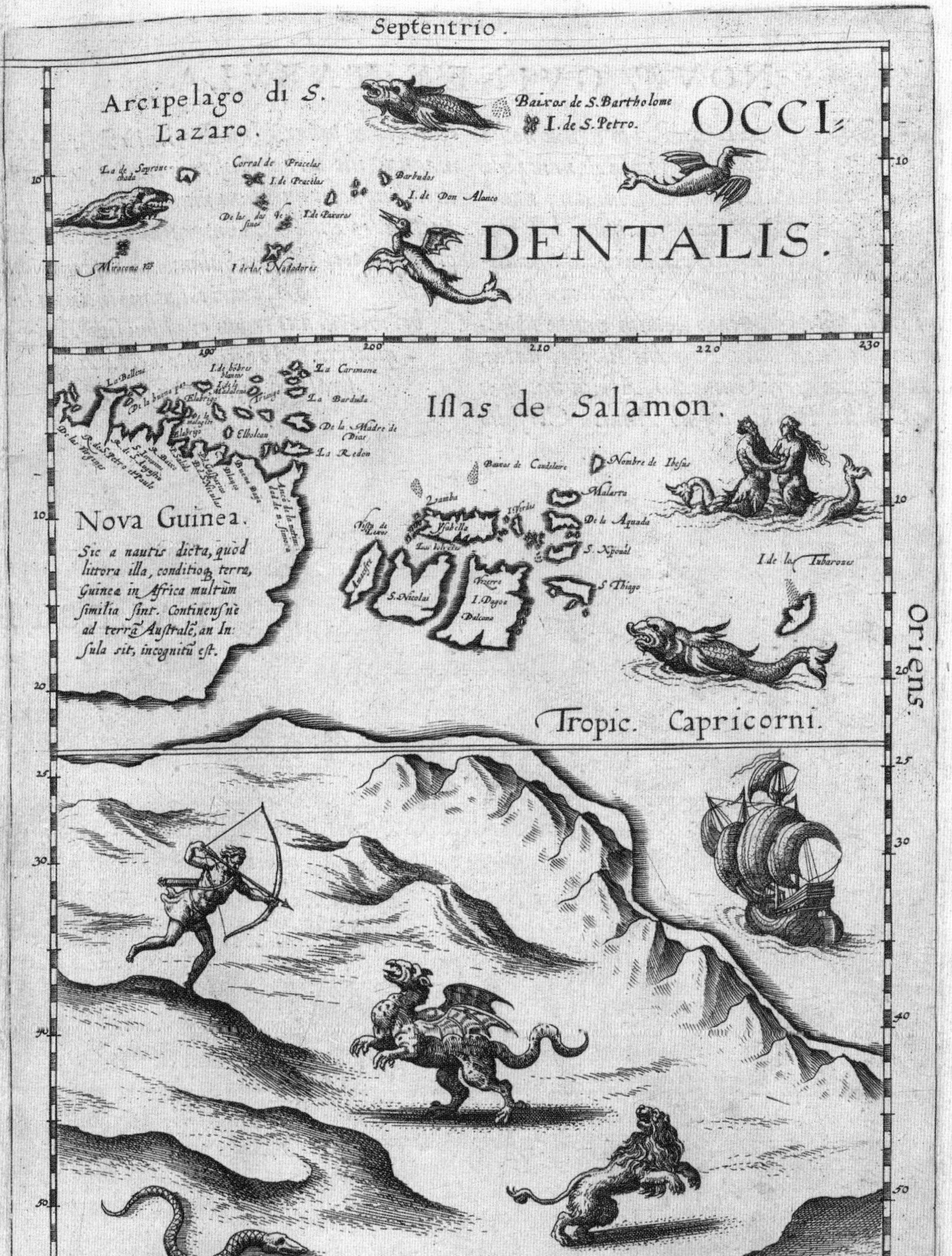

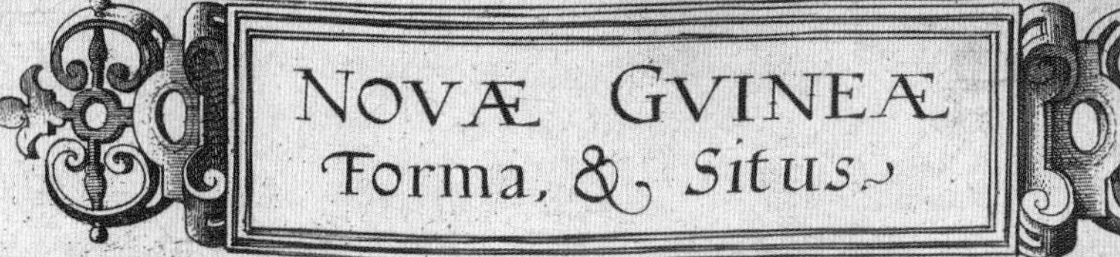

Septentrio.
Arcipelago di S. Lazaro.
Baixos de S. Bartholome
I. de S. Petro.
OCCIDENTALIS.
Islas de Salamon.
Nova Guinea.
Sic a nautis dicta, quod littora illa, conditioq; terra, Guineæ in Africa multum similia sint. Continens'ne ad terram Australem, an Insula sit, incognitum est.
Nombre de Ihesus
I. de los Tiburones
Oriens.
Tropic. Capricorni.
Meridies.
NOVÆ GVINEÆ
Forma, & Situs.

LEO
t' AMSTERDAM
bij Hugo Allardt

EARLY PUBLISHED MAPS

(1598–1760)

The Lion of the Low Countries, 1598

Johannes van Deutecum Jr, Amsterdam

'Leo Belgicus'

Copper engraving; border: 436 × 554 mm (17 1/8 × 21 3/4 in)

ONE OF the best known of all zoomorphic maps is the 'Leo Belgicus' – the Lion of the Low Countries. The Low Countries, or the XVII Provinces, were the Flemish and Dutch territories now known as Belgium and Holland. As well as naturally fitting the shape of the region, the lion was frequently found in the arms of the individual provinces and cities, and so could be readily accepted as symbolic of the XVII Provinces.

The appearance of the first 'Leo' map coincided with a period of great turmoil in the Low Countries. Through accident of dynasty and history the Low Countries had come to be part of the Habsburg Empire, ruled and controlled by the kings of Spain. The so-called Dutch Revolt, a war of independence and self-determination in which the provinces sought to gain freedom from their Spanish masters, to throw off 'the Spanish yoke', began in 1568. The first 'Leo', although conceived by the Austrian author Michael von Eitzing, was engraved by Frans Hogenberg, a supporter of the revolt. He conceived the lion as an emblem of strength and courage, master of its own destiny, roaring to put fear into its enemies. By bringing all the different provinces together within the 'body' of the lion, he was also encouraging the provinces to unite together in the face of their common overlord, the Spanish king.

Hogenberg's map was published in von Eitzing's *De leone Belgico* (1583), a history of the war of independence to that date. The revolt continued until a truce was declared in 1609; when the truce expired, the fighting began again, finally ending with the northern, Dutch, provinces securing their independence.

The 'Leo' symbol must have struck a particular chord with the Dutch public, and a large number of different versions were prepared between 1583 and 1648, while the struggle for independence continued. These continued to be printed long after the fighting had ended. The example illustrated here was first published by Johannes van Deutecum Jr in 1598. This version is an altogether more elaborate construction than Hogenberg's original. The most notable addition van Deutecum made was the insertion of the elaborate borders on three sides, with portraits of the successive governors in the side panels, and the Dutch Stadtholders in the lower border. Text in the two lower corners, in Dutch and French, explains the significance of the 'Leo' map. Two insets depict the seats of government: Brussels ('Palatium Bruxellen Sie') and the Palace of the Court of Holland ('Palatium comitū Holland').

Van Deutecum's plate later passed to Claes Jansz. Visscher (1587–1652), who printed this example in 1650. For this printing, Visscher made a small number of changes to bring the plate up to date, including adding the portraits of Archduke Ferdinand (Spanish Governor from 1634 to 1641) and Prince Frederick Hendrick (Stadtholder from 1625 to 1647).

Artificiosa
et Geographica
tabula sub Leonis
figura XVII. inferi
oris Germaniæ
Provincias repre
sentans, cui ad
dita sunt singula
rum insignia, una cum ordi
naria Præfecturarum distincti
one, Longè elimatius quam
hactenus unquam expressa.
Accesserunt &
icones Guber
natorum Gene
raliũ qui utrin
que Belgium,
Gubernarunt.
Scala Mili: Germanic:
1 2 3 4 5 6 7 8 9 10
Catalogus Civitatum Pago:
rumq, in singulis Provincijs
Civitates. Pagi.
Comitatus Hollandiæ — 33 400
Comitatus Zelandiæ — 10 101
Comitatus Flandriæ — 35 1178
Comitatus Artesiæ — 12 754
Comitatus Hannoniæ — 24 950
Comitatus Namurci — 4 184
Comitatus Zutphaniæ
Ducatus Brabantiæ — 26 700
Ducatus Lutzenburgi 23 1169
Ducatus Limburgi — 5 123
Ducatus Geldriæ — 24 300
Marchionat' Sacri Imp — 1
Dominij Mechliniæ — 1 9
Dominij Ultrajecti — 5 70
Dominij Frisiæ — 11 345
Dominij Transisulaniæ 11 101
Dominij Groningæ — 1 145
GER
MANIÆ
ORIENS
52
51
50
GRONINGA
Groeningen
Embden
Leeuwaerde
FRISLANDIA
DRENT
TRANSISULANA
De Eems
DE ZUYDER ZEE
Enckhuysen
Campen
Coevoerden
Lingen
TWENT
Benthem
Deventer
ZUTPHANIA
GELDRIA
Arnhem
Amersfoort
Wesel
Lippe flu
Rhyn flu
Clivia
Venlo
Keyserswerd
Duyseldorp
Maestricht
Valkeburg
Iuliacens
Ducat
Gulich
Colln
Aken
LIMBURGUM
LEODIUM
Liege
Limburg
Bonn
Poppelsdorp
Godesberg
Coninx Winter
Andernach
Dillenberg

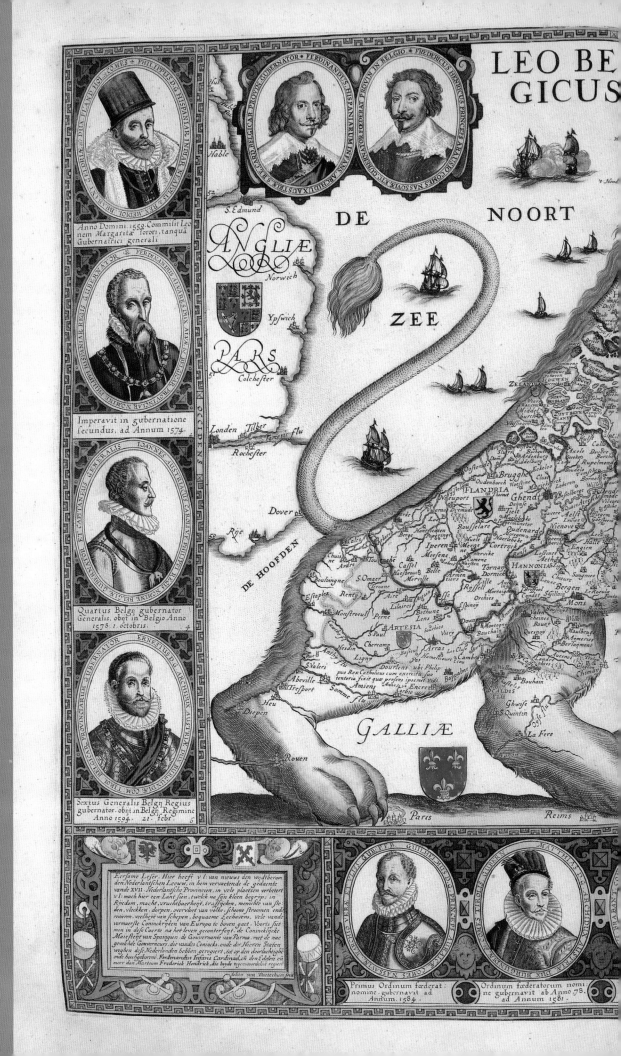

LEO BE
GICUS
DE
NOORT
ZEE
ANGLIÆ
PARS
DE HOOFDEN
GALLIÆ
FLANDRIA
HANNONIA
ARTESIA
OCCIDENS
Anno Domini 1559. Commisit Leonem Margaritæ sorori, tanquā Gubernatrici generali
Imperavit in gubernatione secundus, ad Annum 1574.
Quartus Belgij gubernator Generalis, obijt in Belgio Anno 1578. 1. octobris. 4
Sextus Generalis Belgij Regius gubernator. obijt in Belgij Regimine Anno 1594. 21. febr. 6
Primus Ordinum fœderat: nomine. gubernavit ad Annum. 1584.
Ordinum fœderatorum nomine gubernavit ab Anno 78. ad Annum 1581.
Norwich
Ypswich
Colchester
Londen
Tilber
Rochester
Dover
Rye
S. Edmund
Hable
Zelandia
Brugghe
Ghendt
Dover
Abeville
Tresport
Diepen
Rouen
Paris
Reims
Amiens
Arras
Lille
Bergen
Mons
Cassel
S. Omaer
Boulongne
Monstroeul
Hesdin
Cambray
Somme flu
La Fere
S. Quintin
Oudenarde
Cortrijck
Ipere
Dormick
pus Rex Catholicus cum exercitu suo tentoria fixit quæ præsens ipsemet vidi

Artificiosa et Geographica tabula sub Leonis figura XVII. inferioris Germaniæ Provincias repraesentans, cui addita sunt singularum insignia, una cum ordinaria Præfecturarum distinctione, Longè elimatius quam hactenus unquam expressa. Accesserunt & icones Gubernatorum Generaliū qui utrinque Belgium, Gubernarunt.
Scala Mili: Germanic:
1 2 3 4 5 6 7 8 9 10
Catalogus Civitatum Pagorumq. in singulis Provinciis
Civitates. Pagi.
Comitatus Hollandiæ — 33 400
Comitatus Zelandiæ — 10 101
Comitatus Flandriæ — 35 1178
Comitatus Artesiæ — 12 754
Comitatus Hannoniæ — 24 950
Comitatus Namurci — 4 184
Comitatus Zutphaniæ
Ducatus Brabantiæ — 26 700
Ducatus Lutzenburgi 23 1160
Ducatus Limburgi — 5 123
Ducatus Geldriæ — 24 300
Marchionat Sacri Imp — 1
Dominij Mechliniæ — 1 9
Dominij Ultrajecti — 5 70
Dominij Frisiæ — 11 345
Dominij Transisulaniæ 11 101
Dominij Groningæ — 1 145
GER MANIÆ PARS.
GRONINGA
FRISLANDIA
DRENT
TRANSISULANA
DE ZUYDER ZEE
TWENT
GELDRIA
Groeningen
Leeuwaerde
Enckhuysen
Campen
Deventer
Zutphen
Arnhem
Nieumegen
Wesel
Kessel
Venlo
Keyserswerd
Duyseldorp
Maestricht
Valkeburg
Juliacensis
Gulich
Colln
LEODIUM
Liege
LIMBURGUM
Aken
Bonn
Poppelsdorp
Godesberg
Lintz
Andernach
Engers
Coblens
Nassaw
Dillenberg
Lona flu
Mosella flu
Rhenus flu
Pfaltz
Binge
Mentz
Francfort
Trier
LUTZENBURG
Luxembourg
Arlon
Metz
Excudit 1650
ORIENS
PALATIUM BRUXELLENSE
PALATIUM COMITŪ HOLLAND
MARGARETA AUSTRIACA DUCISSA PARMÆ
Gubernavit Leonem nomine Philippi fratris, ad Annū 1567.
LUDOVICUS REQUESENTIUS MAGNUS COMMENDATOR CASTILIÆ
Tertius in gubernando Leone Belgico, obijt Anno 1576. 5.a Martij. in Belgio.
ALEXANDER FARNESIUS PARMÆ ET PLACENTIÆ DUX GUBERN. ET CAPITAN. GENERALIS REGIONUM BELGICARUM
Quintus Regis Hispaniæ in Belgio Gubernator, gubernavit ad Annum 1592
ALBERTUS CARD. ARCHIEPISC. TOLET. ARCHIDUX AUSTRIÆ
Septimus Regius Belgij Generalis Gubernator gubernavit ad Annum 1621.
ROBERTUS DUDLEIUS COMES LECESTRIÆ
Ordinum fœderatorum nomine gubernavit. ab Anno 1586. ad Annum. 1588.
MAURITIUS D.G. NASSAU PRINCEPS AURAICÆ
Ordinum fœderatorum nomine gubernavit ab Anno 1587 ad Annum 1625
Ami Lecteur. Nous te presentons ici de nouvaeu Le tres-renommé Lion des Pays bas, comprenant en soy La figure des XVII Provinces, corrigé en Lieux infinis. Tu vois ici vn pays, qui au regard de son petit enclos, en richesses puissance, fertilité, traffique, grand nombre de villes, bourgades, et villages, abondance du peuple, et quantité innombrable de bateux, ports maritimes, surmonte plusieurs des plus renommez Royaumes de L'Europe. En oultre peut on veoir dans ceste carte Les pourtraitures de sa Maiesté d'Espagne et de La Gouvernante de Parma & d'aultres Gouverneurs suivans, contrefaits au vif, tant de ceux qui au nom dudit Roy comme aussi sous L'autorité de Mes Seigneurs Les Estats des Provinces Vnies ont gouverné ces Pays bas, iusques au Tresillustre Ferdinand Hisp. Infant, & Le tres-noble, & plus que Martial F. Henric. Lesquels pour Le present en tiennent Le gouvernement.

Untitled Satirical Map of the World, 1605

Joseph Hall, London

Copper engraving; border: 207 × 246 mm (8 ⅛ × 9 ¾ in)

JOSEPH HALL (1574–1656) was a churchman, theological commentator and writer, appointed one of the chaplains of Henry, Prince of Wales in 1608. In 1627 he was made Bishop of Exeter and in 1642 Bishop of Norwich, during the troubled period of the English Civil War.

Hall was a noted satirist, and his most famous work was the *Mundus alter et idem* ('The other and the same World, or Terra Australis up to this time completely unknown, now revealed by long journeys [made] by foreign academics, written by Mercurius Britannicus'), published under a false foreign imprint in 1605, and without author attribution. Despite this, it seems that Hall's authorship was something of an open secret. Although the title page is undated, the book was entered by John Porter in the Registers of the Stationers' Company on 2 July 1605.

The *Mundus alter et idem* is a satire on a corrupt and venal dystopia. The target of his satire was evidently the London society of his own day, but Hall transported this society to Terra Australis, the large southern landmass which geographers believed must exist to counterbalance the landmasses of the northern hemisphere (see p. 26). Hall describes the voyage of the storyteller to Terra Australis, in the ship *Fantasia*. He constructed five maps, probably engraved by William Kip: the map of the world illustrated here, with Terra Australis prominent, and four regional maps of the lands visited on the voyage.

Crapulia (the land of excess) was divided into provinces: 'Pamphagonia' (glutton's paradise) and 'Ivronia' (drunkenness). The inhabitants' lives revolved around food; the land was fertile, but only edible trees and plants were cultivated. The inhabitants exported their superfluous animal hides and wool in order to import more food. Birds that flew into the region ate so much that they could no longer fly, and ended up on the dinner table themselves. Food even served as currency.

Viraginia (land of the Viragoes – here termagants and scolds rather than heroines) was a matriarchal society, with men subservient in every degree. A man was forbidden to be in charge of a household; if a woman died, her spouse would be married off immediately to the maid, or put into service with another female relative. Domestic chores were for the men, who could be beaten at whim. They were forbidden to interrupt women and compelled to give their wives any item of dress they wanted. It was a democratic society in which only women participated: the problem, as described by the narrator in an age-old sexist joke, was that all the women wanted to talk, but not to listen, and all to be obeyed but not to obey.

Moronia was the land of fools. It was very poor, as the inhabitants were not able to perform even basic cultivation. One of their customs, several of which parody Roman Catholic rituals, was to shave their heads, so that the hair did not cause their brains to overheat. The fourth region, Lavernia, was home to thieves and cheats. The citizens of Lavernia preyed upon Moronia, raiding the land and the inhabitants.

Having travelled through all four regions and thoroughly dispirited by what he had seen, the narrator sailed back to England a sadder and a wiser man.

Turchesta
Tenduc
Baiona
Compostella
HISPANIA
Lisboa
Roma
Sardinia
Napels
Sicilia
Constantinopolis
Mare Maior
Candia
ASIA
Cathaia
Indostan
Virginia
C. de la Florida
Acalco
Trugillo
Panama
Saribano
Peru
Brasilia
AFRICA
Anzica
Abassia
Narsinga
Malaca
I. de S. Pedro
AMERICA MERIDIONALIS
PATAGONES
C. de la Victoria
C. de la Crus
Fret. Magellan
Mono motapa
Insula Hermaphroditica
Promontorium Nigrum
Aphrodysia
Noua Gynia
PAMPHAGONIA
IVRONIA
LOCANIA
Amazonia
VIRAGINIA
Phenacia
Larcinia
Codicia
Tryphon flu
PLA MORONIA
MORONIA FELIX
Ploraria
Lingua dotia
MORON FATVA
Baueria
Tuberony Colles
VARIANA
Vel Morion Mobilis
FRVGIONA
Adhuc Incognita
Aspera Moronia
Orgilia
Credulia

The Lion of the Low Countries, 1608

Hessel Gerritsz, Amsterdam

'Leo Belgicus'

Copper engraving; border: 434 × 557 mm (17 ⅛ × 21 ⅞ in)

THE LEONINE depiction of the Low Countries proved a commercial hit with Dutch cartographic publishers and their sympathisers across Protestant Europe; while Michael von Eitzing's original version and its derivatives (see p. 30), remained popular, later publishers continued to experiment with the idea to produce further images.

In about 1608 or 1609, the Dutch cartographer Hessel Gerritsz (1580/1–1632) produced a dramatically different leonine map. Here, the Low Countries are depicted with north at the right, and the lion emblem superimposed, with its head over the southern province of Artois (modern Belgium). Its haunches are formed by the northern provinces, including the Zuider Zee.

The early history of the map is unclear: it seems that Gerritsz prepared two very similar map plates at about the same time, quite possibly for two different publishers. These plates ended up in different hands, and were reprinted by rival publishers, with their own imprints inserted, for many years – the first at least as late as 1705, the second seemingly still available in about 1720. The example illustrated here is a later printing, issued by the Amsterdam mapmaker and publisher Claes Jansz. Visscher, and bearing his imprint dated 1656.

The text under the lion translates as:

> Thus the lion speaks. Just as my huge body has muscular limbs, you can see in my body powerful countries. How great it would be if, united in everlasting peace, each province assisted the others!

This map forms part of a very impressive twenty-four-volume collection of maps, plans and views of towns in the Netherlands, entitled *Germania Inferior, sive XVII Provinciarum geographicae generalis ut et particulares tabulae. Kaert-boeck van de XVII Nederlandtsche Provincien. Door C. Beudeker* ... which is a compilation with, at its core, Willem and Johannes Blaeu's atlas the *Toonneel des Aerdrijcks ofte Nieuwe Atlas*, 1648–1658, and their city-book *Toonneel der Steden van de Vereenighde Nederlanden*, from 1649. This atlas was assembled for Cristoffel Beudeker, a Dutch merchant and landowner, in about 1718, and preserves many rare, separately issued maps and plans, in fine contemporary hand colouring.

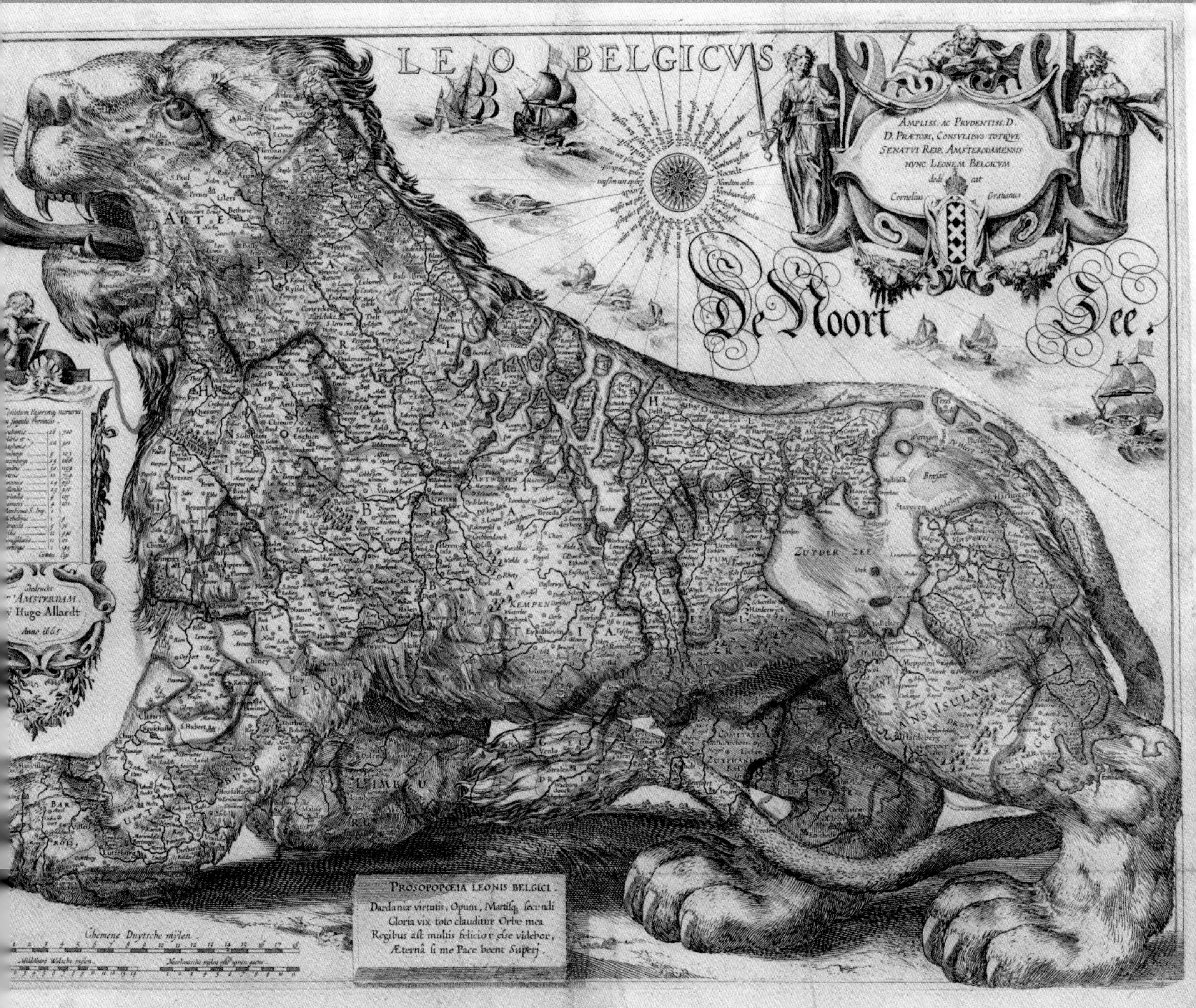
LEO BELGICVS
AMPLISS. AC PRVDENTISS. D. D. PRÆTORI, CONSVLIBVS TOTIQVE SENATVI REIP. AMSTERODAMENSIS HVNC LEONEM BELGICVM dedicat
Cornelius Gratianus
De Noort Zee.
ZUYDER ZEE
Ghedruckt t'AMSTERDAM by Hugo Allardt Anno 1665
PROSOPOPŒIA LEONIS BELGICI.
Dardaniæ virtutis, Opum, Martisq; secundi
Gloria vix toto clauditur Orbe mea
Regibus ast multis felicior esse videbor,
Æternâ si me Pace beent Superi.
Ghemene Duytsche mylen.

The Lion of Holland, 1609

Claes Jansz. Visscher, Amsterdam

'Comitatus Hollandiæ denuo formâ leonis'

Copper engraving; border: 462 × 561 mm (18 ¼ × 22 ⅛ in)

CLAES JANSZ. Visscher began his career as an engraver and etcher, often using the monogram 'CIV' or the punning symbol of a fisherman on his own work, but soon set up his own publishing house. It is best known for its cartographic output, but he also issued topographical prints (particularly panoramas) and a full range of decorative prints. He was succeeded in turn by his son Nicolas (I) (1618–1679), his grandson Nicolas (II) (1649–1672), and then Nicolas (II)'s wife, Elizabeth (fl. 1702–1726). At first overshadowed by the Blaeu and Hondius firms, masters of the Golden Age of Dutch cartography, the Visschers eventually became one of the two leading map publishers in Amsterdam (and thus the world).

This Lion map is of an altogether different type to those shown on pp. 30 and 32: this is the 'Leo Hollandicus', 'the Lion of Holland' rather than the 'Leo Belgicus'. It depicts only the seven northern Dutch provinces in revolt from Spanish control, not the seventeen Provinces forming the whole of the Low Countries.

There is no recorded example of the first state of the map, but there are later printings from 1633 and 1648. It is a matter of debate when the first state was actually printed but, in the context of the events of the Dutch Revolt and the elaborate pictorial border, it seems likely that this image was printed in late 1609.

On 9 April 1609, the Dutch Republic and Spain signed a twenty-one-year truce. It was not the end of the Dutch Revolt, but it was a moment when the Dutch people, and their publishers, could see that Spanish military strength had failed to crush the revolt and sense that independence, recognised *de facto* in the truce, was very much a possibility. In view of the proud and unabashedly patriotic tone of the engraving, it seems certain that the map was put on sale soon after the ceasefire was signed. The central figure of the lion carries a sword over his shoulder with the motto 'Patriae Defensio'; he has successfully defended the independence of the Dutch provinces since the revolt began in 1568.

The map is flanked with twelve finely etched and engraved vignettes containing prospects of the leading Dutch cities – notably Dordrecht, Amsterdam, Rotterdam, Enckhuysen, Haerlem and Leyden – while the arms of thirty-two towns are arranged across the foot of the map.

Across the top are two large pictorial scenes; the left-hand one shows 'Agricolae Hollandiae Australis' (farmers of southern Holland) and 'Nobiles Hollandi' (the nobles) watching an ice-yacht; the right-hand scene depicts 'Mercatores seu Cives' (merchants or citizens) and 'Agricolae Hollandiae Borealis' (farmers from northern Holland) watching two land-yachts. This may well reflect the hope that, with the fighting over, the Dutch could focus again on trade and that peace would see a return of pre-war prosperity. Perhaps too there is a suggestion that the spirit of invention and innovation – represented by the two very unusual yachts – would again propel the economy forwards, not least in the expansion of the Dutch overseas trade as directed by the Vereenigde Oost-Indische Compagnie (VOC or, in English, the 'United East India Company') which had been founded in 1602.

In 1621, the truce lapsed, and the two sides took up arms again. The second phase of the Dutch Revolt formed part of the much wider pan-European Thirty Years' War, which eventually came to an end in 1648. The Treaty of Münster of that year ended the war between Spain and the Dutch Republic, and this proved a most appropriate moment for Visscher to reprint his 'Leo Hollandicus', with the Dutch Republic established, their overseas empire emerging and a new and optimistic golden age dawning for the Seven United Provinces.

t Graeffchap
Hollandt
NOBILES HOLLANDI
MERCATORES SEU CIVES
AGRICOLÆ HOLLANDIÆ BOREALIS
COMITATUS HOLLANDIÆ DENUO FORMÂ LEONIS CURIOSÈ EDITUS A Nicolao Iohannis Visscher Anno 1648.
DE NOORT ZEE MARE GERMANICI PARS
ZUIDER ZEE
DORDRECHT
DELFT
AMSTERDAM
ROTTERDAM
ALCKMAER
ENCKHUYSEN
HAERLEM
LEYDEN
DERGOU
DEN BRIEL
HOORN
T'HOF van HOLLANT
Illustrissimo Ornatissimoq Mauritio Dei gratia, Principi Auransiæ, Comiti de Nassau optimo Patriæ nostræ defensori, hæc tabula dedicatur et offertur. N.I.V.
Edita a Nicolao Iohannis Visscher.

Game of the World, 1645

Pierre Duval, Paris

'Le Jeu du Monde'

Copper engraving; border: 395 × 516 mm (15 ½ × 20 ¼ in); widest: 401 × 516 mm (15 ¾ × 20 ¼ in)

PIERRE DUVAL (or du Val) (1619–1683) is generally credited as being the first mapmaker to combine different elements to produce a board-game suitable for cartographic education through play. Printed games of other kinds had appeared much earlier; in England, the bookseller and publisher John Wolfe, perhaps familar with such things from his early career in Italy, registered his copyright for two games in 1588, while the first pack of English cartographic playing cards was published in 1590.

The most popular game of the time was of the *Jeu de l'Oie* (Game of the Goose) type: a spiral race game played throughout Europe from at least the sixteenth century and which is probably very much older still. The game is known as *Juego de la Oca* in Spanish; *Gioco dell'Oca* in Italian; *Gänsespiel* or *Ganzenbord* in German; *Jogo do Ganso* in Portuguese; and so on. The traditional game had sixty-three squares, some probably with an originally symbolic or cabalistic significance. It was played with a die or teetotum, and the players vied with each other to reach the final square, often called 'game' or 'home'. A game of chance rather than skill, its basic principles are still familiar in games such as Snakes and Ladders: the players throw in turn, and move along the track as directed. Normally the game includes bonus or penalty squares, requiring the player to advance to another square, or to retreat, to pay a forfeit (fine), to throw again or to miss a turn; usually, unlike in Snakes and Ladders, there is also a 'death square', where the unlucky player is simply thrown out of the game.

Duval's *'Jeu du Monde'* is generally acknowledged to be the earliest known cartographic version of the game, dated 1645 in the imprint outside the lower border. It is particularly rare, with most authorities mentioning only three extant examples, including this one from the British Library's collections. However, there is evidence of erasures on these examples, particularly to the text to the left of the Asia inset, suggesting there was an earlier printing that has not survived. Of the many map-games that were produced subsequently, the *Jeu de l'Oie* type was by far the most popular. It appeared in a variety of versions, as different publishers experimented with the basic premise.

In this particular version of the game, there are none of the traditional forfeits: the only penalty is that if a player lands on an occupied square, the person already present has to return to the square the second player started from. The winner had to land exactly on 63 (France); any 'overthrow' would be counted back from 63, and the player would wait to throw again.

To assist the players in identifying the region they landed on, there are small index maps of the four continents in the corners; in the centre is a very sketchy double-hemisphere world map, useful only for locating the four continents.

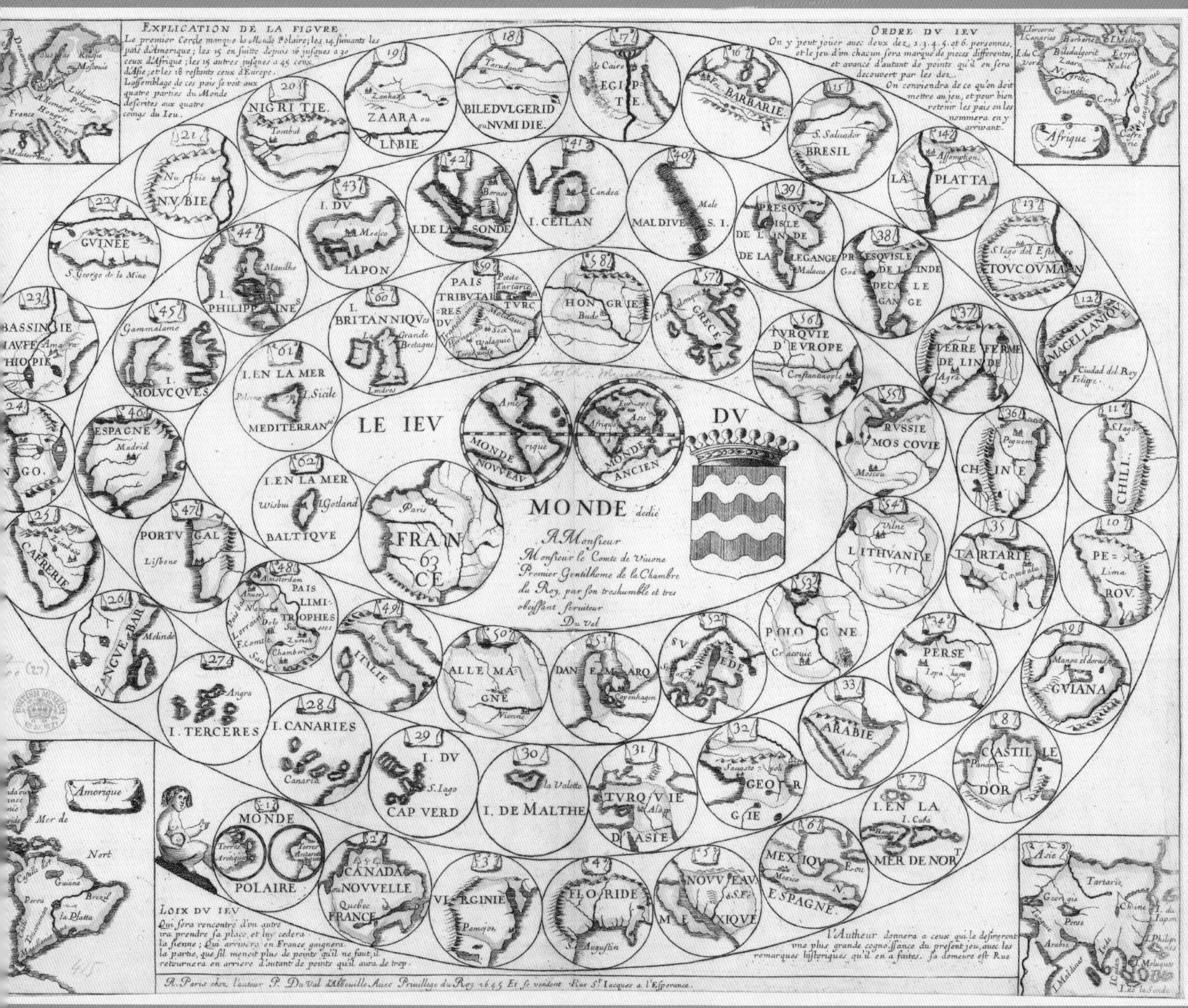
EXPLICATION DE LA FIGVRE
Le premier Cercle marque le Monde Polaire; les 14 suiuants les pais d'Amerique; les 15 en suitte depuis 16 jusques a 30 ceux d'Afrique; les 15 autres jusques a 45 ceux d'Asie; et les 18 restants ceux d'Europe. L'assemblage de ces pais se voit aux quatre parties du Monde descrites aux quatre coings du Ieu.
ORDRE DV IEV
On y peut joüer auec deux dez, 2.3.4.5. et 6. personnes, et le jeu d'vn chacun sera marqué de pieces differentes, et avancé d'autant de points qu'il en sera decouvert par les dez. On conviendra de ce qu'on doit mettre au jeu, et pour bien retenir les pais on les nommera en y arrivant.
LE IEV DV MONDE dedié A Monsieur Monsieur le Comte de Viuone Premier Gentilhome de la Chambre du Roy par son treshumble et tres obeissant seruiteur Du Val
MONDE NOVVEAV
MONDE ANCIEN
MONDE POLAIRE
CANADA NOVVELLE FRANCE
VIRGINIE
FLORIDE
NOVVEAV MEXIQVE
MEXIQVE NOVVELLE ESPAGNE
I. EN LA MER DE NORT
CASTILLE DOR
GVIANA
PEROV
CHILI
MAGELLANIQVE
TOVCOVMAN
LA PLATTA
BRESIL
BARBARIE
EGIPTE
BILEDVLGERID ou NVMIDIE
ZAARA ou LIBIE
NIGRITIE
NVBIE
GVINEE
I. MOLVCQVES
I. PHILIPPINES
IAPON
I. DE LA SONDE
I. CEILAN
MALDIVES I.
PRESQVISLE DE L'INDE DE LA LE GANGE
PRESQVISLE DE L'INDE DECA LE GANGE
TERRE FERME DE L'INDE
CHINE
TARTARIE
PERSE
ARABIE
GEORGIE
TVRQVIE D'ASIE
I. DE MALTHE
I. DV CAP VERD
I. CANARIES
I. TERCERES
ZANGVEBAR
CAFRERIE
ESPAGNE
PORTVGAL
PAIS LIMITROPHES
ITALIE
ALLEMAGNE
DANEMARQ
SVEDE
POLOGNE
LITHVANIE
RVSSIE MOSCOVIE
TVRQVIE D'EVROPE
GRECE
HONGRIE
PAIS TRIBVTAIRES DV TVRC
I. BRITANNIQVES
I. EN LA MER MEDITERRANEE
I. EN LA MER BALTIQVE
FRANCE 63
Afrique
Amerique
Asie
LOIX DV IEV
Qui sera rencontré d'vn autre ira prendre sa place, et luy cedera la sienne; Qui arrivera en France gaignera la partie, que s'il menoit plus de points qu'il ne faut, il retournera en arriere d'autant de points qu'il aura de trop.
L'Autheur donnera a ceux qui le desireront vne plus grande cognoissance du present jeu, auec les remarques historiques qu'il en a faites. Sa demeure est Rue
A Paris chez l'auteur P. Du Val d'Abbeville Auec Priuillege du Roy 1645 Et se vendent Rue St. Iacques a l'Esperance.

‘Map of all the Sea Ports of the World’, Game(?), 1650

Nicolas Berey, Sr, Paris

‘Carte de tous les por[t]s de mer du monde’

Copper engraving; border 343 × 342 mm
(13 ½ × 13 ½ in) diameter (widest surviving)

THIS CIRCULAR map, signed by Nicolas Berey, and probably to be attributed to Nicolas Sr (1610?–1665), is a particularly unusual item. Unfortunately, no intact example survives, to the writer’s knowledge. The Bereys were leading Parisian publishers. The father was active possibly from as early as 1629 (but certainly by 1640) until his death in 1665; Nicolas Jr (1640–1667) succeeded his father but died two years later.

It is a circular engraving, set in three circles, or bands. The inner circle contains the title with Berey’s signature and a small double-hemisphere map of the world; the central band contains maps of the four continents; and the outer ring shows maps of eight European countries or regions, including the British Isles, Spain, France, Italy, Greece and Scandinavia.

Only the central section of a sheet is present; the similarity with Pierre Duval’s game maps (see pp. 40, 48) suggests a common purpose, but presumably the rules were engraved round the outside of the map area, in the four corners of the plate, and were cut off at some point. Without those, it is simply not possible to gauge what sort of game this might have been. This is not a *Jeu de l’Oie* – there is no spiral track. Indeed, there is no track, no numbering, no visible start nor ‘home’ points nor squares – in short, nothing to prove it is a game, except its unusual appearance and its similarity to other map-games of the period. It may simply have been designed as some kind of aide-memoire for a merchant involved with shipping or as a tool for a geography lesson.

All in all, this is a curious survival, but no complete example has been traced, with none found in the online catalogue of the Bibliothèque nationale de France, or other online resources.

CARTE DE TOVS LES PORS DE MER DV MONDE
N. Berey
Ports de Mer des Isles Britaniques
Ports de Mer de Grece
Ports de Mer de Turquie &c.
Ports de Mer d'Espagne
Port de Mer d'Europe
Ports de Mer d'Amerique
Ports de Mer d'Asie
Ports de Mer de France
Ports de Mer d'Afrique
Ports de Mer de Danemarq et Suede
Ports de Mer d'Italie
Ports de Mer d'Allemagne et de Pologne

Draughts Board with Maps of the French Provinces, 1652

Pierre Duval, Paris

'Le Jeu de France pour les Dames'

Copper engraving; maps each approximately 51 × 51 mm (2 × 2 in); sheet at widest: 420 × 402 mm (16 ½ × 15 ⅞ in)

PIERRE DUVAL (1619–1683) followed his map-game of the world with this engraving, a paper sheet forming a board to play the *Jeu de Dames* (chequers or draughts). Unlike a normal board, which has alternating black and white squares, this one replaces the black squares with thirty-two tiny sketch maps of the French provinces, respectively (as given in the map): Bearn; Gascogne; Languedoc; Rovergue; Querci; Perigort; Angoumois; Saintogne; Guienne; Bourgogne; Bresse; Dauphine; Provence; Cevennes; Lyonnois; Auvergne; Bourbonnois; Nivernois; Berri; March; Limosin; Poictou; Tourraine; Anjou; Bretagne; Maine; Perche; Beauce; Champagne; Picardie; Normandie; Isle de France.

The maps are on different scales, adjusted to fit their squares, and in no particular geographical order, but they do cover all of France. For a cartographic publisher, this sort of game sheet could have been an innovative way of introducing people to maps, and interesting them not only in Duval's cartographic games, but also in his general stock of maps of the French provinces, of France itself (from which the sketch maps forming the game are assumed to derive), and of the world beyond.

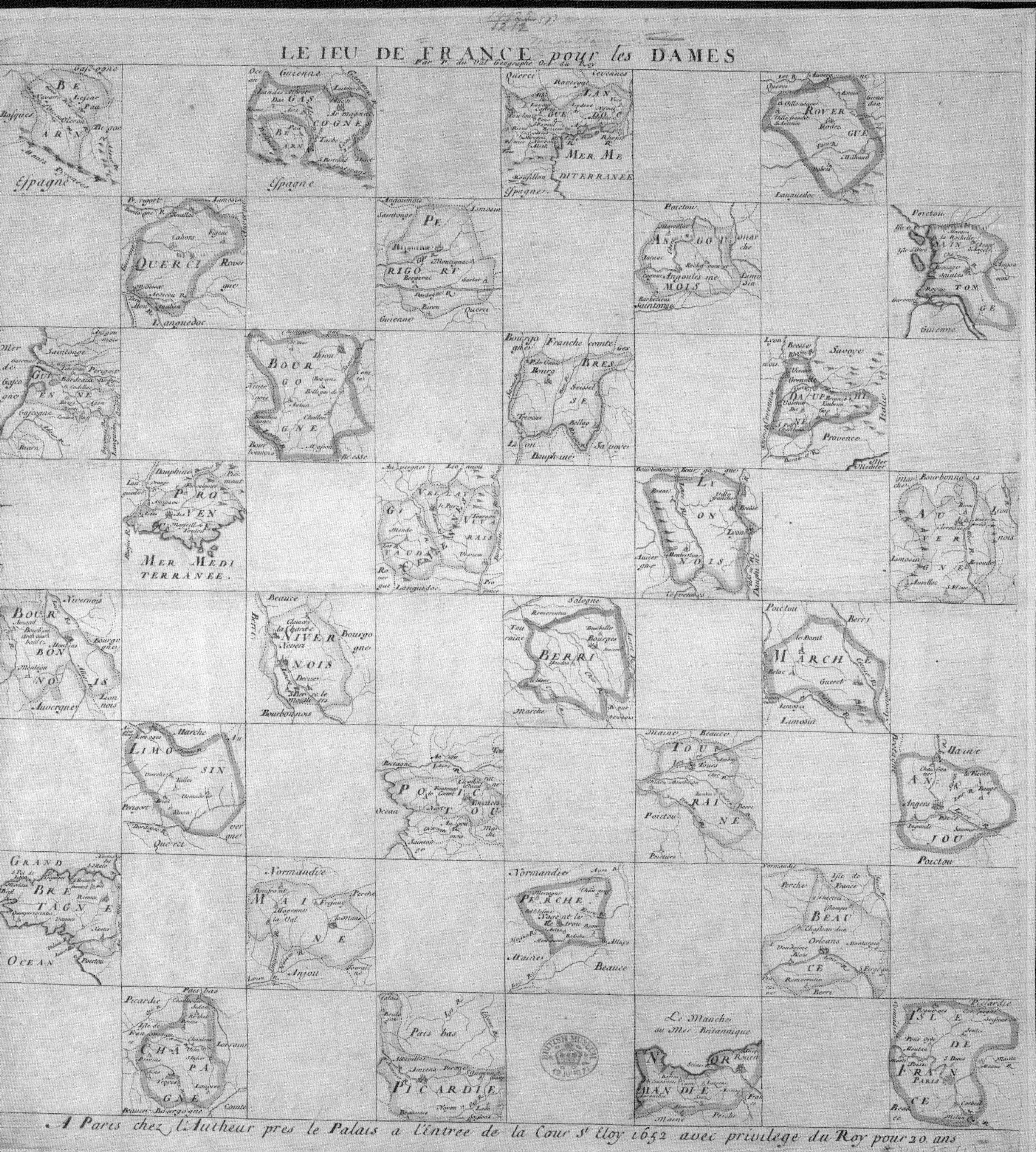
LE IEU DE FRANCE pour les DAMES
Par P. du Val Geographe Ord du Roy
A Paris chez l'Autheur pres le Palais a l'Entree de la Cour St Eloy 1652 avec privilege du Roy pour 20 ans

Untitled Map of the 'Land of Tenderness', 1655

Madeleine de Scudéry, London

Copper engraving; inner border/border: 200 × 295 mm (7 ⅞ × 11 ⅝ in)

THE ROMANCE *Clelia* by Madeleine de Scudéry (1607–1701) tells the story of Aronces and his betrothed Clelia. The day before their wedding, the pair were out walking with her family near Capua when they were separated by a violent earthquake, which gave Horatius, a rival suitor, the chance to seize Clelia and carry her off to Perusia (i.e. modern Perugia). The story is set against the historical struggle between Rome, under Tarquinius Superbus, and the Etruscans, under their king, Lars Porsena. The text refers to 'a Map effectually designed with her hand, which taught us how we might go from New Amity to Tender, and which so resembled a true Map, that there was Seas, Rivers, Mountains, a Lake, Cities and Villages …'

Clelia is based on the character of Cloelia, a Roman heroine. Sent as a hostage to the Etruscans as part of the terms of the peace treaty between Etruria and Rome, she contrived to escape with a group of young Roman girls, swimming the Tiber.

The Land of Tenderness, although it is not named as such in this map, is focused around three rivers, 'Recognaisance F.', "Esteem F.' and 'The River of Inclination', all of which empty into 'The Dangerous Sea', each with its own city, taking its name from the river. The lovers start at 'New Friendship'. The map and the town are divided into two by 'The River of Inclination', so there are four 'roads' they could take. The happy alternatives are both close to the river; on the left-hand side they pass through 'Complacency', 'Submission', 'Small cares', 'Assiduity', 'Empressment' and so on until they reach the town of 'Tender upon R[ecognaisance]'. Hugging the river on the right bank, the route goes through 'Great spirit', 'Pleasing Verses', 'An amorous letter', 'Sincerity', 'A great h[e]art' and on to 'Tender upon Esteem'. Towards the end, the 'Tender upon In[clination]' has a bridge, so the lover can cross over and follow the other route, if desired.

The outer routes are the unhappy outcomes. On the right, the lovers encounter 'Negligence', 'Inequality', leading to 'Lukewarmnes', 'Lightnes', 'Forgetfulnes' before ending at 'The Lake of Indifference'; on the left, the route takes the couple through 'Indiscretion', 'Perfidiousnes', 'Obloquy' and 'Mischief' to 'The Sea of Enmity', where a ship is sinking. If the lovers go too far, they encounter 'The Dangerous Sea', with its many rocks, and beyond it 'Countreys undiscovered'.

Madeleine de Scudéry was a French noblewoman; her many romances are full of characters based on her own circle of friends, and this map originated in the same way, a spatial representation of the stages of love composed on a November day in 1653. It was put into proper form by Paul Pellisson for the first edition of the text published in 1654, and then copied for ,the English editions of 1655–1661 and 1678. The language of the map, therefore, derives from French courtly conceptions of love and courtship, a formal language expressing the philosophical understanding of love rather than the more practical 'landmarks' of the journey associated with later maps of love.

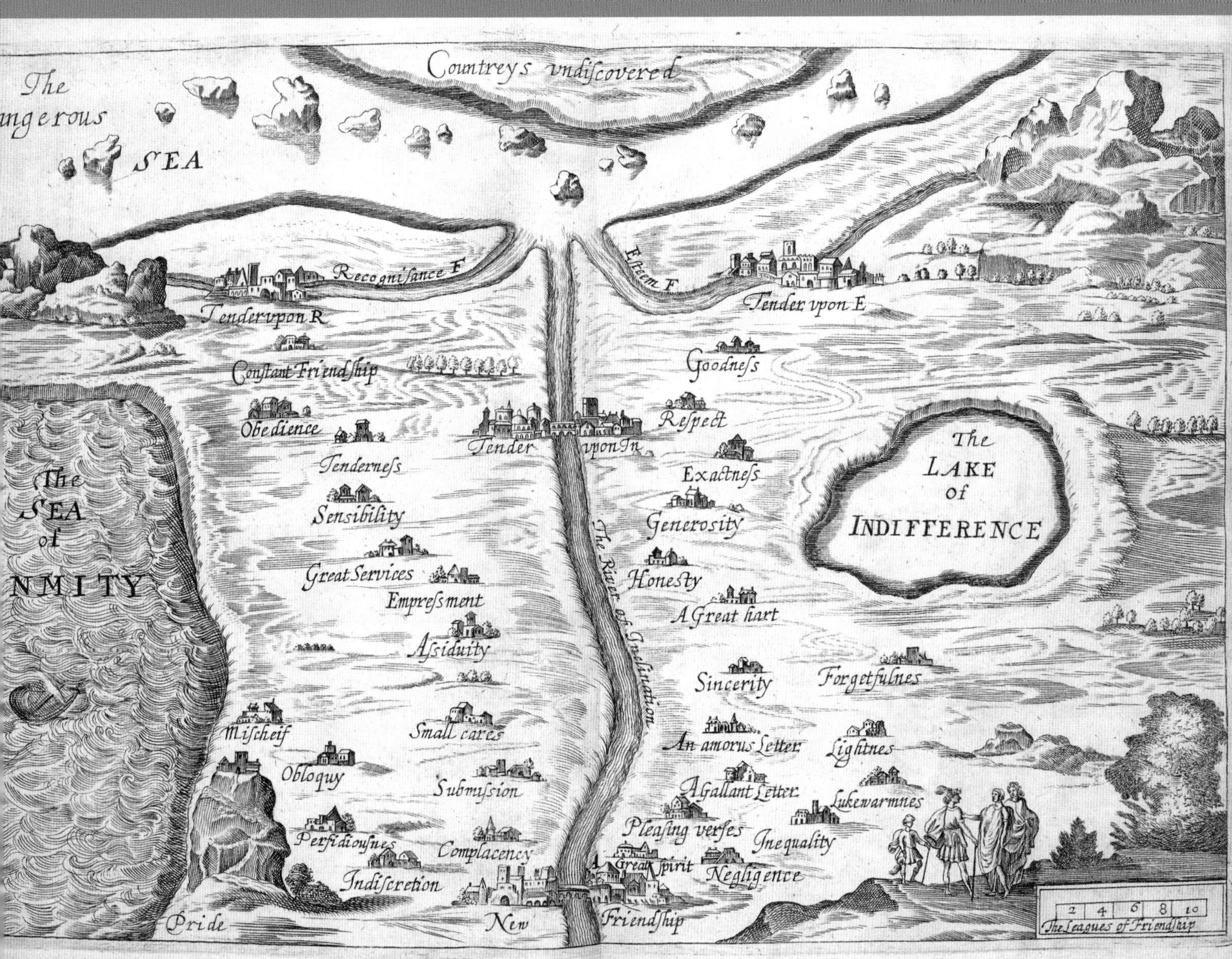
Countreys vndiscovered
The
ngerous
SEA
Recognisance F
Esteem F
Tender vpon R
Tender vpon E
Constant Friendship
Goodness
Obedience
Respect
Tender
vpon In
Tenderness
The LAKE of INDIFFERENCE
Exactness
The SEA of
NMITY
Sensibility
Generosity
Great Services
Honesty
Empressment
A Great hart
Assiduity
The River of Inclination
Sincerity
Forgetfulnes
Mischeif
Small cares
An amorus Letter
Lightnes
Obloquy
Submission
A Gallant Letter
Lukewarmnes
Pleasing verses
Perfidiousnes
Inequality
Complacency
A Great Spirit
Negligence
Indiscretion
Pride
New
Friendship
2 4 6 8 10
The Leagues of Friendship

The Game of France, 1659

Pierre Duval, Paris

'Le Jeu de France par P. du Val Geographe Ordinaire du Roy'

Copper engraving; at widest 383 × 526 mm (15 ⅛ × 20 ¾ in)

THIS MAP-GAME by Pierre Duval (1619–1683) uses France as a cartographic base (compare p. 40), but this time it is re-imagined as a Game of the Goose, with the provinces set in a spiral path leading to the 'game' square, number 63, which is France itself.

Many such games seem to have been primarily intended to be educational, but here Duval has recast the game principally for gambling. Unlike almost every other race game in this book, the *'Jeu de France'* is played not for tokens but for real money, as the rules state (in author's translation):

> The players will agree on a coin to be used as the currency of the game: a denier [one-twelfth of a sol], a sol [later renamed sou], a teston [a silver coin of variable value], a pistole [a gold coin], and all fees, payments, ransoms, fines, imposts and other contributions are to be paid in this coin and placed in the middle of the board (the pot), for the benefit of he who wins the game.

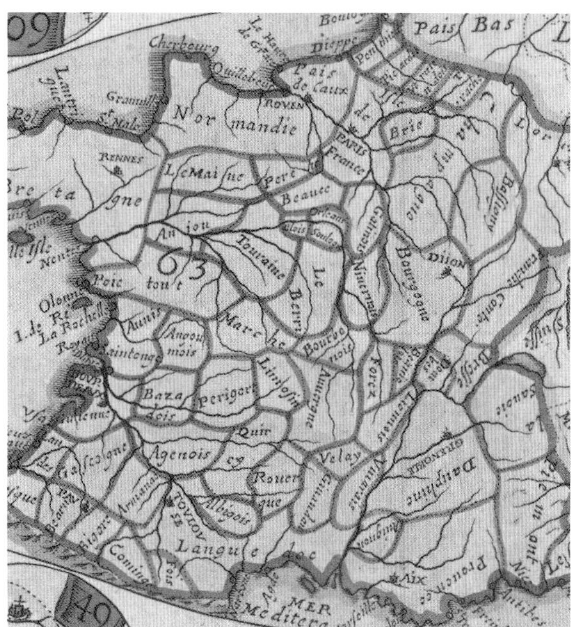

The winner was the player who threw the exact number to land on square 63; any overthrow would be counted back from 63, when the player would have to throw again.

The 'Loix particulieres du jeu' ('Particular rules of the game') engraved outside the spiral, in the bottom left-hand corner of the plate and continued in the bottom right, give additional information about the benefits and forfeits due for landing on particular squares. For example: 'He who goes to Normandy, square 6, will shout "Ha-Rou", summoning to his aid Raoul, the first duke of the country and, for his shares, will receive from all the other players one coin'.

There is no 'death' square, although a player who landed on Touraine (23) had to remain there two turns, while a player who visited Languedoc (52) had to stay there until another player landed on the square. Landing on Provence meant taking a ship to Italy, being captured by pirates and compelled to pay a ransom to continue. Other squares allowed the player to advance to a new square, for the payment of one coin. For example, landing in Brittany (13), the player could take a ship to Bordeaux (44) or from the Basque country (39) proceed to the Isle de Ré (47).

The maps are laid out in approximate geographical groups, starting in the north-east, and then moving south to the Mediterranean littoral, with square 62 being Cannes and its environs. Square 63 is France; it doubles not only as the 'game' square, but also as an over-arching key map, locating the squares in their proper geographical position.

This example is a later printing, retaining the original date of 1659 but also bearing the additional imprint of Antoine de Fer, dated 1671.

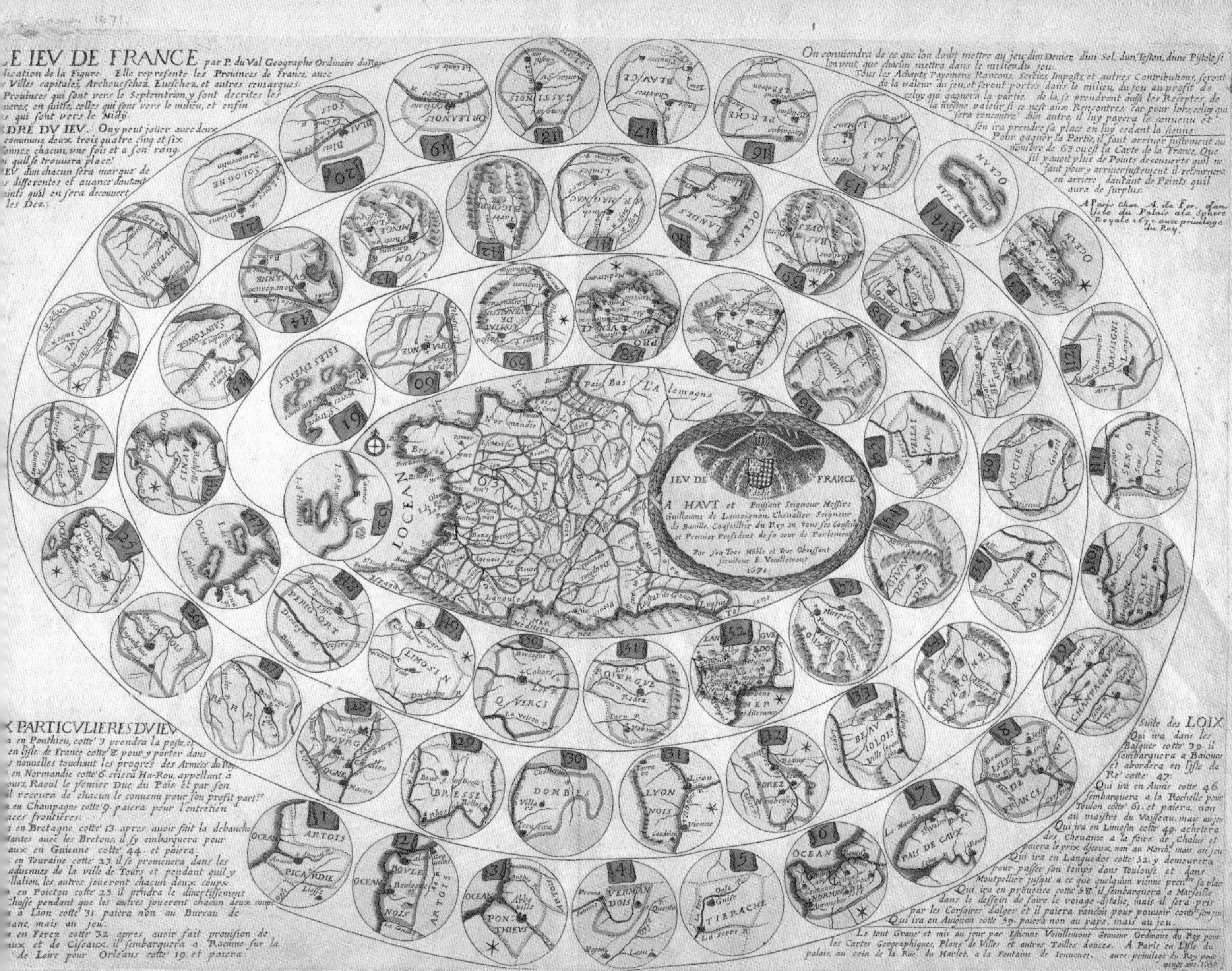
IEV DE FRANCE
OCEAN
MER Mediterra
L'Alemagne
Pais Bas
SOLOGNE
ORLEANOIS
BEAUCE
PERCHE
ARTOIS
PICARDIE
BOVLE NOIS
PONTHIEUS
VERMAN DOIS
TIERACHE
PAIS DE CAUX
DOMBES
BRESSE
LYON NOIS
FOREZ
BEAV IOLOIS
LIMOSIN
QVERCI
ROVERGVE
SAINTONGE
CHAMPAGNE
ISLE DE FRANCE
Suite des LOIX

'Game of the Princes of Europe', 1662

Pierre Duval, Paris

'Les Jeu des Princes de LEurope Par P. Du Val Geographe du Roy'

Copper engraving; border: 404 × 526 mm (15 ⅞ × 20 ¾ in)

THIS MAP-GAME by Pierre Duval (1619–1683) is a further example of a cartographical *Jeu de l'Oie* or Game of the Goose, in this instance another gambling racing game played for money. (Compare other games by Duval on pp. 40, 42, 44, 46.) It features the countries of Europe, with sixty-three roundel maps, and a rectangular map of Europe, which serves as the bank.

To win the game it was necessary to land exactly on square 63, which is France. As Duval put it:

> Pour gagner la partie il faut arriuer justement au nombre de 63, ou est decrite la France l'oeil et la Perle du Monde, et qui est a l'Europe ce que l'Europe est aux autres parties de la Terre. (To win the game it is necessary to land exactly on number 63, which depicts France, the eye and the pearl of the world, which is to Europe what Europe is to the rest of the world.)

The winner not only had the satisfaction of triumphing over his rivals but also scooped the 'bank' which contained all the fines, ransom payments, import duties and so on: 'Celui qui y arrivera se trouvera Maistre des paymens, Rancons, Sorties, Imposts et autres Contributions qui auront esté faits pendant le jeu'.

One element of the game that the players could decide to include was that each player, on landing on a particular region, had to give the name of the place, and its principal cities.

The 'Loix du jeu' ('Rules of the game') are engraved in the upper and lower right-hand corners, giving the specific penalties (or rewards) for landing on particular squares. The most serious penalties are square 5 (Portugal): the player has to remain there, waiting for a ship to take him to the East Indies, until another takes his place, while at square 59 (Crete), the player is imprisoned for plotting against the Turkish rulers and again has to wait for another player to take his place. Anyone landing on 57 ('Petite Tartarie') has to play a ransom and, once free, return to Spain to start the game again. Perhaps the most interesting of the squares is number 6, where the traveller arrives in Holland but is sent on to square 60, to assist in the wedding ceremony of Charles II of England and Catherine of Braganza, the Infanta of Portugal. The nuptials were held on 21 May 1662, the year the map was first published.

This second printing appeared in 1670, with the date revised and the imprint of the new publisher, Alexis-Hubert Jaillot, son-in-law of Nicolas Berey, substituted for that of Berey.

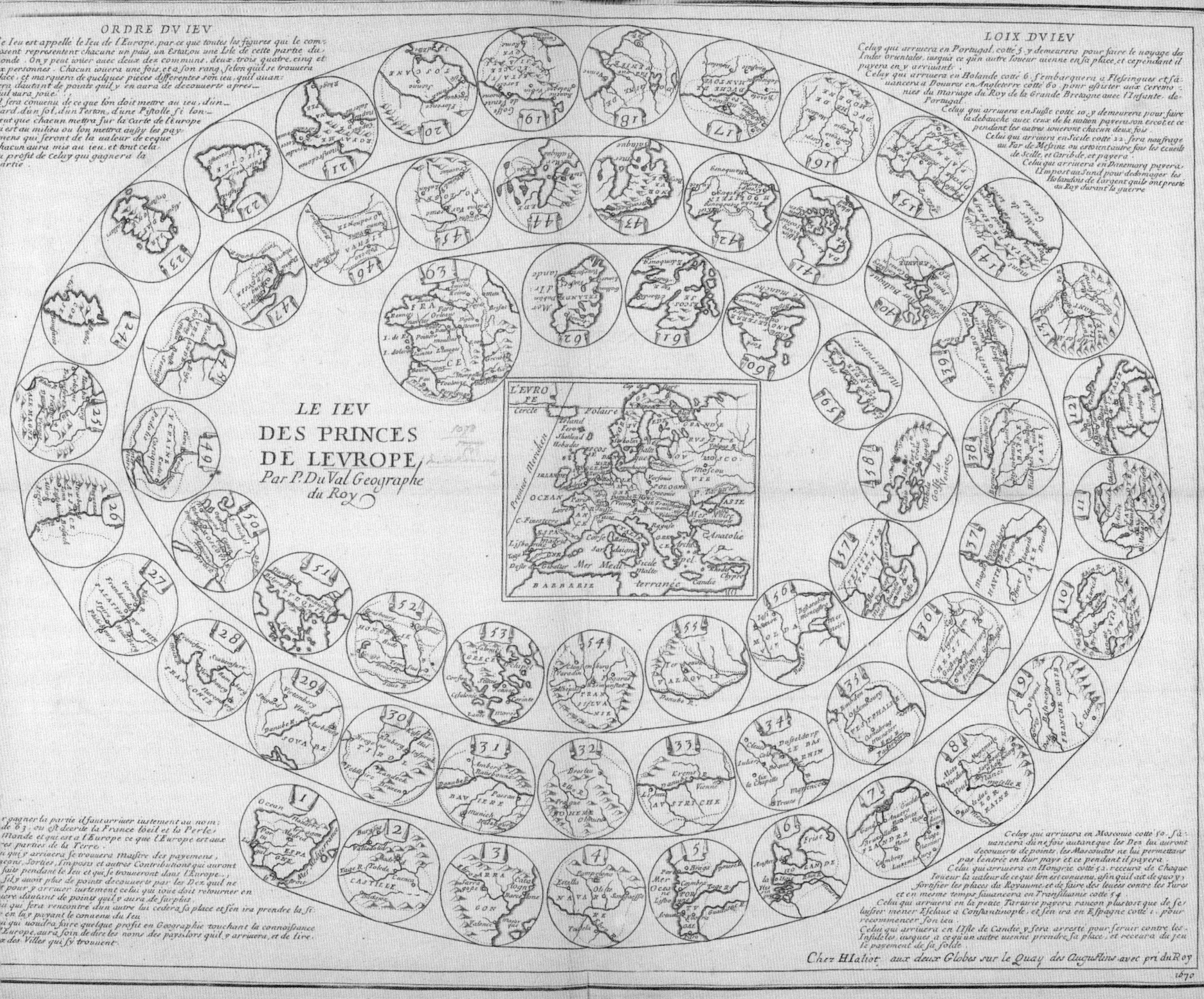
ORDRE DV IEV
LOIX DV IEV
LE IEV
DES PRINCES
DE LEVROPE
Par P. Du Val Geographe
du Roy
Chez H Iaiiot, aux deux Globes sur le Quay des Augustins avec pri du Roy
1670

Geography Reduced to a Game for the Instruction of Young Venetian Nobles [*c.*1665]

Casimir Freschot, Venice

'Geografia ridotta a givoco per instruttione della giovane nobilta Venetiana'

Copper engraving; sheet: 615 × 832 mm (24 1/4 × 32 3/4 in); on two sheets joined

THIS IS QUITE possibly the earliest surviving map board-game published outside France. Casimir Freschot (*c.*1640–1720) was a Frenchman employed by Nicolò Michiel, a Senator in Venice, as a tutor for his three sons, Angelo, Constantino and Hieronimo. The precise dates of his sojourn in Venice are not known, but were almost certainly in the early part of his life.

As part of his duties, Freschot drew this large-format game as an educational tool in geography for the instruction of his charges, who are listed in the dedication found on the map. The game is formed of 153 miniature maps; the final destination, Venice, is depicted at the centre, in a larger format as befits its status. The map also contains the dedication, the 'Regole del giuoco' ('Rules of the game') and maps of each of the four continents (Asia, Europa, Africa and America) engraved along the upper border.

Although Freschot became a prolific author, with some fifty books to his name, this is his only known map and the geography of the individual maps is imprecise. It seems unlikely that Freschot created it from scratch; it is more plausible that it is copied directly or indirectly from earlier map-games, perhaps by Pierre Duval, or from more general maps of the continents, which would explain the rather crude outlines for many of the regions shown. That the map made the transition from manuscript to print is presumably due to the patronage of Nicolò Michiel. As with other contemporary maps of this type, it may well be that clarity and adequacy for a limited educational purpose were deemed more important than scientific geographical accuracy.

The plates were engraved by Antonio-Francesco Lucini (1610–*c.*1665), one of the most famous Italian engravers, best known for his engraving of the sea-charts for Sir Robert Dudley's *Dell'Arcano del Mare*, published in Florence in 1646–1647. The signature, 'Il Cav:re AF: Lucini Fecio', is a form he used at the end of his working life, hence the ascribed dating of this map.

The map is the board for a 'Game of the Goose' type geographical race. The maps, although laid out in a rectangular frame, form a spiral path working in from the outer border and leading to Venice. The miniature maps are engraved in such a way as to be readable to the player seated on that side of the map. This was obviously a sensible step as the board itself is so large, but it was also perhaps easier for the engraver, who could rotate the printing plate as he worked. However, this does have the rather curious effect that, when viewed normally (with the text the right way up) the uppermost maps are upside-down.

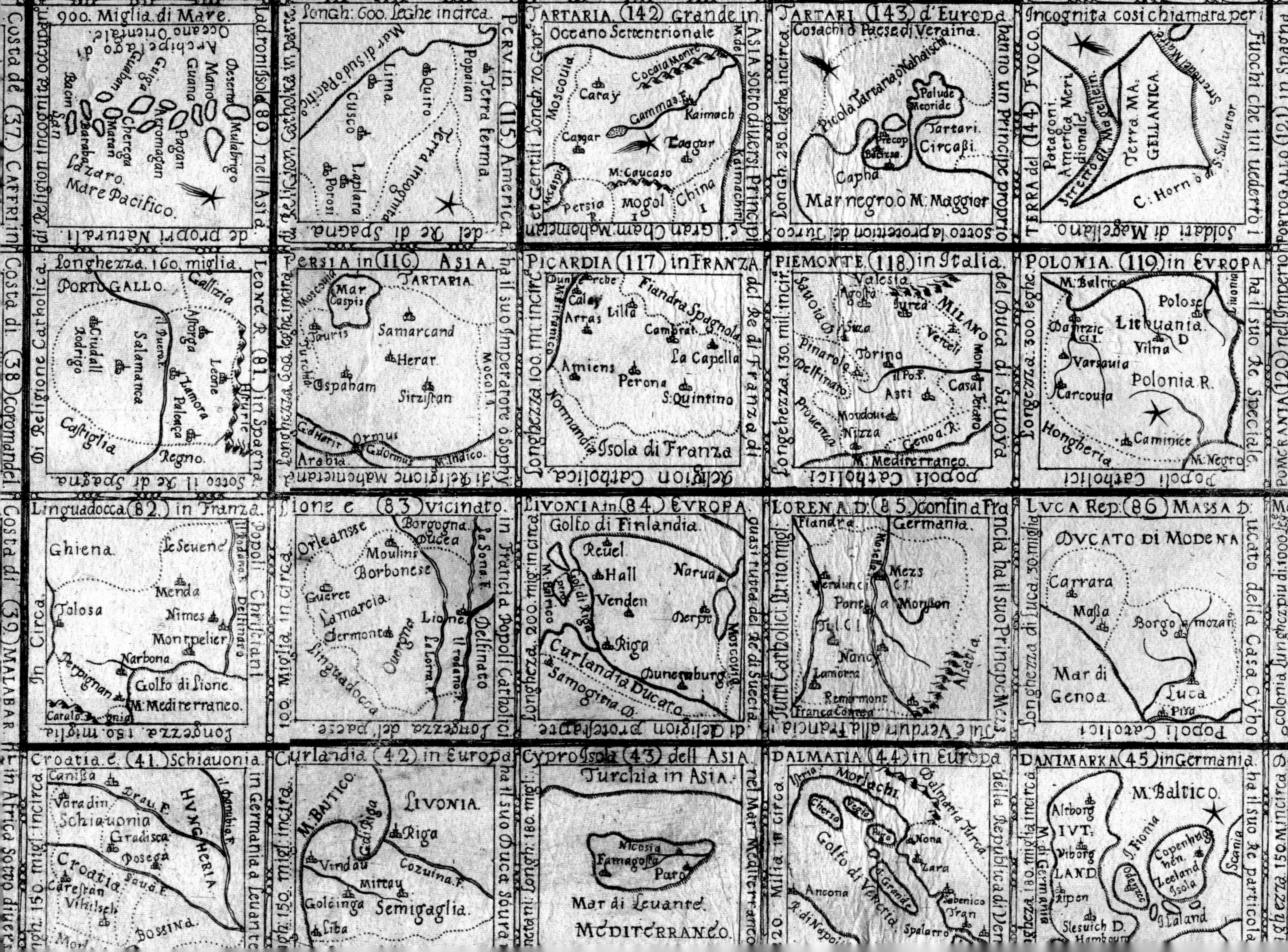

900. Miglia di Mare.
Ladroni Isola (80) nell Asia
de propri Naturali
di Religion incognita occupan
Mare Pacifico.
Longh: 600. leghe incirca.
Peru in (115) America
del Re di Spagna.
di Religion Catholica in parte
Lima
Quito
Cusco
Laplata
Porosi
Terra ferma.
Popaian
Terra incognita
Tartaria (142) grande in
Asia sotto diuersi Principi
e Gran Cham Mahometan
et Gentili Longh. 70. Gior.
Oceano Settentrionale
Moscouia
Catay
Cascar
M. Caucaso
Persia
Mogol
China
Tartari (143) d'Europa
hanno un Principe proprio
Sotto la protettion del Turco.
Longh. 280. leghe incirca.
Cosachi ò Paese di Veraina.
Picola Tartaria o Nahaischi
Palude Meoride
Precop
Tartari.
Circaßi.
Capha
Mar negro ò M. Maggior
Incognita così chiamata per i
Fuochi che iui uederno i
Soldati di Magellano.
Terra del (144) Fuoco.
Patagoni
America Meridionale.
Streto di Magellen.
Terra MA. GELLANICA.
C. Horn
S. Saluator
Portogallo (121) in Spagna
Longhezza. 160. miglia.
Leone R. (81) in Spagna
Sotto il Re di Spagna.
Di Religione Catholica.
Portogallo.
Salamanca
Ciudall Rodrigo
Castiglia
Leone
Regno.
Persia in (116) Asia
ha il suo Imperatore ò Sophy
di Religione Mahometana
Longhezza. 600. leghe incirca
Tartaria
Mar Caspis
Samarcand
Herar
Espaham
Sitzistan
Ormus
Arabia
Picardia (117) in Franza
del Re di Franza di
Religion Catholica.
Longhezza. 100. mi. incirc.
Fiandra Spagnola
Arras
Lilla
Cambrai
La Capella
Amiens
Perona
S. Quintino
Normandia
Isola di Franza
Piemonte (118) in Italia.
del Duca di Sauoya
Popoli Catholici
Longhezza. 130. mil. incir.
Valesia
Milano
Torino
Pinarolo
Asti
Casal
Nizza
Prouenza
M. Mediterraneo.
Polonia (119) in Europa
ha il suo Re Speciale
Popoli Catholici
Longezza. 300. leghe.
M. Baltico
Lithuania
Vilna
Varsauia
Carcouia
Polonia R.
Hongheria
M. Negro
Pomerania (120) nell Imperio
Linguadocca (82) in Franza.
Popoli Christiani
Longezza. 150. miglia
In Circa.
Ghiena
Tolosa
Nimes
Montpelier
Narbona
Golfo di Lione.
M. Mediterraneo.
Lione e (83) Vicinato.
in Francia Popoli Catholici
Longezza del paese
100. Miglia in circa.
Orleansse
Borgogna
Moulins
Borbonese
Lamarcia
Lione.
Delfinato
Linguadocca
Livonia in (84) Europa
quasi tutta del Re di Suecia
di Religion protestante
Longhezza. 200. mig. in circa
Golfo di Finlandia.
Reuel
Hall
Venden
Narua
Riga
Curlandia Ducato
Samogitia
Lorena D. (85) confina Fra
ncia ha il suo Principe
Tutti Catholici Strutto migl.
Fiandra
Germania
Mezs
Nancy
Remirmont
Alsatia
Luca Rep. (86) Massa D.
ucato della Casa Cybo
Popoli Catolici
Longhezza di Luca. 30. miglia
Ducato di Modena
Carrara
Maßa
Mar di Genoa
Luca
Pisa
colonia lunghezza piu di 1000. leg.
Costa de (37) Caffri in
Costa di (38) Coromandel
Costa di (39) Malabar
in Africa sotto diuers
Croatia. e. (41). Schiauonia.
in Germania a Leuante
Longh. 150. migl. incirca.
Schiauonia
Croatia
Hungheria
Bossina
Curlandia (42) in Europa
ha il suo Duca Soura
Longh. 150. migl. incirca.
M. Baltico
Livonia.
Riga
Semigaglia.
Cypro Isola (43) dell Asia
nel Mar Mediterraneo
Longh. 180. migl.
Turchia in Asia.
Nicosia
Famagosta
Mar di Leuante.
Mediterraneo
Dalmatia (44) in Europa
della Republica di Ven
120. Milia in circa.
Morlachi
Golfo di Venetia
Dalmatia Turca
Ancona
Danimarka (45) in Germania.
ha il suo Re particola
Longhezza 180. miglia incirca
M. Baltico.
Iutland
Copenhaghen
Scania
Longhezza. 150. M. incirca

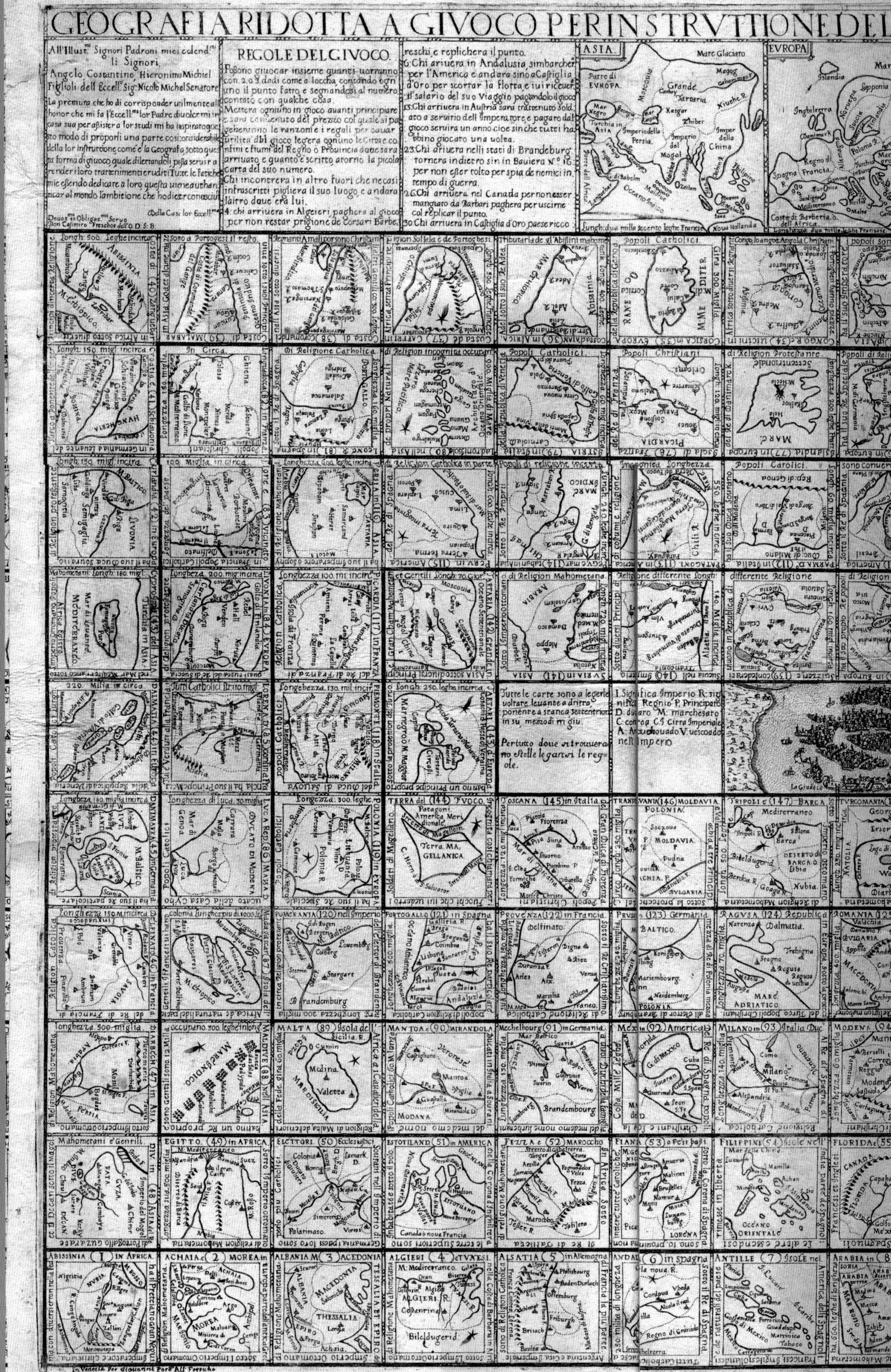
GEOGRAFIA RIDOTTA A GIVOCO PER INSTRVTTIONE DELL
All'Illust.mi Signori Padroni miei colend.mi li Signori.
Angelo Costantino, Hieronimo Michiel
Figlioli dell Eccell.o Sig: Nicolò Michel Senatore
La premura che ho di corrisponder utilmente all'honor che mi fa l'Eccell.mo lor Padre di volermi in casa sua per aßister a lor studi mi ha inspirato questo modo di proporli una parte così considerabile della lor instruttione come è la Geografia sotto questa forma di giuoco, quale dilettandoli poßa servir a render i loro trattenimenti eruditi. Tutte le fatiche mie eßendo dedicate a loro, questa viene a authenticar al mondo l'ambitione che ho di esser conosciuto
Della Casa lor Eccell.ma
Devot.mo et Obligat.mo Servo
Don Casimiro Freschot dell'O. D. S. B.
REGOLE DEL GIVOCO.
Poßono giuocar insieme quanti vorranno con 2. o 3. dadi come a loccha, contando ognuno il punto fatto e segnandosi al numero contato con qualche cosa.
Mettera ogniuno in gioco avanti principiare e sara convenuto del prezzo col quale si pagheranno le ranzoni e i regali per cavar utilità del gioco legera ognuno le Città e confini e fiumi del Regno o Provincia dove sara arrivato e quanto è scritto atorno la picola carta del suo numero.
Chi incontrera in altro fuori che ne casi infrascritti piglierà il suo luogo, e andara l'altro dove era lui.
4. chi arrivera in Algeieri paghera al gioco per non restar prigione de Corsari Barbereschi, e replichera il punto.
6. Chi arivera in Andalusia, s'imbarcherà per l'America e andara sino a Castiglia d'Oro per scortar la Flotta, e ivi riceverà il salario del suo Viaggio pagandolo il gioco
13. Chi arrivera in Austria sara trattenuto soldato a servitio dell Imperatore, e pagato dal gioco servira un anno cioe sin che tutti habbino giocato una volta.
23. Chi arivera nelli stati di Brandeburg tornera indietro sin in Baviera N.o 16 per non eßer tolto per spia de nemici in tempo di guerra.
26. Chi arrivera nel Canada per non esser mangiato da Barbari paghera per uscirne col replicar il punto.
30. Chi arrivera in Castiglia d'Oro paese ricco
ASIA
EVROPA
Tutte le carte sono a legerle voltare levante a dritta, ponente a stanca settentrion in su, mezzodi in giu.
Pertutto dove si trovveranno Stelle leggansi le regole.
I. Significa Imperio R: significa Regnio P. Principato D: ducato M: marchesato C: contea C.I. Città Imperiale A: Arcivescovado V: vescovado nell Imperio
In Venetia Per Giovanni Parè All' Fortuna

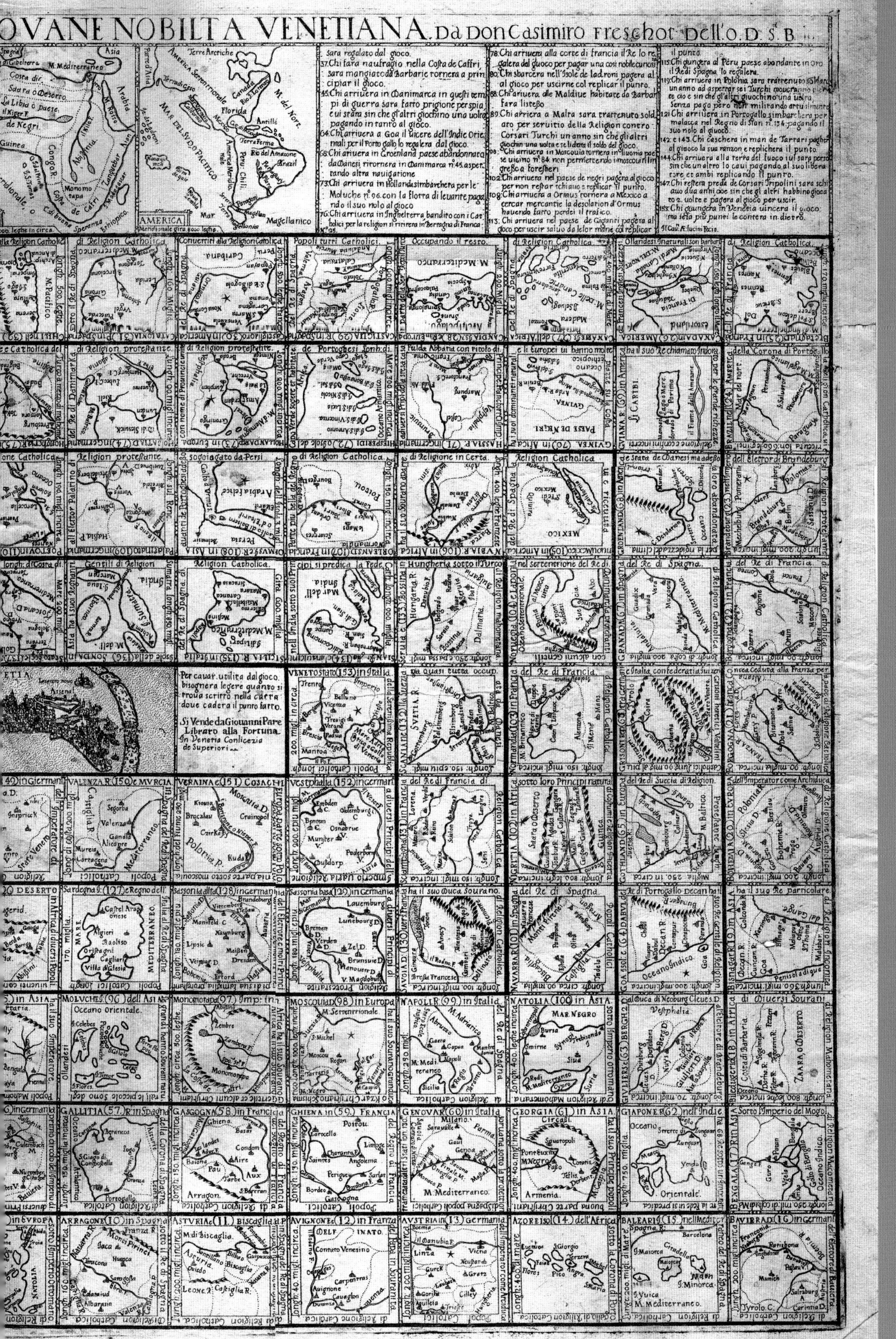
OVANE NOBILTA VENETIANA. Da Don Casimiro Freschot Dell'O. D. S. B.
sara regalato dal gioco.
37. Chi fara naufragio nella Costa de Caffri, sara mangiato da Barbarie tornera a principiar il gioco.
45. Chi arriuera in Danimarca in questi tempi di guerra sara fatto prigione per spia, e ui stara sin che gl'altri giochino una uolta pagando in tanto al gioco.
64. Chi arriuera a Goa il uicere dell'Indie Orientali per il Porto gallo lo regalera dal gioco.
68. Chi ariuera in Groenland paese abandonnata da Danesi, ritornera in Danimarca n° 45 aspettando altra nauigatione
73. Chi arriuera in Hollanda simbarchera per le Moluche n° 96 con la flotta di leuante pagando il suo nolo al gioco
76. Chi arriuera in Inghelterra bandito con i Catholici per la religion si ritirera in Bertagna di Franca N° 25.
78. Chi arriuera alla corte di francia il Re lo regalera dal guoco per pagar una cosi nobile curiosita
80. Chi sbarcera nell'Isole de ladroni pagera al al gioco per uscirne col replicar il punto.
88. Chi ariuera alle Maldiue habitate da Barbari fara listeßo
89. Chi arriera a Malta sara trattenuto soldato per seruitio della Religion contro Corsari Turchi un anno sin che gli altri giochin una uolta e se lidara il soldo del gioco.
98. Chi arriuera in Moscouia tornera in liuonia paese uicino n° 84 non permettendo i moscouiti l'ingreßo a foresteri
102. Chi arriuera nel paese de negri pagera al gioco per non restar schiauo e replicar il punto.
108. Chi arriuera a Ormus tornera a Mexico a cercar mercantie la desolation d'Ormus hauendo fatto perdei il trafico.
113. Chi arriuera nel paese de Giganti pagera al gioco per uscir saluo da le lor mani col replicar il punto.
115. Chi gungera al Peru paese abondante in oro il Re di Spagna lo regalera.
119. Chi arriuera in Polonia sara trattenuto soldato un anno ad aspettar se i Turchi mouerano niente cio e sin che gl'altri giuochino una uolta, senza paga pero non militando attualmente
121 Chi arriuera in Portogallo simbarchera per malacca nel Regno di Siam n° 134 pagando il suo nolo al giuoco.
142 e 143. Chi caschera in man de Tartari paghera al gioco la sua ranzon e replichera il punto.
144. Chi arriuera alla terra del fuoco iui sara perso sin che un altro lo caui pagando al suo liberatore et ambi replicando il punto.
147. Chi restera preda de Corsari Tripolini sara schiauo dua anni cioe sin che gl altri habbino giocato 2. uolte e pagera al gioco per uscir.
153. Chi giungera in Venetia uincera il gioco: ma seta piu punti le contera in dietro.
Il Cau. A. lucini Fecit.
Per cauar utilita dal gioco. bisognera legere quanto si troua scritto nella carta doue cadera il punto fatto.
Si Vende da Giouanni Pare Libraro alla Fortuna In Venetia Con licezia de Superiori
AMERICA
VENETO stato (153) in Italia
VALENZA R. (150) e MURCIA
VCRAINA e (151) COSACHI
Vestphalia (152) in Germania
SICILIA R. (135) in Italia
Moscouia (98) in Europa
Napoli R. (99) in Italia
Natolia (100) in Asia
Gallitia (57) R. in Spagna
Gascogna (58) in Francia
Ghiena in (59) Francia
Genoua R. (60) in Italia
Georgia (61) in Asia
Giapone (62) nell'Indie
Arragone (10) in Spagna
Asturia (11) Biscaglia
Auignone (12) in Franza
Austria in (13) Germania
Azore isol (14) dell'Africa
Baleari (15) nel Mediter.
Bauiera (16) in Germania

Symbolic Rose Map of Bohemia, 1677

Christoph Vetter, Prague

'Bohemiæ Rosa Omnibus sæculis cruenta in qua plura quàm 80. magna prælia commissa sunt, nunc primum hâc formâ excusa'

Copper engraving; border: 388 × 263 mm (15 ¼ × 10 ⅜ in);

THE LATIN title of this symbolic map of Bohemia, now the western part of the Czech Republic, translates as 'The Rose of Bohemia, bloody for all the centuries, where more than eighty great battles were fought, now drawn in this shape for the first time'. The map was drawn by Christoph Vetter (1575–1650) and engraved in 1668 by Wolfgang Kilian, a member of a dynasty of engravers at Augsburg; it was published in Bohuslaus Aloysius Balbinus' *Epitome historica rerum Bohemicarum* – a history of Bohemia which appeared in 1677.

The rose symbol was closely associated with southern Bohemia. Two of the most powerful noble families of the region used the rose as their emblem: the Rozmberks a red rose, and the lords of Hradec a black. The imagery would have been familiar to any Bohemian, particularly as Prague is shown prominently at the centre. Spreading out from the centre, the regions of Bohemia are each shown as a petal, with eighteen administrative districts (as listed in the key): Prague itself, fourteen regions, and Loket, Cheb and Hlad, areas given specific legal and local privileges by their Habsburg rulers.

The map is an overt propaganda piece for the benefits of Habsburg rule. The motto at the top of the sheet, 'Iustitia et Pietate' ('Through Justice and Piety'), is that of Leopold I (1640–1705), Holy Roman Emperor, Archduke of Austria and, among his numerous other titles, also King of Bohemia. The roots of the rose are firmly planted in Vienna, the capital and seat of power of the Habsburg dynasty, which had held sway over Bohemia since 1526.

Prague itself had been for a time the capital of this empire, established as such by the Holy Roman Emperor Rudolph II in 1609, the same year in which he guaranteed the long-held Bohemian principle of religious freedom and toleration in his *Maiestas Rudolphina*. His successors, Matthias I and Ferdinand II, were not so tolerant, and the ill-treatment of Bohemian Protestants spurred an attempt to replace Ferdinand as King of Bohemia with the elected Frederick V – the Protestant Winter King, married to Elizabeth Stuart, daughter of the King of England and Scotland, James I & VI (the Winter Queen or the Queen of Hearts). This Bohemian Revolt (1618–1620) was crushed at the Battle of White Mountain in 1620, but proved the spark for the pan-European Thirty Years' War (1618–1648), one of the most appalling and brutal conflicts in European history, a bloody war both religious and political, leading to widespread carnage, famine and devastation.

Vetter's message in the poem at bottom left reinforces the allusion to the benefits (for Bohemia) of Habsburg rule, not least from the advent of peace to the region. It can be translated as:

> A most graceful Rose grew in the Bohemian woods, and an armed lion stood guard next to her. This Rose grew out of the blood of Mars, not of Venus. Here Rhodes [*sic*]. Here the woods and the country was created. Do not fear anything, lovely Rose! He [ie. the Austrian] comes from the south to the graceful gardens. Let horrible war cease under the reticent Rose!

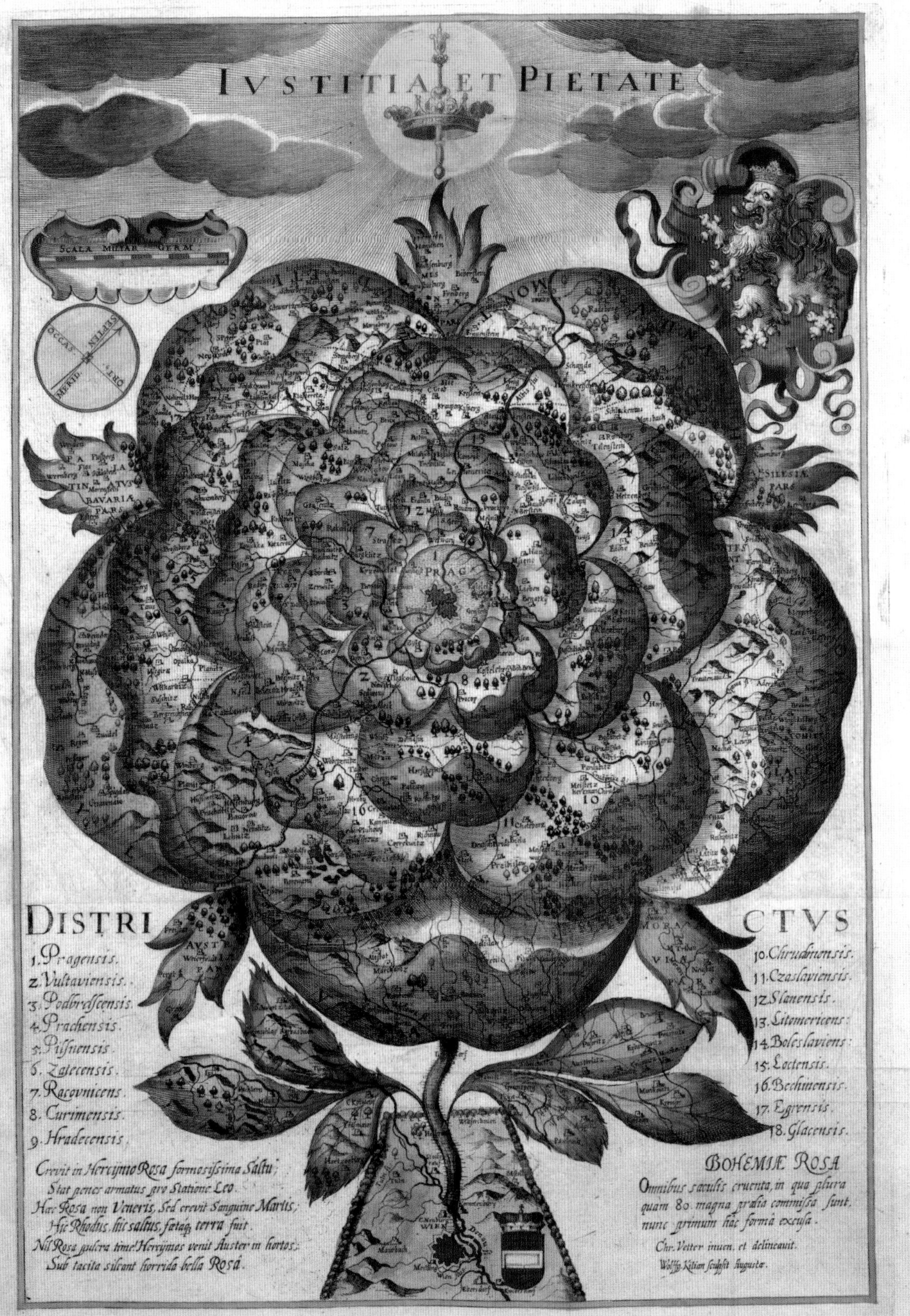
IVSTITIA ET PIETATE
SCALA MILITAR GERM
DISTRI CTVS
1. Pragensis.
2. Vultaviensis.
3. Podbrdscensis.
4. Prachensis.
5. Pilsnensis.
6. Zatecensis.
7. Racovnicens.
8. Curimensis.
9. Hradecensis.
10. Chrudimensis.
11. Czaslaviensis.
12. Slanensis.
13. Litomericens.
14. Boleslaviens.
15. Loctensis.
16. Bechinensis.
17. Egrensis.
18. Glacensis.
Crevit in Hercynio Rosa formosissima Saltu;
Stat pener armatus pro Statione Leo.
Hæc Rosa non Veneris, Sed crevit Sanguine Martis;
Hic Rhodus, hic saltus, fataq terra fuit.
Nil Rosa sudera time! Hercynnes venit Auster in hortos;
Sub tacita sileant horrida bella Rosa.
BOHEMIÆ ROSA
Omnibus saeculis cruenta, in qua plura quam 80 magna prælia commissa sunt, nunc primum hac forma excusa.
Chr. Vetter inven. et delineavit.
PRAG

Allegorical Map of the Baltic Sea as Charon, 1701

Olof Rudbeck Jr, Uppsala

Copper engraving; 200 × 170 mm (7 ⅞ × 6 ¾ in)

OLOF RUDBECK JR (1660–1740) was a famous Swedish scientist, botanist and author. He was also the tutor of Carl Linnaeus, founder of the modern system of botanical nomenclature. His *Olavi Rudbeckii filii Nora Samolad sive Laponia illustrata* is a personal diary and narrative of a journey through Lapland and its environs in the 1690s. Describing his crossing of the River Dalälven, which empties into the Gulf of Bothnia, Rudbeck mused about the parallels between his passage and the classical story of Charon, who ferried the souls of the newly dead across the Styx. He compares the shape of the region with Charon's form as described by the ancients:

> He put me in mind of what has been affirm'd by some Modern Authors of the Shape or whole extent of the Baltick Sea to represent the posture of a Gyant, which, if taken with some grains of allowance, may perhaps challenge the same probability, as the Representations made by some Geographers of other Countries; as of Europe, like a Virgin; of Holland, like a lion, &c. Then take this vast tract of our Baltick Sea ... you will find it to represent in an exact Map, the shape of an old Gyant bending his head forward, with a crooked Back ... (cited in Gillian Hill, *Cartographical Curiosities*, British Library, 1978).

That comparison inspired this most unusual map of the Baltic with Charon superimposed – although, to get the proper effect, Rudbeck had to invert the Baltic, so that the Skagerrak, the strait above the northern tip of Denmark, is at the top of the page, with the Danish Sond islands forming Charon's head, while the Gulf of Bothnia between Sweden and Finland is positioned at the foot of the page, serving as his right leg, and the Gulf of Finland as his left.

Rudbeck, also an etymologist, came to believe that the ancient Greek underworld, Hades, was a real place, located north of the Arctic Circle, although it is not clear how he thought the Mediterranean Greeks knew of the region. He suggested that Charon's name originated from the Swedish word *bårin* or *barin*, a type of funeral barge. He also found other elements of the story of the Greek underworld paralleled in Norse legend; he equated Cerberus, the three-headed dog that guarded Hades, with Garm, a hound that guarded Hel, the Norse underworld.

ÖSTER SJÖN
SKAGER RAK
BOTHNE
FINNE WIKEN
Kamben seu Kimbernäs
Anhout
Lesö
Kullen
Kullby
Kattenborg
Stora Belt
Fyn
Lilla Belt
Bardt
Rugön
Bornholm
Scåne
Öresund
Seland
Dantzig
Königsberg
Brösterort
Carlscrona
Christianopel
Calmar
Bork
Liba seu Lise
Narvik
Gothland
Riga
Öset
Dagerört
Stockholm
Tyrin
Kervea
Akerön
Gefle
Åland
Kokör
Jungfrusund
Åbo
Söderhamn
Hudvikswall
Hermonar
Raiseborg
Reise Sund
Biornborg
Sundswall
Christinestad
Hernösand
Hälsingfors
Borgo
Wasa
Ny Karby
Kuerken
Jacobstad
Umeå
Gamla Karby
EURO PE
Pitheå
Luleå
Brahestad
Uleå
Kimi
Torneå

Untitled Map of the World, 1718

Jean Crepy, Paris

'Nouvelle Methode de Geographie ou Voiage du Monde par les Villes les plus Considerables de la Terre ou par un Jeu On apprend la situation des païs & de Villes, leur dependance & la Religion des peuples avec une Mappemonde ou les routes de ce Voyage sont marquées.'

Copper engraving; world map 167 × 340 (6 ⅝ × 13 ⅜ in); sheet 432 × 560 mm (17 × 22 in)

THE CREPY family were leading map publishers, active in Paris in the eighteenth century. Members of the family tended to sign themselves simply with the imprint 'Chez Crepy' so it can be difficult to assign particular maps to the various individuals (creating a cartobibliographic nightmare for the modern researcher). This map is variously credited to Jean Crepy (c.1650–1739), founder of the business, or to Jean-Baptiste Crepy Sr, presumed to be his son, but it is possible it could be by Louis Crepy, who may or may not be the same person as Etienne-Louis Crepy.

Regardless of the difficulty of attribution, this is an interesting version of the race game, comprising a spiral track with seventy-eight steps in which the travellers journey from Brest on a voyage round the world, the first one to arrive safely at Paris being the winner. For the use of the players, the central part of the sheet is a double-hemisphere map of the world, marking the track that they must follow; for the more crowded parts of Europe there are three small inset maps: 'Carte de Grece', 'Carte d'Italie' and 'Carte de France, d'Allemagne, Païs-Bas, Loraine &c.'

The seventy-eight stops are generally cities within Europe, as noted in the title: 'A new method of geography or voyage round the world through the most important cities of the earth ...'. However, square 38 is the 'I. de Candie' (Crete), and some other regions are shown even though they are without principal cities, such as 'Terres Magellaniques' (Tierra del Fuego).

The game begins in the bottom left-hand corner at Brest, with the instruction 'en Commancent met 1 au Jeu pour l'Embarquement' – each player had to pay one coin, of the value agreed by the players, into the pot to start the game, and then proceed round the board, following the engraved instructions ('Regles de Jeu') on either side of the hemispheres, and obeying the instructions appended to each square. Often these instructions would order the player to place a coin on a particular square; anyone landing on that square would keep the money, but the winner would take the pot and the uncollected money remaining on the squares at the end of the game. The game also required a 'guide', a judge in case of dispute but also a question master, because one of the rules was that any player could avoid laying down money if he was able to answer a geographical question posed by the 'guide'. In this way the game could justify the claim in the title to be educational.

Square 38, Crete, was one of the penalty stops, as described in rule 6: 'Celuy qui tombe à 38. y demeure à cause du Labyrinthe de Crete jusqu'à cequ'un autre y vienne qui reste à sa place, le premier va à 24. sans rien payer' ('The one who lands on 38 remains there because [he is lost in] the labyrinth of Crete until another comes there and takes his place; the former moves to 24, without penalty').

The final square, Paris, has the banner caption 'Omnes quidem currunt sed unus accipit bravium', a biblical quotation from Corinthians – 'they which run in a race run all, but one receiveth the prize'.

c dez, chacun sa marque, et
t sur les cases.
t premiere case pour revenir à Paris
e qui est au Jeu et sur les cases, dont
point, et seulement le quart, s'il Jouë.
s de trois, lequel est obligé de
s contenus dans la case d'où il
lieux sont situez, s'il y manque, il
orés luy sous même peine.
où il y a un ⚓ marque de port de
mais si c'est ⚓ on ne paye rien, en
dans d'autres païs.

Nouvelle Methode de Geographie ou VOIAGE du MONDE

par Les Villes les plus Considerables de la Terre ou par un

JEU

On apprend la situation des païs & des Villes, leur dependance & la Religion des peuples, avec une Mappemonde où les routes de ce Voyage sont marquées.

des cases les enleve, mais en
6. Celuy qui tombe à 37. y dem
Jusqu'à cequ'un autre y vien
va à 24. sans rien payer.
7. Où l'on est rencontré, on
8. Quand le lieu où l'on reto
de jettons, on se met a l
9. Celuy qui amene de
et quand ce sont deu
10. Le reste suivant ce
11. Celui qui sera obligé de
indiquant sur la mappemond
sans Chifres

Carte de France, d'Allemagne, Païs-Bas, Loraine &c.

NON SED UNUS ACQUIT BRAVIUM

78

PARIS Ville Capit. du FRANCE.

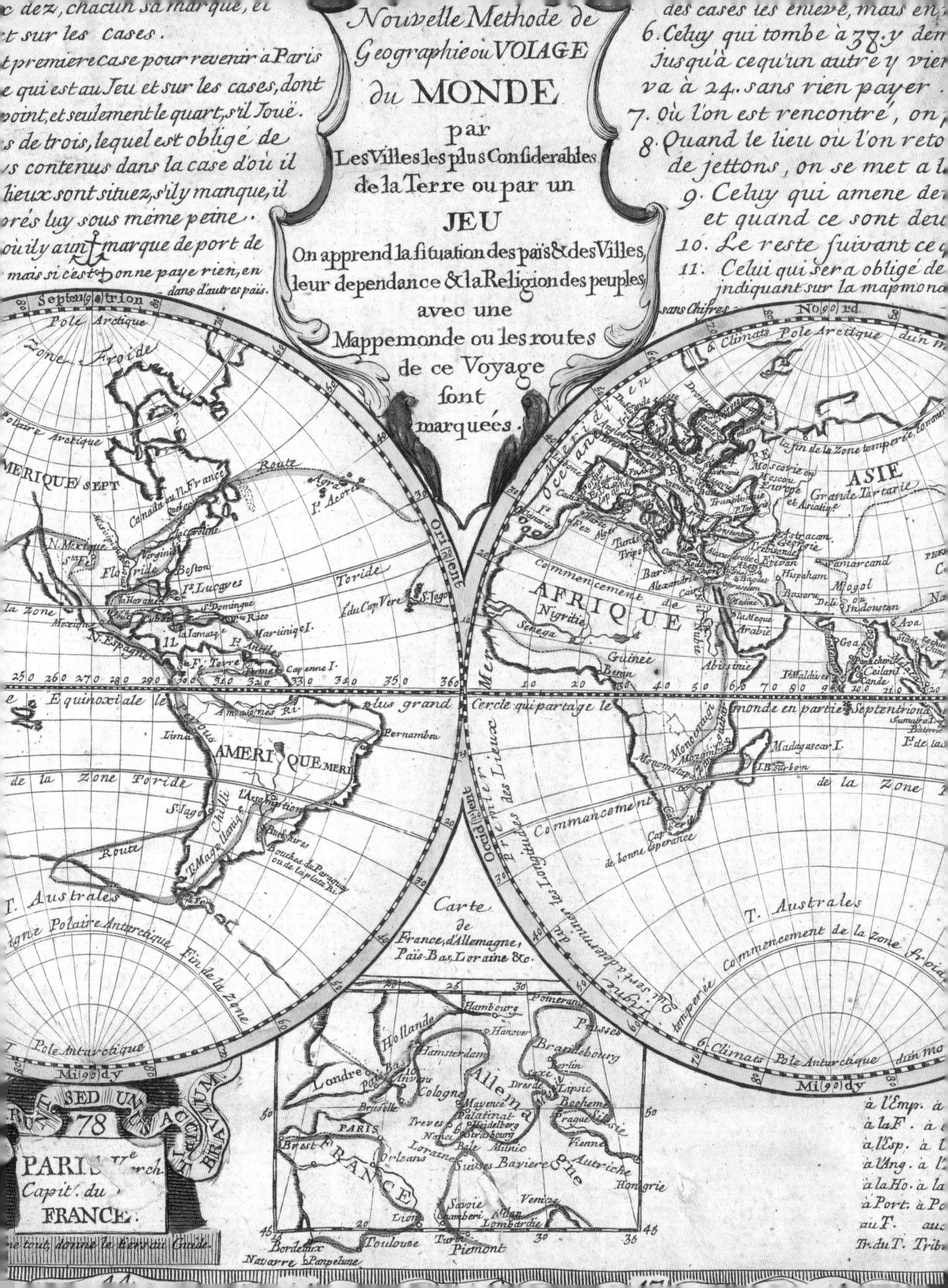

ne tout, donne le tiers au Guide.

à l'Emp. à
à la F. à
à l'Esp. à
à l'Ang. à
à la Ho. à la
à Port. à Po
au T. au
Tr. du T. Trib

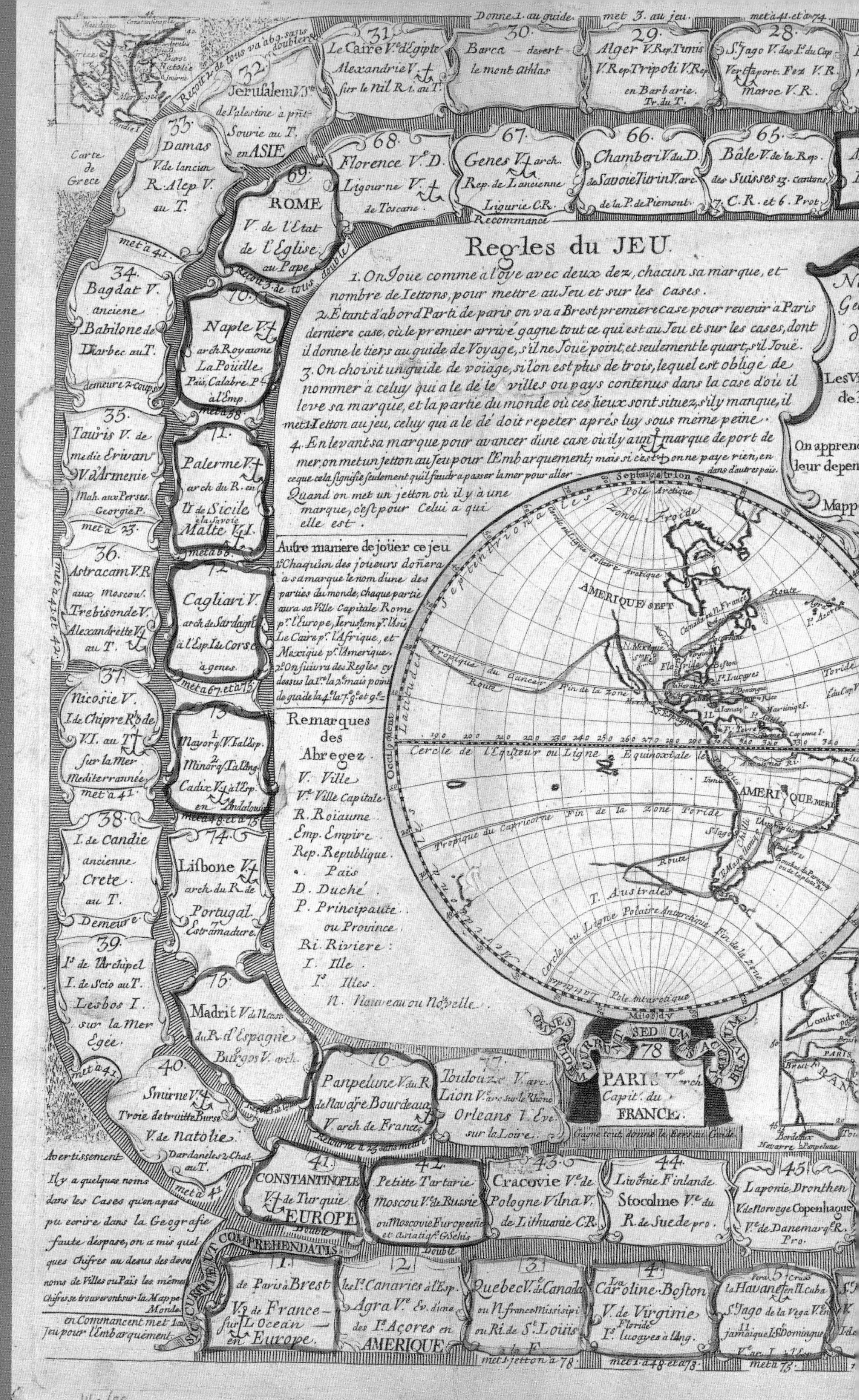
Regles du JEU
1. On Joue comme a l'oye avec deux dez, chacun sa marque, et nombre de Jettons, pour mettre au Jeu et sur les cases.
2. Etant d'abord Parti de paris on va a Brest premiere case pour revenir a Paris derniere case, où le premier arrivé gagne tout ce qui est au Jeu et sur les cases, dont il donne le tiers au guide de Voyage, s'il ne Jouë point, et seulement le quart, s'il Jouë.
3. On choisit un guide de voiage, si l'on est plus de trois, lequel est obligé de nommer a celuy qui a le dé le villes ou pays contenus dans la case d'où il leve sa marque, et la partie du monde où ces lieux sont situez, s'il y manque, il met 1 Jetton au jeu, celuy qui a le dé doit repeter aprés luy sous même peine.
4. En levant sa marque pour avancer d'une case où il y a une marque de port de mer, on met un jetton au Jeu pour l'Embarquement; mais si c'est on ne paye rien, en ce que cela signifie seulement qu'il faudra passer la mer pour aller dans d'autres païs.
Quand on met un jetton où il y a une marque, c'est pour Celui a qui elle est.
Autre maniere de joüer ce jeu
1° Chaqu'un des joueurs doñera a sa marque le nom d'une des parties du monde, chaque partie aura sa Ville Capitale Rome pr. l'Europe, Jerusalem pr. l'Asie, Le Caire pr. l'Afrique, et Mexique pr. l'Amerique.
2° On suivra des Regles cy dessus la 1re. la 2e. mais point de guide la 4e. la 7e. 8e. et 9e.
Remarques des Abregez.
V. Ville
Ve. Ville Capitale.
R. Roïaume.
Emp. Empire.
Rep. Republique.
. Pais
D. Duché.
P. Principauté ou Province.
Ri. Riviere:
I. Ille.
Is. Illes.
N. Nouveau ou Nouvelle.
Avertissement
Il y a quelques noms dans les Cases qu'on a pas pu ecrire dans la Geografie faute d'espace, on a mis quelques Chifres au desus des deux noms de Villes ou Païs les mêmes Chifres se trouveront sur la Mappe-Monde.
en Commancent met 1 au Jeu pour l'Embarquement.
1. de Paris à Brest Ve. de France sur L'Ocean en Europe.
2. les Is. Canaries à l'Esp. Agra Ve. Ev. dione des Is. Açores en AMERIQUE
3. Quebec Ve. de Canada ou N. france Misisipi ou Ri. de St. Loüis
4. La Caroline Boston V. de Virginie Is. Floride Lucayes a l'Ang.
5. la Havane en l'I. cuba St. Jago de la Vega Ve. en Jamaique I. St. Domingue
28. St. Jago V. des Is. du Cap Verd. port. Fez V. R. Maroc V. R.
29. Alger V. Rep. Tunis V. Rep. Tripoli V. Rep. en Barbarie. Tr. du T.
30. Barca desert le mont athlas
31. Le Caire Ve. d'Egipte Alexandrie V. sur le Nil R. au T.
32. Jerusalem Ve. de Palestine a pnt Sourie au T. en ASIE
33. Damas V. de lancien R. Alep V. au T.
34. Bagdat V. anciene Babilone de Diarbec au T.
35. Tauris V. de medie Erivan V. d'Armenie Mah. aux Perses. Georgie P.
36. Astracam V. R. aux Moscou. Trebisonde V. Alexandrette V. au T.
37. Nicosie V. I. de Chipre Rhode V. I. au T. sur la Mer Mediterranée
38. I. de Candie ancienne Crete au T.
39. Is. de l'Archipel I. de Scio au T. Lesbos I. sur la Mer Egée.
40. Smirne V. Troie detruitte Burse V. de natolie
41. CONSTANTINOPLE Ve. de Turquie en EUROPE
42. Petitte Tartarie Moscou Ve. de Russie ou Moscovie Europeene et asiatique G. Schis
43. Cracovie Ve. de Pologne Vilna V. de Lithuanie C.R.
44. Livonie Finlande Stocolme Ve. du R. de Suede pro.
45. Laponie Dronthen V. de Norvege Copenhague Ve. de Danemarc R. Pro.
65. Bâle V. de la Rep. des Suisses 13. cantons 7. C.R. et 6. Prot.
66. Chamberi V. du D. de Savoie Turin V. arc. de la P. de Piemont.
67. Genes V. arch. Rep. de L'ancienne Ligurie C.R.
68. Florence Ve. D. Ligourne V. de Toscane
69. ROME V. de l'Etat de l'Eglise au Pape
70. Naple V. arch. Royaume La Poüille Païs, Calabre P. à l'Emp.
71. Palerme V. arch. du R. en I. de Sicile à la Savoie Malte V. I.
72. Cagliari V. arch. de Sardaigne à l'Esp. I. de Corse à genes.
73. Mayorq. V. I. a l'Esp. Minorq. I. a l'Ang. Cadix V. à l'Esp. en Andalousie
74. Lisbone V. arch. du R. de Portugal. Estramadure.
75. Madrit V. de N. caste du R. d'Espagne Burgos V. arch.
76. Panpelune V. du R. de Navarre Bourdeaux V. arch. de France.
77. Toulouze V. arc. Lion V. arc. sur le Rhone Orleans V. Eve. sur la Loire.
78. PARIS Ve. arch. Capit. du FRANCE.
Gagne tout, donne le tiers au Guide
OMNES VIDEMUS CURRUNT SED UNUS ACCIPIT BRAVIUM
SIC CURRITE UT COMPREHENDATIS
Carte de Grece
AMERIQUE SEPT.
AMERIQUE MERI.
Cercle de l'Equateur ou Ligne Equinoxiale
Tropique du Cancer
Tropique du Capricorne
T. Australes
Pole Arctique
Pole antarctique
Zone Froide
Route
140/88

26. Monomotapa V.R. Monomugi R. C. de bone esperance en Cafrerie a la Hol.
25. Nubie, Abissinie & c. Mozambique V. a Port. Zanguebar Madagascar I. a la F. AFRIQUE
24. Medine V. la Mecque V. d'Arabie Aden V. de l'Arabie Heureuse
23. Hispahan V.e du Roiaume Basora V. de Perse Mah. Péche des Perles
22. Goa V. arc. côte de Malabar Id. Mah. et C.R.
21. Maldives R.e 12000 I.s Mah.
63. Venize V. patriarchat Republique C.R. d'Italie
62. Duraz V. d'Albanie Spalatre V.e arch. de Dalmatie
61. Setine Ancienne Athenes V. de Grece Ministra V. de Morée
60. Salonichi V. de Macedoine I. de Negrepont
59. Bude V. du R. d'Hongrie Transilvanie Belgrade V.e a l'Emp.
20. Pondicheri V.e Côte de Coromandel a la F. Cande V.e en Ceilan I.
Carte d'Italie
Regles du JEU
5. Celuy qui en avançant sa marque rencontre des jettons sur des cases les enleve, mais en retournant, non, il passe pardessus.
6. Celuy qui tombe à 37. y demeure à cause du Labyrinthe de crete Jusqu'à cequ'un autre y vienne qui reste à sa place, le premier va à 24. sans rien payer.
7. Où l'on est rencontré, on paye un jetton et non autre chose.
8. Quand le lieu où l'on retourne est occupé d'une marque ou de jettons, on se met a la 1.er case vacante d'ensuite.
9. Celuy qui amene deux dez de même nombre, double et quand ce sont deux six, double deux fois.
10. Le reste suivant ce qui est marqué autour des cases.
11. Celui qui sera obligé de payer au jeu ou ailleurs en sera quitte en indiquant sur la mapmonde les païs Villes Isles &c. sans noms Ecrits et que le guide lui proposera, savoir autant de lieux differens qu'on devra donner ou mettre de jettons.
3.o Celui qui posera sa Marque sur la Capital de son nom gagnera la partie.
4.o Toutes les fois quon posera sa marque sur les Cases de la partie d'un autre on lui doñera 1 Jetton mais si c'est sur la Capitale on le fera sortir du jeu.
On n'aura point d'egard a ce qui est Ecrit autour des Cases.
Remarques
Ev. Eveché.
arc. Archevêché.
C.R. Catolique Romain.
Gr. Sch. Grec Schismatique.
pr. Protestans.
mah. Mahometans.
Id. Idolâtres.
Va a. Signifie mettre sa marque a la Case Indiquée.
Met a. Signifie mettre un jetton sur chaqu'une des Cases indiquée.
Le Guide se doit mettre au milieu pour pouvoir tout lire Aisément.
à l'Emp. à l'Empereur.
à la F. à la France.
à l'Esp. à l'Espagne.
à l'Ang. à l'Angleterre.
à la Ho. à la Hollande.
à Port. à Portugal.
au T. aux Turcs.
Tr. du T. Tributaire du Turc.
ASIE
AFRIQUE
T. Australes
58. VIENNE V. Eveché d'Autriche C.R. a l'Emp.
19. Deli V. de l'Emp. du Mogol ou Indoustan Id. et Mah. Mines de Diamans.
57. Silesie P. Prague V. arc. du R. de Boheme C.R. a l'Emp.
18. Samarcand V. de la grande Tartarie Païs divisé par ordes. Idol.
56. Pomeranie. Prusses Berlin V.e de l'Elect. de Brandebourg.
17. PEKIN V.e de l'Emp. de la Chine Namquin V. Idol.
55. Dresde V. de l'Elect. de Saxe Lipsic V.e de misnie.
16. Malaca V. a la Ho. Siam V.R. Ava V.R. Cochinchine Roïaume.
54. Heidelberg V. de l'Elect. Palatin pr. Munich V.e de Baviere C.R.
15. Borneo I. Batavie V.e de Java a la Hol. I. de Sumatra apelée les I.s de la Sonde
53. Nanci V. du Du. de Loraine Strasbourg V.e Ev. D'Alsace a la F.
14. Celebes I. des Moluques Carpentarie N. Guinée a la Hol. N. Hollande
52. Mayence V. Arc. Electorat Treves V. Arc. Elect.
13. Manille V.e arc. des I.s Philipines S.t Jean des I.s Mariannes a l'Esp.
51. Hanover V. Electorat pr. Cologne V.e arc. Elect. d'Allemagne
12. Yedo V. Meaco V.e des I.s du Japon en ASIE
47. Edimbourg V.e du R. D'écosse Dublin V. du R. d'Irlande
48. Principauté de Galle Londre V.e Ev. du R. d'angleterre
49. Anvers V. Ev. Bruselle V.e des Païs-Bas C.R. a l'Emp.
50. Amsterdam V.e de la Rep. de Hollande pr. 10. Zelande P.
7. S.t Fe V. arc. de Terre ferme. amazones Ri. Pernambouc V.e de Bresil a Port.
8. Buen-aires V.e de paraguay ou la plata Ri. assomption V. a l'Esp.
9. Terres magelaniq. T. de Feu T. Australes
10. S.t Jago V. Ev. de Chili Lima V.e arc. de Perou a l'Esp.
11. Mexique V.R. de la N. Espagne S.t Fe V. arc. du N. Mexique a l'Esp.
A Paris Chez Crepy rue S.t Jacques au Lion d'Argent. 1718.

Map of France Game, 1718

Jean Crepy, Paris

'Nouvell Methode de Geographie ou Voiage Curieux par les Villes le plus Considerables et les principaux Pais...'

Copper engraving; at widest: 438 × 570mm (17 ¼ × 22 ½ in)

IN 1718 Jean Crepy (*c*.1650–1739) published a geographical map-game focused on France as a companion to the one shown on p. 60. Both are race games of the 'Game of the Goose' type, set on a spiral course; here the course goes around a map of France, highlighting its regions and cities. There are 109 squares, commencing with the province of 'L'Orleannois' and proceeding to the winning square, 109, which is Paris, the square surmounted by a bust of the French king: 'Louis XV. Roi de France et Navarre'.

The main set of rules is engraved on the left of the map; a second and shorter set of rules, for an alternative version of the game in which each player takes the name of a particular province and wins when his or her token reaches its capital, is to be found to the right of the title, along with the key to the symbols, found both in the map and the individual squares of the game. The map contains an engraved track that replicates the sequence of squares around it.

In the description for square 106, Compiègne, the tiny settlement of Crépy (Crépy-en-Valois) is surprisingly listed, presumably because it was the ancestral home of the mapmaker's family.

The map contains a brief text extolling the virtues of France and describing briefly its government and geographical setting:

> La France, le meilleur et l'un des plus Abondans Pais de l'Europe, est un Roiaume Hereditaire pour le Roi en qui Reside toutte l'Auctorité du Gouvernemente; elle est située entre le 13. et le 26.e degré de Longitude et les 42. et 52 degré 15.m de Latitud. septentrionalle. elle à l'Orient l'Italie, la Savoie, la Suisse et l'Allemagne; a l'Occident l'Ocean et l'Espagne; au Septentrion l'Angleterre et les Pais-Bas; au Midy la Mer Mediterranée. &c.
>
> (France, the best and one of the wealthiest countries of Europe, is a hereditary kingdom with a king in whom resides all authority of government; it is situated between 13 and 26 degrees Longitude, and between 42 and 52 degrees Latitude. She has to the East Italy, Savoy, Switzerland and Germany; on the West the Atlantic and Spain, to the North England and the Low Countries and in the South the Mediterranean Sea.)

As with most games of this nature, this one has a 'death' square: any player who lands on number 38 (Limousin – it being described as 'bad country') has to 'sort le jeu' ('leave the game'), while those landing on the penultimate square, number 108, must 'Recomene', presumably an error for 'Recommence', that is, start again. Otherwise the principal penalty squares are those with an anchor, each representing a *port de mer* (sea port); the player landing on one of these had to pay one token for each anchor, with two being the forfeit for visiting La Rochelle.

Each square has a potted description of the town or region. The game has a strong educational undercurrent; according to rule 3, the 'guide' (umpire) had to name the town and region where each player landed, or pay a fine of one token; the player concerned had to repeat this accurately or pay the same penalty.

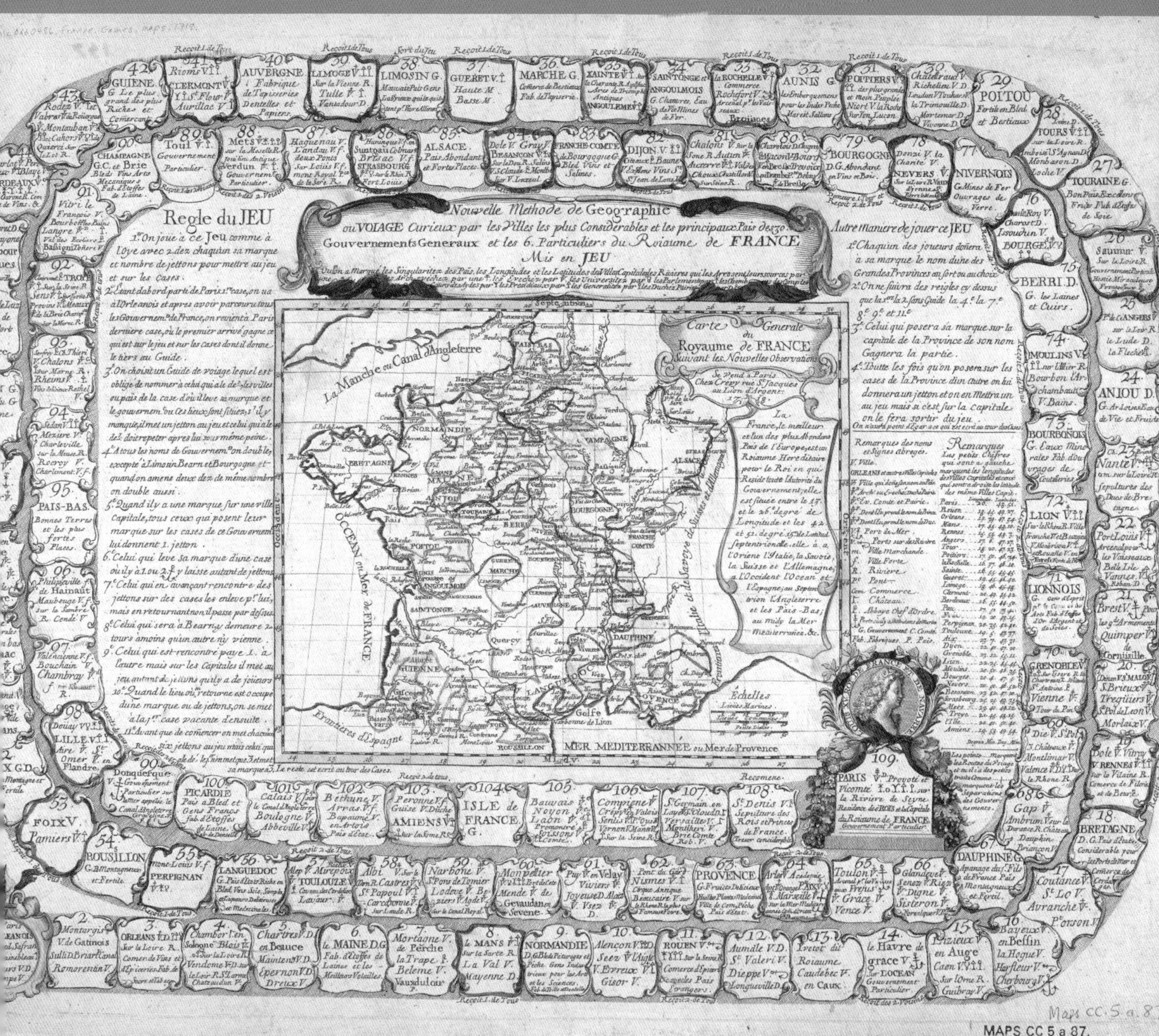
Nouvelle Methode de Geographie
ou VOIAGE Curieux par les Villes les plus Considerables et les principaux Pais des 30 Gouvernements Generaux et les 6. Particuliers du Roiaume de FRANCE
Mis en JEU
Regle du JEU
Autre Maniere de jouer ce JEU
Carte Generale du Royaume de FRANCE Suivant les Nouvelles Observations
La Manche ou Canal d'Angleterre
OCEAN ou Mer de FRANCE
MER MEDITERRANNÉE ou Mer de Provence
Frontieres d'Espagne
Remarques
PARIS
ISLE de FRANCE
MAPS CC 5 a 87.

'An Accurate Map of Utopia, Which is the Newly Discovered World of Fools', 1720

Anonymous [Nuremberg]

'Accurata Utopiae Tabula Das ist Der Neu entdeckten Schalck Welt'

Copper engraving; border: 482 × 561 mm (19 × 22 ⅛ in)

ALTHOUGH THE map is entitled 'Utopia', in fact it depicts a dystopia, a society in which the moral duty of the individual to act for the benefit of society as a whole has been corrupted by greed. This map, like that on p. 68, refers to the events of the Mississippi and South Sea Bubbles, in which a pan-European wave of stock-market greed brought the countries of Europe to their knees and led to widespread financial losses. The title and labels in the map point to a German maker, perhaps Johann Baptist Homann, but the map is frequently found in Dutch collections, which may suggest a Dutch publisher with German connections, such as Pieter Schenk.

The title is framed by the front of a beer barrel; astride the barrel is a gambler, with money pouring out of his hat – 'Der uber fluss' ('the overflow'). A figure behind him has grabbed too greedily and lost his balance as he falls to his fate. To the left is an amorous couple, she half-dressed, but while he is intent on amour, she too seems to be reaching for the cascading coins. At his feet sprawl two figures; one grasps the barrel as he reaches over to get his cap under the flow of 'Bier flu' (best translated, in the circumstances, as 'river of beer'), ignoring 'Meth. flu' and 'Vein flu'; his drinking companion, however, has imbibed to excess, and vomits into 'Sau flu'.

Schlarraffenland, or the Land of Cockaigne as it was known in other cultures, was a realm of vice, made popular in contemporary fiction. The manner in which Schlarraffenland is geared for a life of pleasure and leisure is exemplified by the image of the man presenting the pie containing a heron-like bird, cooked feathers, beak and all: it was believed that the animals of the region were ready-cooked so that the hungry had to divert little or no time from their life of indolence to preparing their food. This map, also known in versions by Johann Baptist Homann and Georg Matthaüs Seutter, is an elaborate recasting of the Schlarraffenland metaphor in cartographic form. The country is divided, with all the appearance of a genuine map, into nineteen regions, each devoted to a particular vice: 'Mammonia' (Mammon, or financial greed), 'Superbia' (arrogance), 'Stomachi' (gluttony), 'Bibonia' (drunkenness), 'Lusoria' (sexual dissipation), 'Bacchanalia' (uninhibited partying) and so on.

On the northern boundary lies 'Terra Sancta Incognita' ('the Undiscovered Holy Land') hemmed in by mountains, so that only those willing to make the special effort to get there could cross over. To the south lies 'Tartaria', in which is 'Da Hollische Reiche' ('the kingdom of Hell'), into which all the inhabitants of Schlarraffenland will one day descend. If Utopia was conceived as a paradise for the educated upper classes, Schlarraffenland was conceived as a paradise for the uneducated lower classes, and many of the allusions and puns in the map range from the merely bawdy to the downright crude.

Accurata
UTOPIÆ TABULA
Das ist
Der Neu entdeckten
SCHALCK WELT,
oder des so offt benanten und doch nie erkanten
SCHLARRAFFENLANDES
Neu erfundene lächerliche Land tabell
Der FROMMEN des Zeitlichen AUFF und UNTERGANGS
auch ewigen VERDERBENS Regionen
samt einer erklerung
TERRA SANCTA INCOGNITA
UNBEKANTE LAN DER DER FROMMEN
PRODIGALIA REGN
MAMMONIA
SCHLARAFFEN
DAS TRAURIGE MEER
DAS TOBACK LUDER MEER
DAS SUSSE GOLFO DI FEIN NÆRRISCHE VENEREA MEER
DAS VERSOFFENE MEER
EBRIUM
UND VOLLE MEER
STOLIDUM
DAS NARRISCHE MEER
LUSORIA REGN
DOLE
MARE
INFERUM

‘The Very Famous Island of Mad-head, Lying in the Sea of Shares’, 1720

[Henri-Abraham Chatelain?], Amsterdam

‘Afbeeldinghe van’t zeer vermaerde Eiland Geks-Kop. geligen in de Acti-ze ontdekt door Mons.r Lau-rens, werende bewoond door een verzameling van alderhande Volkeren. die men dezen generalen Naam (Actionisten) geest’.

Copperplate engraving; image 160 × 225 mm (6 ¼ × 8 ⅞ in)

THIS FAMOUS cartographical political satire was published in the aftermath of the collapse of the French Compagnie de la Louisiane ou d’Occident, and of similar English and Dutch overseas companies. The full title translates as ‘Representation of the very famous island of Mad-head, lying in the sea of shares, discovered by Mr. Law-rens, and inhabited by a collection of all kinds of people, to whom are given the general name shareholders’.

The Compagnie was established in August 1717 by John Law, a Scottish banker and financier, at a time of fiscal crisis in France caused by her expensive wars. Law claimed that by exploiting the French possessions in America, he could pay off the French national debt. Accordingly the Compagnie was granted control of Louisiana. Laws’ grand plan to exploit the resources of the region – the ‘Mississippi Scheme’ – captured the popular imagination, as did the wild claims of profits to be made, and people rushed to invest. The shares were originally priced at 500 *livres*, but rose to 18,000 *livres* (as a comparison a high-ranking official might then have earned 3,600 *livres* per annum). Such levels were unsustainable; wiser investors, and hordes of speculators, sold their shares, causing a run on the share price. The company’s capital was drained and the Compagnie went bankrupt. Many investors all over Europe were ruined. After the disastrous failure of the Mississippi Scheme, parallel schemes also failed, notably the English South Sea Company in the ‘South Sea Bubble’, but also a number of smaller Dutch companies. A new term in finance emerged from the devastation: ‘bubble’ came to describe any of the periodic and unsustainable rises in the price of shares in which promises of instant wealth ultimately prove illusory.

This engraving (from a volume of satires relating to the ‘Mississippi Bubble’, sometimes attributed to Henri-Abraham Chatelain [1684–1743], a Dutch clergyman and geographical writer and publisher) is full of punning references to these ‘bubbles’, focusing on the way the investors’ foolishness and greed had seduced them, and the consequences for those who had lost everything. The scene at bottom right is entitled ‘Vlugt De Inwoonde von’t Eilant Geks-Kop’ (‘Flight of the Inhabitants of Mad-head Island’), with an investor fleeing from his angry creditors in a land-yacht, an allusion to some of the bizarre inventions being proposed at the time. He is fleeing *na viane* – to Vianen – home to an infamous lunatic asylum. At top left, another investor, with his money-chest and money-bags empty, complains ‘O Jammer en Elend Hoe Kom Ik Hier Te Pas’ (‘Oh Pity and Misery! How have I come to this?’).

The scene on the left, labelled ‘Quinquempoix Beplystert’ (‘Quinquempoix Besieged’) depicts the headquarters of the Compagnie (located in the Rue Quinquempoix) under siege by unhappy investors. The headquarters also give their name to the capital of the island in the central map. The island of Mad-head is shaped like a man’s head, with the ears of a jackass, wearing a fool’s cap with a bell. The island’s rivers, the ‘R: de Seine’, ‘R. de Teems’ (Thames) and ‘R. de Maas’ (Meuse), are named after the principal rivers of the capital cities of the countries involved in the scheme: Paris, London and Amsterdam. ‘Z.Z. have’ (‘South Sea Haven’), located at the island’s ‘mouth’, alludes to the English South Sea scheme.

Deez' Schets vertoond het vreemd gewest
Van Gekskop, 't geen men op het lest
Door Missipse en Bubbel–winden,
En Zuidzé stormen kwam te vinden,
Maar menig die van 't vaste land
Zyn heil ging zoeken op dat strand
Vind zig te dérelyk bedrógen,
Eerst blonk het alles schoon voor oogen;
Nu is 't vol giftig ongediert,
't Geen door de scherpe Distels zwierd
En doorns, die dat land omvangen,
Vol Schorpioenen, Spinnen, Slangen,
Waar by de Kat en Nagtuil voegt,
Die 't ligt verägten, om vernoegt
In duisterheid van roof te léven.
Wat vrugten kan dat land tog géven
't Geen 't allermeest bestaat in schyn,
Als Valsheid, Droefheid, en Fenyn?
Terwyl de Gekheid daar als Koning
In Quinquenpoix houd hof en woning,

Regerende door kwaá Praktyk,
Zyn nieuw gebakke Koninghryk,
Maár ach! waar zal die Vorst belanden?
Het eiland beeft, en op de stranden
Bruld een verwoede zuide–wind,
Waar door zig elk verlégen vind;
En roept het eiland krygt dé stuipen,
Zo dat hy die niet wil verzuipen
Moet denken op een snelle vlugt.
Fluks maakt men op dit droef gerugt
Veel wagens naar de nieuwste moden,
Daar de Acties zyn tot 't zeil van nóden,
Om dus van Gekskops malle strand
Te ráken in een ander land,
Van Wanhoop, Droefheid, Armoed', Schanden,
Of liever naar de Nederlanden,
In Kuilenburg, of Ysselstyn,
Of in Vianen, zo 't kon zyn,
Om daar als Uilen 't hoofd te buigen;
Wyl Quinquenpoix tog leid in duigen.

‘An Astronomicall and Chronologicall Clock, Shewing All the Most Usefull Parts of an Almanack’, 1725

John Naylor, London

Copper engraving; at widest: 629 × 385 mm (24 ¾ × 15 ⅛ in)

JOHN NAYLOR was one of a group of clockmakers based in Nantwich, Cheshire. He is first recorded in about 1725. By 1740, he had certainly relocated to London, where he is thought to have died in 1751.

In about 1726 he began work on a astronomical calendar table clock, with an elaborately engraved metal face; as part of the design, he included a map of the northern hemisphere south to ‘Biledulgerid’ (North Africa), ‘Ormus’ (Hormuz) and ‘Surat’. An example of the clock can be found in the collections of the British Museum. The design of the map is very similar to an earlier clock face made by John Carte, a leading London clock-maker active at the start of the century. At the same time, Naylor also prepared a printing plate with a copy of the clock face that was to be used to print advertising or promotional flyers for the clock; this is an example. The British Museum has the first state of this promotional engraving, with the text at the top commencing ‘The Explanation March the first 1725/6 ...’, acquired along with their example of the clock.

The British Museum also has a second state of the complete promotional engraving, with the text at the top revised and now commencing ‘The Explanation March the first 1750/1 ...’; it was reprinted to coincide with the British changeover from the Julian to Gregorian calendar. The British Library also has an example, but the sheet has been cut down, so shows only the clock face, and lacks the engraving of the Sun God (Helios or Apollo) and the explanatory text.

The example illustrated here is a third state; John Naylor’s name has been replaced by that of his heir Joseph Naylor. Joseph was almost certainly not a clock-maker, but rather inherited the finished clock. In the fashion of the time, he planned a lottery to dispose of the clock and maximise the cash value of his inheritance, for which undertaking he prepared an accompanying promotional booklet, entitled ‘An explanation of an Astronomical Clock the Workmanship of Jos. Naylor, Joyner; near Namptwich in Cheshire, [which] is to be disposed of by One hundred Chances, at Two Guineas each, which are to be decided by a Machine ... The Clock will be fixed at a convenient Apartment near St. James’s, whereof notice will be given and the winning chance decided April, 1751.’

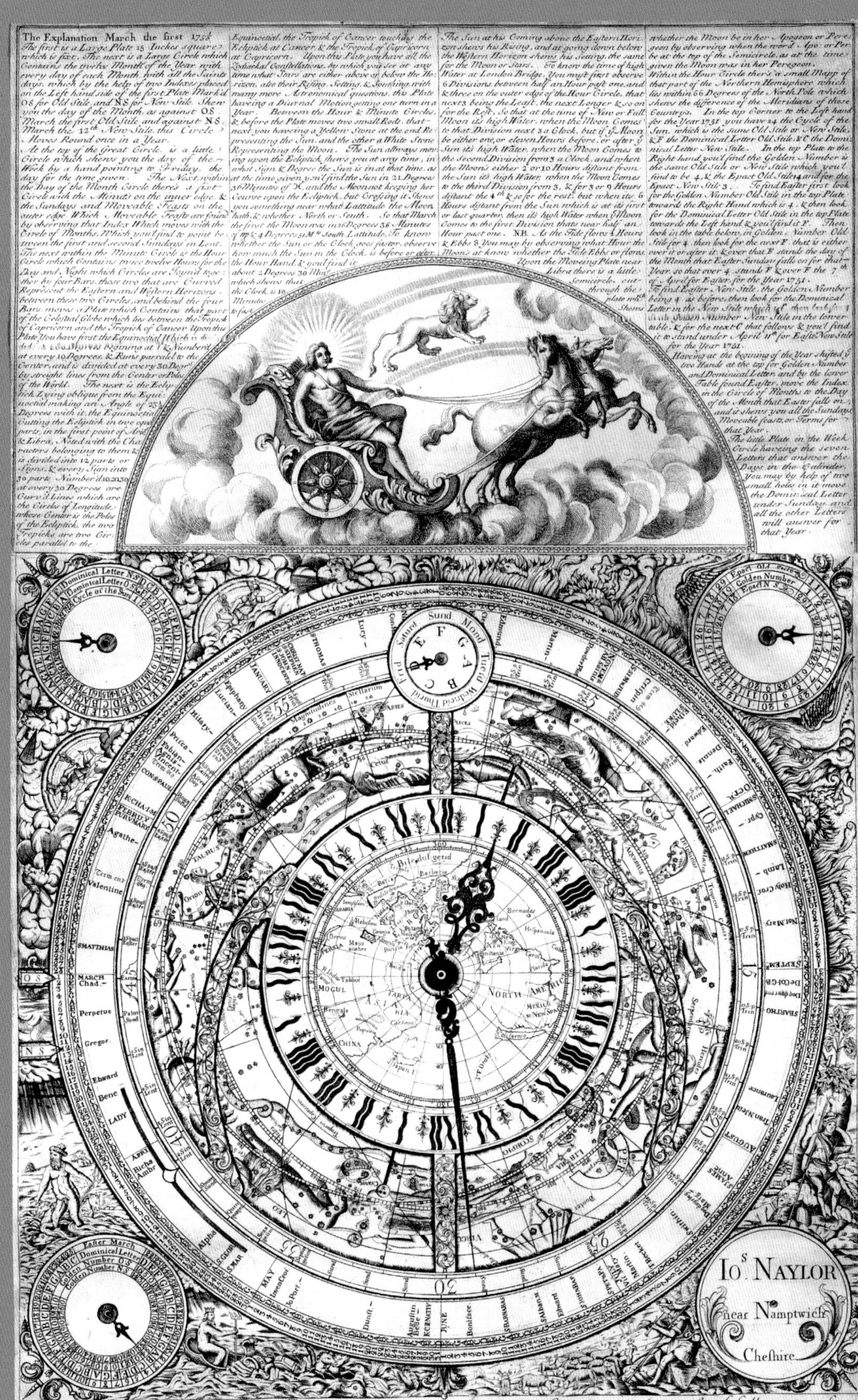

The Explanation March the first 1751
Ios. Naylor
near Namptwich
Cheshire
An Astronomicall and Cronologicall Clock shewing all the most usefull parts of an Almanack

Untitled Map of 'Brobdingnag', North America and 'New Albion', 1726

Jonathan Swift, London

Copper engraving; inner border/border: 158 × 100 mm (6 ¼ × 4 in)

JONATHAN SWIFT (1667–1745) was an Anglo-Irish clergyman, satirist, political pamphleteer and author, active in the inner circles of the Tory Government in London between 1710 and 1714. His *Travels into Several Remote Nations of the World*, better known as *Gulliver's Travels*, is one of the most famous works of travel fiction, incorporating political and moral satire, but also reflecting Swift's general disillusionment with the world. This sense of disillusionment grew increasingly evident as the four books were written. The first book is by far the best known: this is the account of Gulliver's time in Lilliput, where he is a giant to the tiny inhabitants. The next three books revolve around a juxtaposition of circumstances; in book two, 'A Voyage to Brobdingnag', Gulliver finds himself the midget and the inhabitants the giants. In book three, 'A Voyage to Laputa, Balnibarbi, Glubbdubdrib, Luggnagg and Japan' he finds a country obsessed with the pursuit of science and scientific reason, but without regard to the purpose of the quest or its applications. In the 'Voyage to the Country of the Houyhnhnms' he encounters the Houyhnhnms, a breed of horses who have attained perfection in nature but at a cost of their 'humanity' – having lost their feelings in blind pursuit of absolute reason and logic.

Swift blurs the distinction between fact and fiction by creating a realistic setting for each of his parables. Gulliver notes (in book two):

> I now intend to give the reader a short description of this country, as far as I travelled in it, which was not above two thousand miles round Lorbrulgrud, the metropolis. For the Queen, whom I always attended, never went farther when she accompanied the king in his progresses, and there staid till his Majesty returned from viewing his frontiers. The whole extent of this prince's dominions reaches about six thousand miles in length, and from three to five in breadth. From whence I cannot but conclude, that our geographers of Europe are in a great error, by supposing nothing but sea between Japan and California; for it was ever my opinion, that there must be a balance of earth to counterpoise the great continent of Tartary; and therefore they ought to correct their maps and charts, by joining this vast tract of land to the north-west parts of America, wherein I shall be ready to lend them my assistance.

Each of Swift's countries is depicted on a map; one of the interesting features of the *Travels* is that it is produced in a fashion identical to that of a contemporary travel account, so an uninformed reader might take it for a work of fact, not fiction. The map depicted here shows Brobdingnag. Swift (or his engraver, possibly Herman Moll but just as possibly John Sturt and Robert Sheppard, who engraved the portrait frontispiece) has blended Swiftian invention with geographical fact. Brobdingnag is depicted on the northernmost coast of America, above the 'Streights of Annian', 'New Albion' (northernmost California, as named by Sir Francis Drake) and Monterey. The features are recognisable from Moll's atlas maps, with California at the time depicted as an island – Moll claiming to have known mariners who had sailed round it.

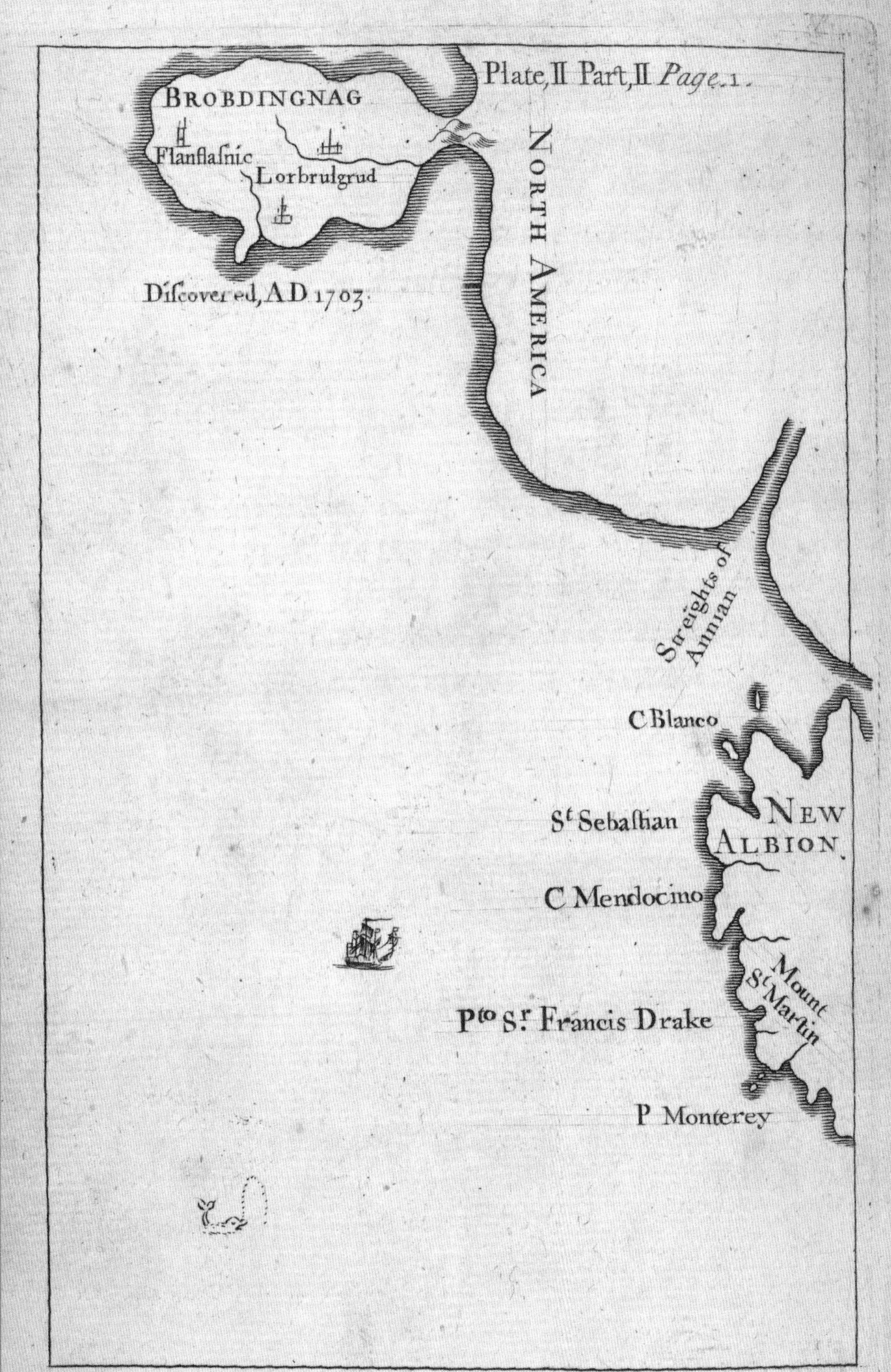
Plate, II Part, II Page. 1.
BROBDINGNAG
Flanflasnic
Lorbrulgrud
Discovered, A.D. 1703
NORTH AMERICA
Sreights of Annian
C Blanco
St Sebastian
NEW ALBION
C Mendocino
Mount St Martin
Pto Sr Francis Drake
P Monterey

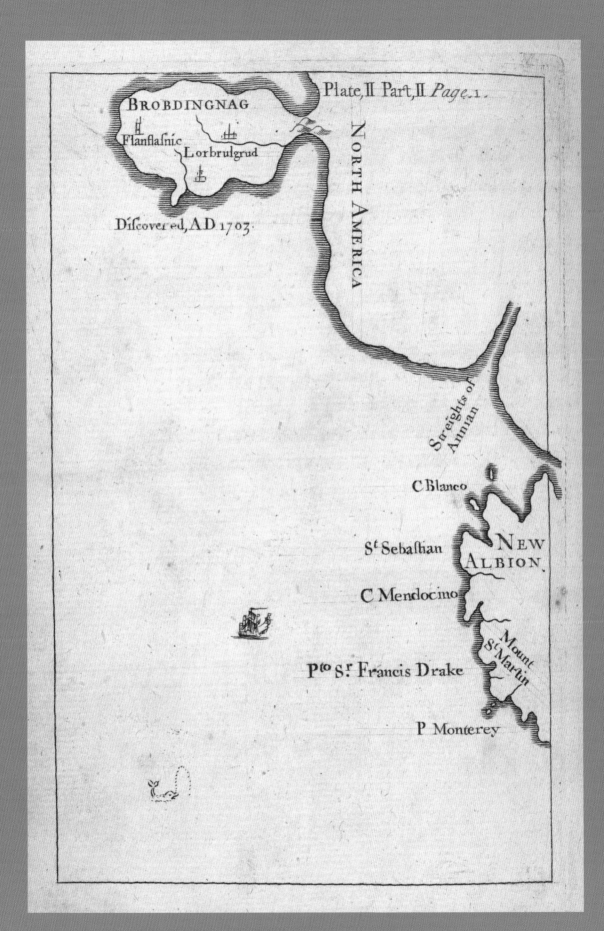

Allegorical Map of the Siege of the Castle of Love, 1735

Georg Matthaüs Seutter Sr, Augsburg

'Representation Sÿmbolique et ingenieuse projettée en Siege et en Bombardement comme il faut empecher prudemment les attaques de L'Amour'

Copper engraving; border: 489 × 568 mm (19 ¼ × 22 ⅜ in)

GEORG MATTHAÜS SEUTTER (1678–1757) founded one of the two most prominent German publishing houses of the eighteenth century; he was apprenticed to Johann Baptist Homann, founder of the other publishing house, in 1697, before setting up his own firm.

In this map, Seutter has recast typical contemporary town plans into an allegorical map of the siege of the Castle of Love. Unusually for a map of love, the plan is drawn from a male perspective: the Castle of Love is the male heart, while it is the women who are shown trying to break through the male's outer defences.

The male heart is shown in the keep of a fortified town. This is reminiscent of the siege maps from the War of the Spanish Succession (1701–1714), which would have been very familiar to a contemporary audience. The keep is surrounded by a moat, labelled 'La Mer Glacee Sans Passion' ('Frozen Sea without Passion'). Like the title, the key alludes to the many (military) stratagems that the (male) defender might employ to defend his heart against the advances of (female) attackers ('Methode pour defendre et conserver son coeur contre les attaques de l'amour'). Around the castle are arrayed the female forces of love, represented by the artillery batteries on the mainland and the naval forces on the glacial sea, bombarding the defences with such surprises as 'regards languissant' ('languishing looks'). Each of the guns has an individual female charm: 'Enchantment', 'Tendresse', 'Un certain je ne Sais quoi', 'Surprises', and 'Charmes' (a full list is given in the key) while the defenders shelter in their bastions – among them 'Precaution', 'Prudence', 'Experience', 'Indifference' and 'Resolution'. The attacking forces are commanded from their headquarters at bottom left, labelled the 'Camp de l'Amour', wherein are the tents of its 'General Cupido', settled in with his forces until the defences are breached and the besieged brought to surrender.

As the defences are slowly breached by female wiles, the defender is forced to retreat out of the keep through the Gates of Wisdom, to successive hideouts along the lake: 'Conseil des fideles amis' ('Counsel of faithful friends'), 'Deliberation' and 'Inspiration de nos propres sens' ('Inspiration of our right senses') onward to the 'Jardin de Plaisir ...' where the first pleasant meeting takes place. Then finally, in desperation, he takes flight by an underground way to 'Le Palais de L'Amour' ('the Palace of Love'), lured by the song of the Sirens who inhabit the lake, where the final surrender to the lady takes place. From here there is no return without loss of liberty.

All this is expanded in the detailed key along the lower border of the map, while Venus (or Aphrodite in Greek mythology), the goddess of love, in her chariot, looks down on the campaign from the finely engraved title cartouche at top right.

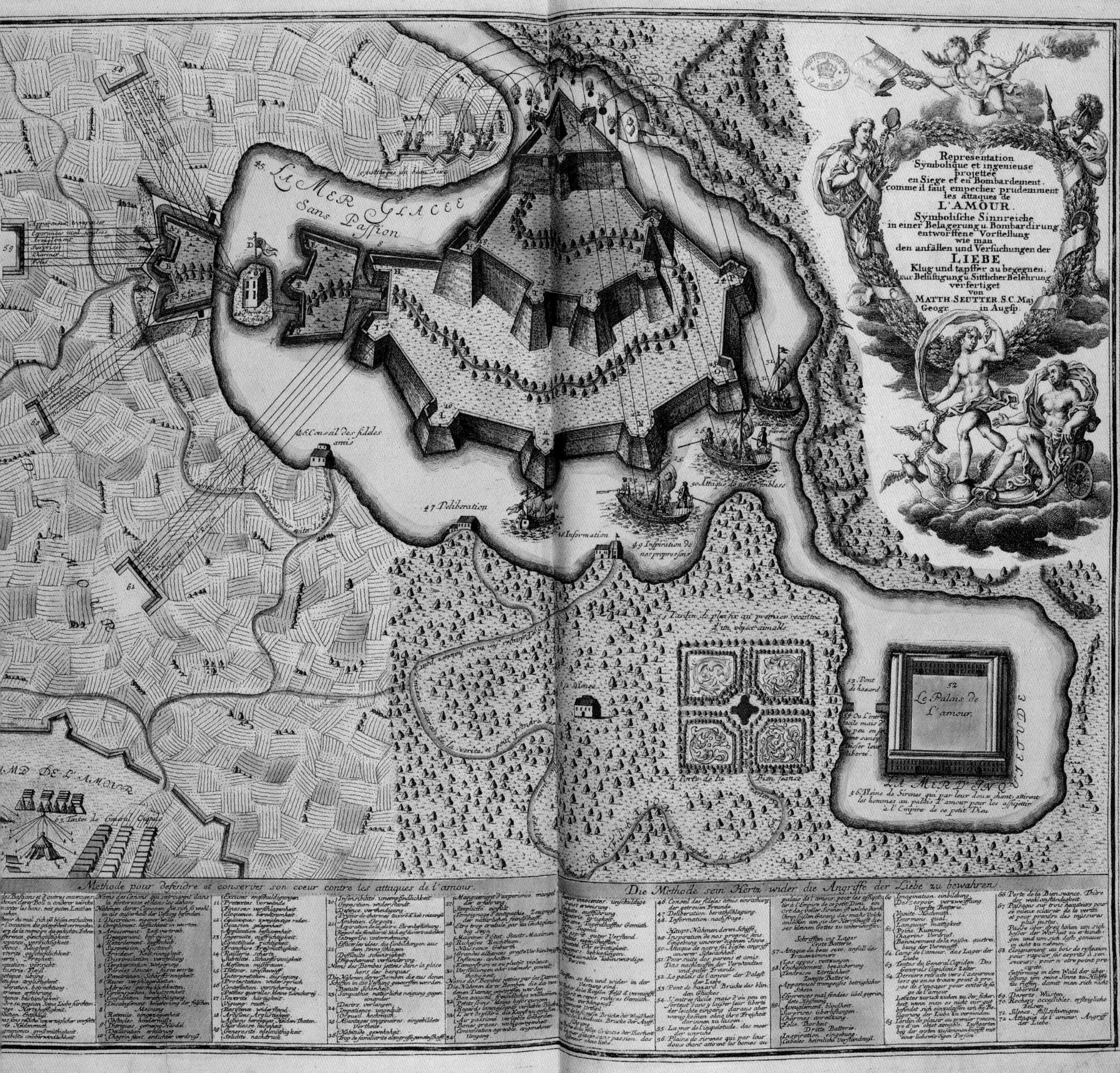
Representation
Symbolique et ingenieuse
projettée
en Siege et en Bombardement,
comme il faut empecher prudemment
les attaques de
L'AMOUR.
Symbolische Sinnreiche
in einer Belagerung u. Bombardirung
entworffene Vorstellung
wie man
den anfällen und Versuchungen der
LIEBE
Klug und tapffer zu begegnen.
zur Belustigung u. Sittlicher Belehrung
verfertiget
von
MATTH. SEUTTER. S.C. Maj.
Geogr. in Augsp.
La Mer Glacée
Sans Passion
46 Conseil des fidels amis
47 Deliberation
48 Information
52 Le Palais de l'amour
Methode pour defendre et conserver son coeur contre les attaques de l'amour.
Die Methode sein Hertz wider die Angriffe der Liebe zu bewahren.

The Harz Mountains, Germany, as from the Air, 1749

L.S. Bestehorn, Nuremberg

'Perspectivische Vorstellung des brühmten blocken oder Blokenbers mit der jenigen Gegens...'

Copper engraving; border: 473 × 551 mm (18 5/8 × 21 3/4 in)

THIS ATTRACTIVE bird's-eye view was compiled by the German surveyor L. S. Bestehorn in 1732; it depicts the region of the Harz Mountains, a wooded and hilly area of Germany centred on Blocksberg Mountain. The map was published in 1749 by the leading Nuremberg mapmakers and publishers, Homann Heirs – the heirs of Johann Baptist Homann (1664–1724), Geographer to the Holy Roman Emperor.

The map would be an otherwise unremarkable product of that celebrated cartographic workshop, were it not for one element. Someone involved in the production of the printed version – the draughtsman, the engraver or the publishers themselves – drew two small witches dancing on the crest of the mountain and six witches in flight, circling the peak, two on broomsticks, two on pitchforks and two astride goats. This is a reference to old legends in which the mountain is a centre of Devil worship, 'der fabulose Hexen Platz' ('the fabled witches' place'). The images relate to Walpurgisnacht, the eve of 1 May, the feast-day of St Walpurga (an English missionary), and one of the most enthusiastically celebrated of the traditional festivals of northern and central Europe. In German tradition it was the night on which witches were thought to gather on the top of the mountain to dance for the Devil. He would select the most beautiful one for his own pleasure for the year ahead. The others would then disperse, to seek new recruits into their number before reassembling the next year.

In the key at top left, the various features are identified. They include an altar and a fountain with a copper ladle with which to drink the water.

This may seem to modern viewers a harmless bit of cartographic fun, inventive licence, but in the eighteenth century not everyone agreed. The publishers must have received complaints: a second printing of the map, from 1752, bore an additional panel of text, an apology from the publishers to anyone offended, saying that the drawings were simply the harmless invention of the engraver. However, the apology is somewhat tongue-in-cheek, as this text is dated Walpurgis-day 1751.

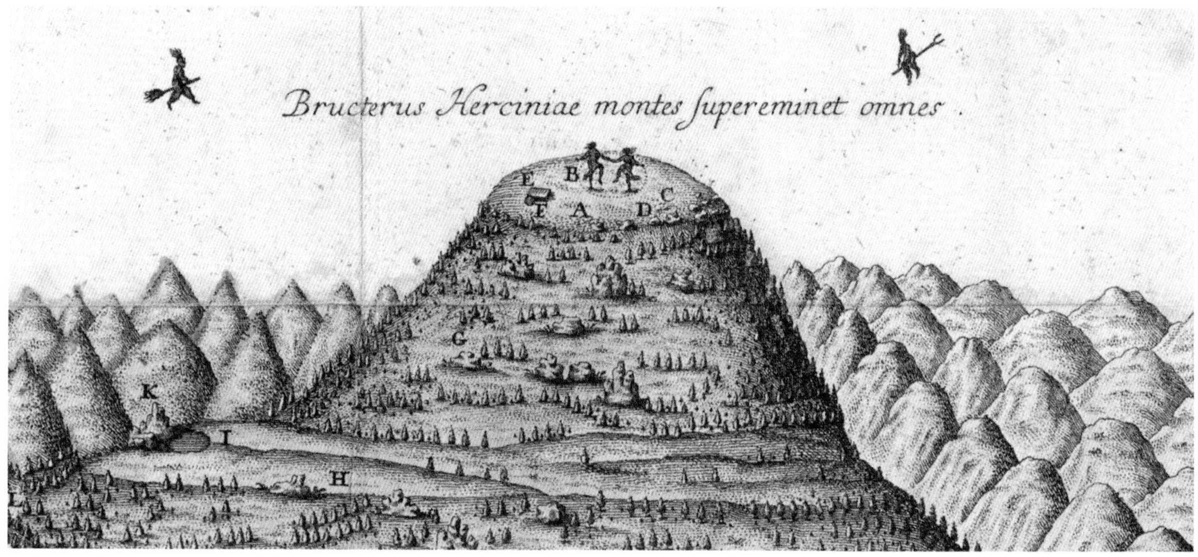

Vue de la Montagne de BROKEN située dans le Territoire du Comté de Wernigerode, qui est dans les forêts de Hartz.
Erklaerung der Buchstaben.
PERSPECTIVISCHE VORSTELLVNG
des berühmten
BLOCKEN ODER BLOKENBERGS
mit der jenigen Gegend, so weit solche
von dem, der auf der Spitze des Berges stehet,
gesehen werden kan.
Gezeichnet A.o 1732. von L. S. Bestehorn.
herausgegeben von Homænn. Erben. C.P.S.C.M. 1749.
Nota: Joh. Prætorius u. der Ilfeldische Insp. Albr. Ritter haben Nachrichten von dem Bloksberg geschrieben. Jener hat 1653. 5 Jul. mit seiner Gesellschafft von Ilsenburg aus bis an die Spitze des Bergs 4 Stund, wobey er halben Weg geritten, herunter aber 2 Stunden zugebracht.
Erklaerung der Ziffern.
1. die Klippe, der Ilsenstein.
2. die Ilse, Fluss.
3. der Westenberg.
4. der Rinckenberg.
5. die Ecker, Fluss.
6. der Schimmel Wald.
7. die ahls Burg.
8. der Funff Eichen Platz.
9. der kahle Königs Berg
10. das alte Schloss zu Hartzeburg.
Nota. Von dem Flecken Ilsenburg ist der Brocken noch eine kleine Meile entfernet.
Bructerus Herciniae montes supereminet omnes.
Hohe Geist
Hasselfelt
Wernigerode
Ilsenburg
Derenburg
Halberstadt
Clausthal
Zellerfeld

Map Screen, 1749

John Bowles, London

'A Map of the World or Terrestrial Globe in two planispheres, laid down from the Observations of the Royal Academy of Sciences, Wherein as an Introduction to the study of Geography are inserted several things relating to the Doctrine of the Earthly Globe'

180 × 256 cm (70 ⅞ × 100 ¾ in)

MAP SCREENS, by their nature, are particularly fragile objects. They were constructed of a wooden frame with the maps glued to a coarse canvas backing stretched, in the manner of a painting, across the frame. As such they were extremely vulnerable to damage but also to extremes and variations of temperature. Heat makes paper and canvas expand or contract at different speeds, which tends to lead to the paper tearing and flaking off the backing. As a consequence, perhaps only four British map screens are recorded today: two in a private collection and two in the British Library.

George Willdey (1676?–1737) was the first London mapseller to advertise map screens, at first obliquely, but in a catalogue of 1738 his heir Thomas boasted of selling

> the best large Map of the World ever yet done in England, and the most useful of any yet seen; the Two Hemispheres are Six Foot long, and Three Foot deep; round it is generally placed a Sett of 20 Sheet Maps of the Kingdoms and chief Dominions in Europe, being those before mentioned, beginning with the Northern Hemisphere and ending with the Map of Flanders, which makes it ornamental and useful. This Map made in a Screen (as a large quantity of them are now used for) is the most diverting and instructive Screen yet seen; it may also be fixed up in different Manners and Sizes, for Rooms, Passages, Stair-cases, Halls, &c.

The screen illustrated here was assembled by John Bowles (1701–1779) in about 1746, the date found on the central four-sheet map of the world. It is composed of a series of maps arranged around a central world map. Five of the outer maps come from Bowles' older stock: the two maps flanking the world map were published by Herman Moll in 1715 and 1717 respectively and three of the maps along the base were published by William Berry in the 1680s. All these plates had subsequently come into the possession of Bowles and, as all the maps were by now rather out of date, they made excellent candidates for being pasted on a screen in this fashion. They were both decorative and cheap compared to more up to date maps from Bowles' current stock. There are also ten views of London, or London buildings.

This screen was probably made for someone with connections to the British colonies in America, which are depicted in two maps: Moll's 'A New and Exact Map of the Dominions of the King of Great Britain on ye Continent of North America, Containing Newfoundland, New Scotland, New England, New York, New Jersey, Pensilvania, Maryland, Virginia and Carolina ....' (the so-called 'Beaver Map', first published in 1715) and 'A Map of the British Empire in America with the French and Spanish Settlements adjacent thereto ...', the index map to Henry Popple's twenty-one-sheet 1733 wall map of the British possessions in North American and the West Indies.

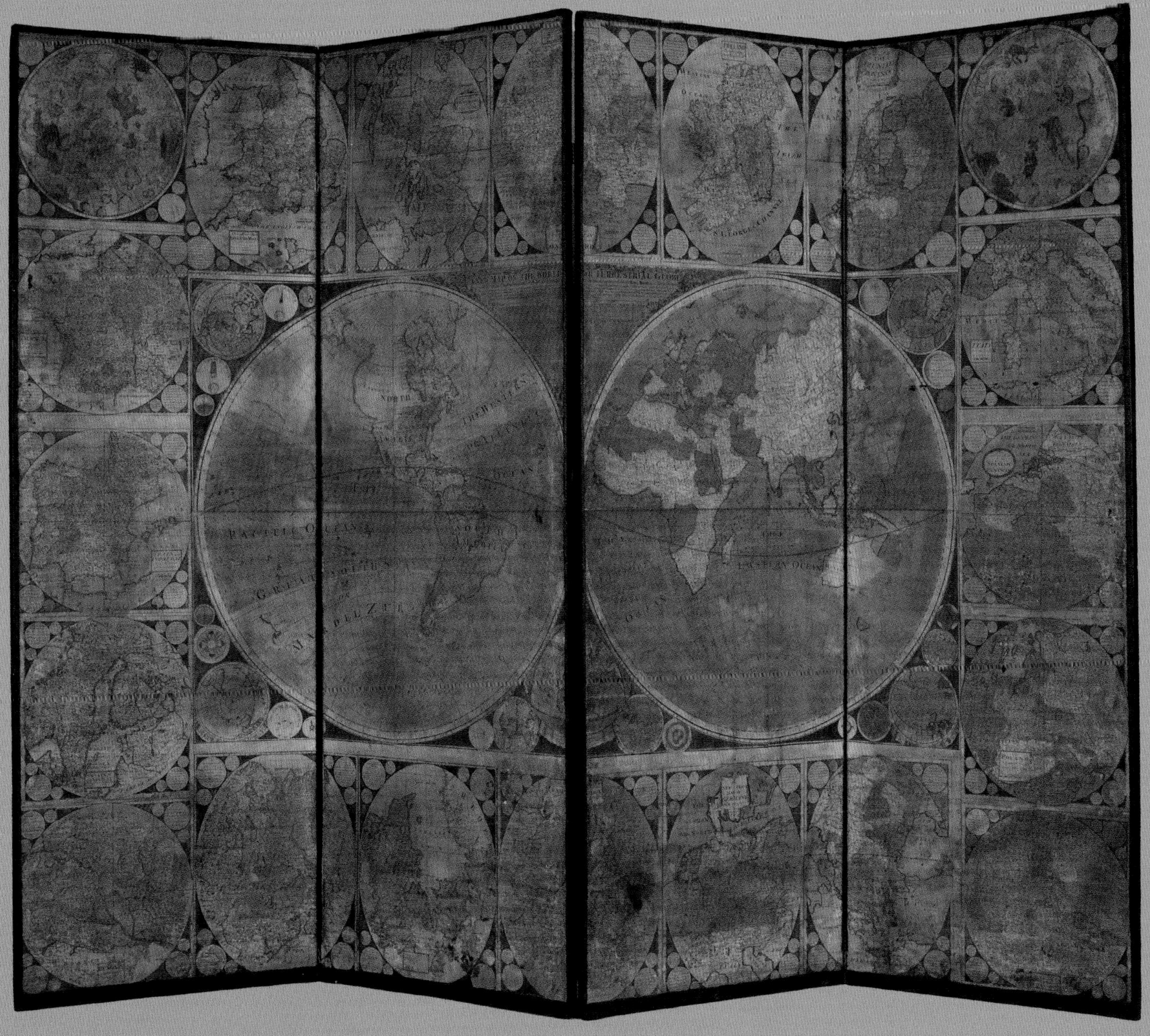

'A New & Correct Plan of London, Including All Ye New Buildings &c', Map-Fan, *c*.1760

Richard Bennett, London

Copper engraving; the map mounted on silk as a fan, with ivory and bone telescopic sticks.
Map radius 149 mm (5 ⅞ in), outer circumference 680 mm (26 ¾ in),
length of sticks 240 mm (9 ½ in)

ILLUSTRATED FOLDING fans were a fashionable accessory throughout the eighteenth century. While many earlier fans had been imported from overseas, often from as far afield as India and China, fans of English manufacture were starting to become more common. The Worshipful Company of Fan Makers received its charter as one of the Livery Companies of the City of London in 1709. The popularity of the pictorial fan is attested by a war in the advertisement columns of the newspapers in 1710 over rival fans depicting the fiery preacher Henry Sacheverell, whose thundering denunciations of Catholics and Non-Conformists had made him the current darling of the High Church Tories, especially after his ill-advised impeachment and trial by the Whig administration (he was barred from preaching for three years in March 1710).

The earliest advertiser to offer an English-made pictorial fan seems to have been Lucy Beardwell, who announced:

> Emblematical Fans, with the true Effigies of the Reverend Dr. Henry Sacheverell done to the Life, and several curious Hieroglyphicks in Honour of the Church of England, finely painted, and mounted on extraordinary genteel Sticks, are to be sold only at Mrs. Beardwell's, Printer of this Paper, next the Red-Cross-Tavern, in Black-Friars. (*Post Boy*, 22–24 August 1710)

A rival fan-maker, Mr Humphrey, quickly responded:

> Whereas in Yesterday's [25 August] and Monday's [21 August] Supplement and Thursday's [24 August] Post-Boy, there was this Advertisement, viz. Emblematical Fans with the true Effigies of the Reverend Dr. Sacheverell, &c. Sold by Mrs. Beardwell; and some Buyers having, by Mistake, taken the said Fan to be the same with the New One sold at the House of Mr. Humphrey in Lovell's Court in Pater-Noster-Row, which, besides it being the most beautiful Landskip both in blacks and whites and Painting, on Account of its great variety of Figures, does really contain, together with the Effigies of the Doctor and 6 Bishops, the noblest Emblems of the Church of England, as also Emblematically sets forth the base Attempts and Practices of her secret and open Enemies; all adapted to the present State of things: All Buyers therefore are requested to take a View of this Fan before they buy that or any other. N.B. Ladies may have these Fans mounted upon any Sticks of their own. (*Daily Courant*, 26 August 1710).

Fans with maps on them were a natural enough development and Richard Bennett produced this fine fan-map of London in about 1760, covering the area from 'Sadlers Wells' south to the River Thames and from 'Green Park' east to 'Shadwell Dock'. Geographically, it follows the work of John Rocque, who carried out the great mid-eighteenth-century survey of London. Such fans were evidently intended as a talking-points and conversation pieces, but this particular fan came with an in-built utility: a table of hackney carriage fares for journeys across town, so that travellers, making their way home, perhaps from dinner or the theatre, would know the driver was charging the correct fare.

For a fan, a special map had to be engraved, which curved to open as the sticks were spread apart; unfortunately, because of the fragility of the paper, and the wear and tear from the repeated opening and closing, antique fans are of considerable scarcity. Indeed, this is one of only two recorded examples of this particular map-fan of London.

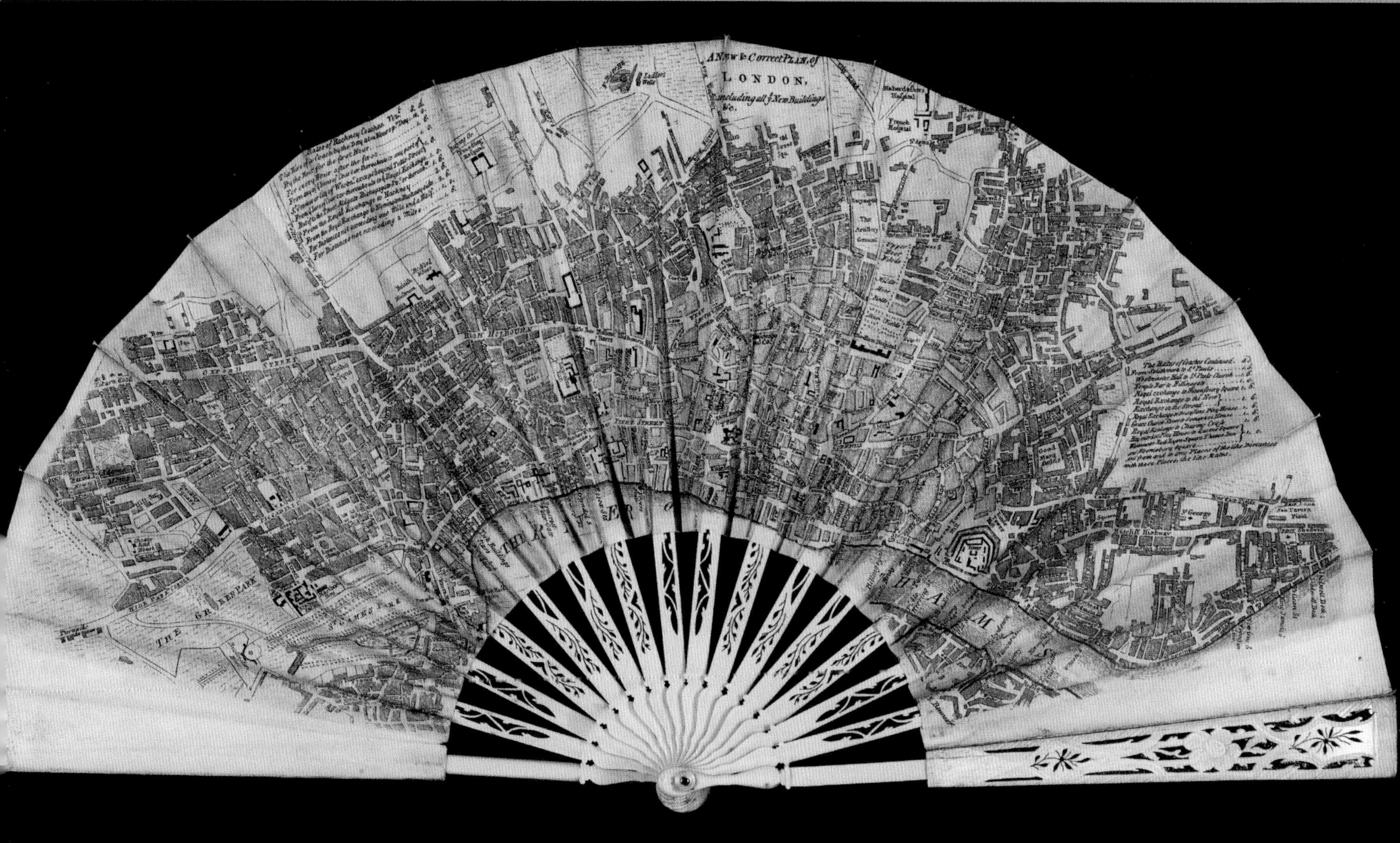
A New & Correct PLAN of
LONDON,
including all ye New Buildings
&c.

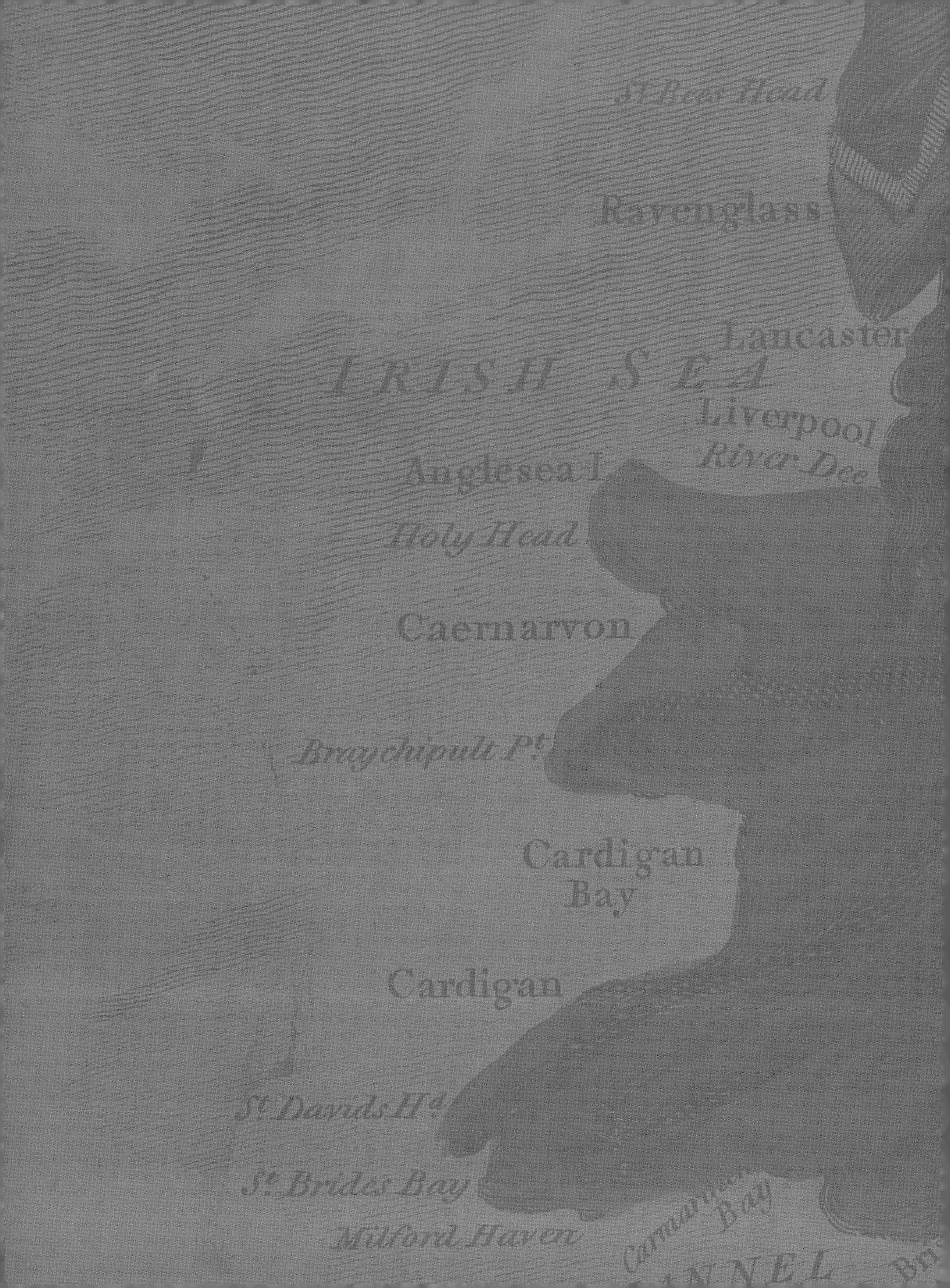

St. Bees Head

Ravenglass

Lancaster

IRISH SEA

Liverpool

River Dee

Anglesea I.

Holy Head

Caernarvon

Braychipult Pt.

Cardigan Bay

Cardigan

St. Davids Hd.

St. Brides Bay

Milford Haven

COMMERCIAL CARTOGRAPHY AND EDUCATION

(1761–1848)

Allegorical Map of Spain's Overseas Possessions, in the Form of a Queen, 1761

Vicente de Memije, Manilla

'Aspecto symbolico del mundo Hispanico, puntualmente arreglado al geografico ...'

Copper engraving; border: 1000 × 635 mm (39 ⅜ × 25 in), on two sheets joined

THIS MAP was drawn by a Jesuit priest, Vicente de Memije, as one of a pair. The first, the 'Aspecto Geographico', was a factual map of the Americas; this companion map, the 'Aspecto Symbolico', is an allegorical depiction, intended to show all of Spain's possessions, both at home and overseas, drawn into one orbit and united by the queen. The empire is a physical unity, both anointed and protected by the grace of (the Catholic) God, and with the duty of spreading the word of God to the 'heathens' within the Spanish possessions.

From the start of the great age of discovery the two Iberian naval powers, Spain and Portugal, had vied for control of the wider world beyond Europe. To avoid continuing conflict, the two countries agreed the Treaty of Tordesillas (1494) and the subsequent Treaty of Zaragossa (1529), which divided the world outside Europe between them along a meridian (a line of longitude) 370 leagues west of the Canaries. By this Spain was granted most of North and South America, although Brazil still fell within the Portuguese orbit, as did Africa and India.

The Philippines clearly belonged to the Portuguese 'half' of the globe, but were soon settled by the Spanish. The two countries sought to regularise matters by the Treaty of Madrid (1750). Portugal was granted larger borders in Brazil, while the Philippines were confirmed as Spanish. Memije was based in the Philippines, and he clearly intended that his 'Aspecto Symbolico' map should emphasise to the dedicatee, Carlos III, King of Spain, that the Philippines were an integral, and important, part of his overseas empire. The map depicts the Spanish overseas possessions in the guise of a queen. The engraving is a little pale, but the map is oriented with West Africa and the Atlantic Ocean at the top; the queen's shoulders and upper torso are formed by the Americas, with the sweep of the South American continent forming part of her robe. The Pacific region is shown as her dress, with the tracks followed by the Spanish fleets as the folds in the dress, and her feet rest on the Philippine Islands 'I. Luzon' and 'I. Mindanao'. At left are the coasts of Tartary, with Japan and Korea. Just to the right can be discerned the coast of Australia, and part of the southern continent, 'Isla de Borneo'.

The queen's crown, labelled 'Espana', bears the names of the Spanish provinces; the jewel at her throat is actually a compass-rose, with the names of the winds, supported on a band composed of treasure galleons, the source of Spain's wealth. One hand is reaching for a flaming sword that is being passed to her by an angel, symbolising Spain's pre-ordained destiny to rule the world; in the other hand she holds the Equator, which serves as a flagpole for the Spanish standard.

Above her head, the sun's rays originate from Rome, with the Holy Spirit depicted as a dove (also a symbol of peace) at its centre, and with Catholic motifs and Bible quotations. God's light is shining on Spain and the empire.

AFRI
CA
ISLA S. SANTIAGO
NEVLTRA
AMERICA SEPTENTRIONAL
SENO MEXI-CANO
AMERICA MERIDIONAL
PLVS
ULTRA
REYNO
DE
EL
AUSTRO
ASPECTO SYMBOLICO DEL MUNDO HISPANICO,
PUNTUALMENTE ARREGLADO AL GEOGRAFICO.
QUE A SU GLORIOSO CATHOLICO REY D. CARLOS TERCERO EL MAGNANIMO
DEDICA, Y CONSAGRA D. VICENTE DE MEMIJE CON IX THESES, Y XC PROPOSICIONES, QUE ACERCA DE EL DEFIENDE:
PRESIDIENDO EL R.P. PASQUAL FERNANDEZ, PUBLICO PROFESSOR DE MATHEMATICAS
EN LA UNIVERSIDAD DE MANILA DE LA COMPAÑIA DE IESVS AÑO DE 1761.

‘North and South America in its Principal Divisions’, Jigsaw, 1767

John Spilsbury, London

Copper engraving, mounted on wood and cut as a jigsaw; border: 446 × 481 mm (17 ½ × 19 in)

THERE HAS been considerable debate over the origin of British cartographic jigsaw puzzles. In 1763, a London trade directory listed one Leprince, who claimed to be the ‘Inventor of the dissection of maps on wood’, at a Marylebone Street address. It is not clear whether this refers to Mme Jeanne-Marie le Prince de Beaumont (1711–1780), the French writer and tutor who ran a school in Henrietta Street in the 1750s, but who returned to France in 1762; or perhaps her half-brother, the engraver Jean-Baptiste le Prince (1734–1781), who worked both in Russia (1757–1762) and in England and is said to have invented the art of aquatint; or perhaps most likely, another half-brother, Jean-Robert le Prince, reportedly a geographer who died in London in or about 1762. There is certainly evidence that Jeanne-Marie le Prince used maps mounted on wood in her teaching in the 1750s, the implication being that these were dissected. She is thought to have supplied Lady Charlotte Finch, governess of the royal nursery, with sixteen dissected maps for the use of the children of George III.

There is little doubt, however, that John Spilsbury (1739?–1769), the maker of this jigsaw and the one shown on p. 88, was the first commercial publisher in England of these educational toys. His earliest jigsaw was made in 1762 and he advertised himself in the following year as an ‘Engraver and Map Dissector in Wood, in order to facilitate the Teaching of Geography’ (Mortimer’s *Universal Director*, 1763), a statement repeated on his trade cards and in newspaper advertisements placed by himself and then, after his early death in 1769, by his wife.

This jigsaw is one of a set of the four traditional continents – Australia was not at that time classified as being a continent. The map itself is a good contemporary delineation of the Americas, but it has been adjusted for the particular needs of dissection – most notably it is simplified, with few place names.

Compared to a modern jigsaw, it is notable how few pieces there actually are. The principal map area is cut into seventeen pieces, with the sea, Europe and Africa on nineteen larger pieces. It may be that the process of cutting the wood was a both expensive and difficult task, with the attendant risk of damaging the pieces, so the maker made as few incisions as possible. For example, the English colonies in North America are all on one piece.

NORTH and SOUTH
AMERICA
in its
PRINCIPAL DIVISIONS
BY
J Spilsbury
1767
HUDSONS BAY
GULF OF MEXICO
ATLANTIC OCEAN
THE GREAT SOUTH SEA
PACIFIC OCEAN
CANADA
NEW MEXICO
AMAZONIA
BRASIL
PARAGUAY
EUROPE
AFRICA

'Asia in its Principal Divisions', Jigsaw, 1767

John Spilsbury, London

Copper engraving, mounted on wood and cut as a jigsaw; border: 439 × 469 mm (17 ¼ × 18 ½ in)

THIS MAP of Asia is another of the maps of the continents issued as jigsaws by John Spilsbury in 1766 and 1767. Like the Americas example on p. 86, the map is simplified, with only the most important names included. Often maps that were intended to be made into jigsaw puzzles were printed from purpose-prepared maps, which lacked the detail of contemporary atlas maps, but had more sharply defined, if less precise, boundaries to help the dissector in the onerous task of cutting out the individual pieces.

As with the companion map of the Americas, the individual pieces here are relatively large: the map is composed of twenty-one pieces, most of generous size, such as Independent Tartary and Chinese Tartary, with the sea-pieces included as an optional extra.

The dissecting process was carried out by hand. The cutter has not attempted to follow the contours of the shore exactly, but has cut a simplified line along the coast. Internal boundaries have been followed more exactly but still are relatively simplified.

It is worth noting that, unlike modern jigsaws, the pieces do not have interlocking plugs (sometimes called tabs or tongues); instead, the pieces butt up against each other rather than joining. It was only with mechanisation of the cutting that smaller, more complicated or intricate shapes could be created. The name 'jigsaw' was itself unknown before the middle of the nineteenth century (although puzzles of this date appear to have been made with a fret-saw rather than the slightly later and slightly different jigsaw, and the name may be altogether a misnomer). Nevertheless, considering that this jigsaw (and its companions) are among the earliest surviving examples of the genre, the workmanship is something to be admired.

A S I A
in its
Principal Divisions,
By
J. Spilsbury,
1767
FROZEN OCEAN
Nova Zemlia
EUROPE
RUSSIA
ASIA
BLACK SEA
TURKEY IN ASIA
PERSIA
ARABIA
ARABIAN SEA
AFRICA
INDIA
BAY OR GULF OF BENGAL
CEYLON
ISLES MALDIVIAS
INDEPENDENT TARTARY
CHINESE TARTARY
CHINA
COREA
THE GREAT SOUTH SEA
ISLES OF JAPAN
PHILIPINE ISLES
NEW GUINEA
Tropic of Cancer
Spilsbury
ENGRAVER
MAP & PRINT Seller
in Russel Court
Covent Garden
LONDON

'The Royal Geographical Pastime or the Complete Tour of Europe', Map-Game, 1768

Thomas Jefferys Sr, London

Copper engraving; border: 484 × 478 mm (19 × 18 ⅞ in); widest with text: 488 × 661 mm (19 ¼ × 26 in)

THE FIRST extant English map-game was published in 1759 by John Jefferys Jr, a teacher of writing and mathematics who lived in Westminster. That map is particularly rare; there is no recorded example of the first printing from 1759, and only a single exemplar of a later printing bearing the publication line of Carington Bowles (1724–1793), the son of John Bowles, who traded independently under his own name from about 1762 onwards.

Thomas Jefferys Sr (1719–1771) – seemingly unrelated to John Jefferys – was the leading English mapmaker and publisher of the period, with retail premises close to Charing Cross in London. He was appointed Geographer to Frederick, Prince of Wales in 1746, and to George, Prince of Wales in 1757 and then Geographer to the King when George acceded to the throne. John Spilsbury (maker of the two jigsaw maps on pp. 86 and 88) was apprenticed to Jefferys for seven years in 1753.

It is dangerous to argue a theory when there is so little evidence, but it would seem likely that Thomas Jefferys was aware of John Jefferys' map-game of Europe; the map trade in London of the day was an extremely small circle and Jefferys routinely advertised his maps in the newspapers. Jefferys would certainly have been aware of Spilsbury's jigsaws, which were circulating from at least 1762, and quite probably Mme le Prince de Beaumont's dissected maps used in the royal nursery (see p. 86); we should therefore understand that Jefferys was aware of rival mapmakers' forays into the market for cartographical games.

Yet it was apparently not until 1768, with the publication of this map, that Jefferys followed the same route. Indeed, this is only the second recorded English cartographical game – nine years after the first. While it has been suggested that Jefferys was slow to appreciate the commercial possibilities, one might guess that as Geographer to the King, Jefferys saw his role as creating and selling maps for more serious purposes. With the cartographic demands of the Seven Years' War (1756–1763) and then his ambitious domestic surveying programme of the English counties, which commenced immediately after the war, Jefferys had enough else on his mind. It was perhaps his bankruptcy in 1766 that turned his attention to smaller projects that could be produced and sold quickly.

The map illustrated here is complete with the rules, printed from letterpress and pasted on either side of the map; these contain an explanation and directions for play, with a description of each of the 103 steps of the game, ending with 103, London. The 'totum' (spinning die), pillars and counters called for in the rules are no longer present.

Although the map was published in 1768, the latest date found in the text is 1758, against St Malo (85), Cherburg (88) and Carrickfergus (92).

This is another example of the 'Game of the Goose' type, based on a race around Europe, with detours to Algiers, Egypt and through the Near East. Here the 'bonus' squares are the capital cities of Europe, including Dublin by virtue of the residence of the Lord Lieutenant. The winner is the first to land on London; as in other such games, the player has to throw the exact number to win. If he or she overthrows, rather than counting the excess backwards, the player has to return in ignominy to Paris (83). The penalty squares are, most notably, Algiers (68) where the player is imprisoned by Corsair pirates until another player comes along, and the Scilly Islands (89), a 'death' square where the player 'loses his chance of the game', presumably being shipwrecked and drowned.

THE ROYAL GEOGRAPHICAL PASTIME
OR THE COMPLETE TOUR OF EUROPE
By Thomas Jefferys
GEOGRAPHER to the KING
THE FROZEN OCEAN
EXPLANATION
ICELAND
SWEDEN
FINLAND
RUSSIA
LITHUANIA
GERMAN OCEAN
THE ENGLISH CHANNEL
BAY OF BISCAY
FRANCE
SPAIN
GERMANY
POLAND
HUNGARY
LITTLE TARTARY
GULF OF VENICE
TURKEY IN EUROPE
BLACK SEA
ASIA MINOR
SYRIA
HOLY LAND
EGYPT
THE MEDITERRANEAN SEA
THE LEVANT
PART OF AFRICA
THE ATLANTIC OCEAN
To His Royal Highness
GEORGE PRINCE OF WALES,
DUKE OF CORNWALL, &c. &c.
and Knight of the Most Noble Order of the Garter.

‘A Complete Tour Round the World’, Map-Game, 1770

Thomas Jefferys Sr, London

‘The Royal Geographical Pastime exhibiting a complete tour round the world in which are delineated the North East and North West Passages into the South Sea, and other modern Discoveries, By Thomas Jefferys, Geographer to the King’

Copper engraving; border: 484 × 472 mm (19 × 18 ⅝ in); with the side panels of text: 505 × 695 mm (19 ⅞ × 27 ⅜ in)

THE MAPMAKER Thomas Jefferys Sr (1719–1771) was initially slow to add cartographic games to his repertoire of stock (see p. 90), but once he did, he quickly introduced three game maps in three years. This is the second and the most cartographically interesting of the three. The map is drawn on an unusual (to our eyes anyway) projection, in which all the world is visible in one plane, although the map extends outside the circular frame of the hemisphere. As described in the map itself, ‘Note This Map is drawn upon a Stereographic projection, of which London is the Centre. The Horizon is greatly extended, to exhibit the distant parts of the World at one View ....’ This may well be a response to rival French mapmakers who frequently constructed their world maps with France at the centre. Notwithstanding its purpose as a game, it is one of the best world maps to originate in England at the time. Another interesting feature is that this one of the earliest printed maps to use red as the colour for the British Empire, as noted in the key: ‘The Tour round the World is coloured ... Blue. The British Empire in America ... Red. The Russian Empire with their Discoveries in America. Brown.’

The ‘Tour round the World’ is a ‘Game of the Goose’ style race; the side panels contain an explanation of the theme of the game, the manner in which is played, the rules, and descriptions of the stops that each traveller could or would make in their passage round the board.

The winning stage is ‘103. Land’s End – being the first pleasant place in England which is seen by mariners in their return from long voyages, and is equally wished for by the players – is the Game’, but ‘The players attending to the instructions hereafter given, will proceed regularly towards No. 103, which is at the Land’s End; and he who is fortunate enough to gain this number, wins the game: But as the chances of the play will oftener carry him beyond than exactly to it, he is then to return back to No. 89, which is at Oronoko-River: where he must remain till it comes to his turn to spin and try his fortune again; and this method is to be pursued by all the players, till one of them hits the lucky number’.

The descriptions of the different destinations give an abbreviated but still interesting account and history of each place. For example, ‘45. Manilla – the capital of the Philippine-Islands, was taken by the English from the Spaniards in 1762, who ransomed it from plunder. The ransom, however, is not yet paid.’ In the Pacific North-West is ‘58. The North West Passage into the South Seas – discovered by Nich. Shapley, from New-England, through Hudson’s Bay and by Admiral de Fonte, a Spanish admiral, from the South Sea, through the Archipelago of St. Lazarus, in 1640. The traveller, who has been so fortunate as to find his way through this Passage, shall be removed to Cape Horn, no. 79.’

Some of the descriptions are perhaps not entirely accurate, but would nonetheless appeal to a young audience conscious of Britain’s dominant role across the world, highlighting the great victories that won the empire.

As always, there are the hazard stops, notably ‘99. Bahama Islands – noted for shipwrecks, and often fatal to the Spanish galleons; and to the traveller, who will be shipwrecked on these islands, and lose his chance for the game’. Perhaps the most entertaining-sounding stop is ‘102. Newfoundland – famous for its cod-fishery, which is reputed the greatest yet known in the world. Here the traveller must stay one turn, to eat a dish of chauder*, and drink a cup of black strap†’. (The symbols refer to footnotes: ‘* A large fresh cod, boiled with a piece of fat salt pork. † Spruce-beer and rum, sweetened with treacle’).

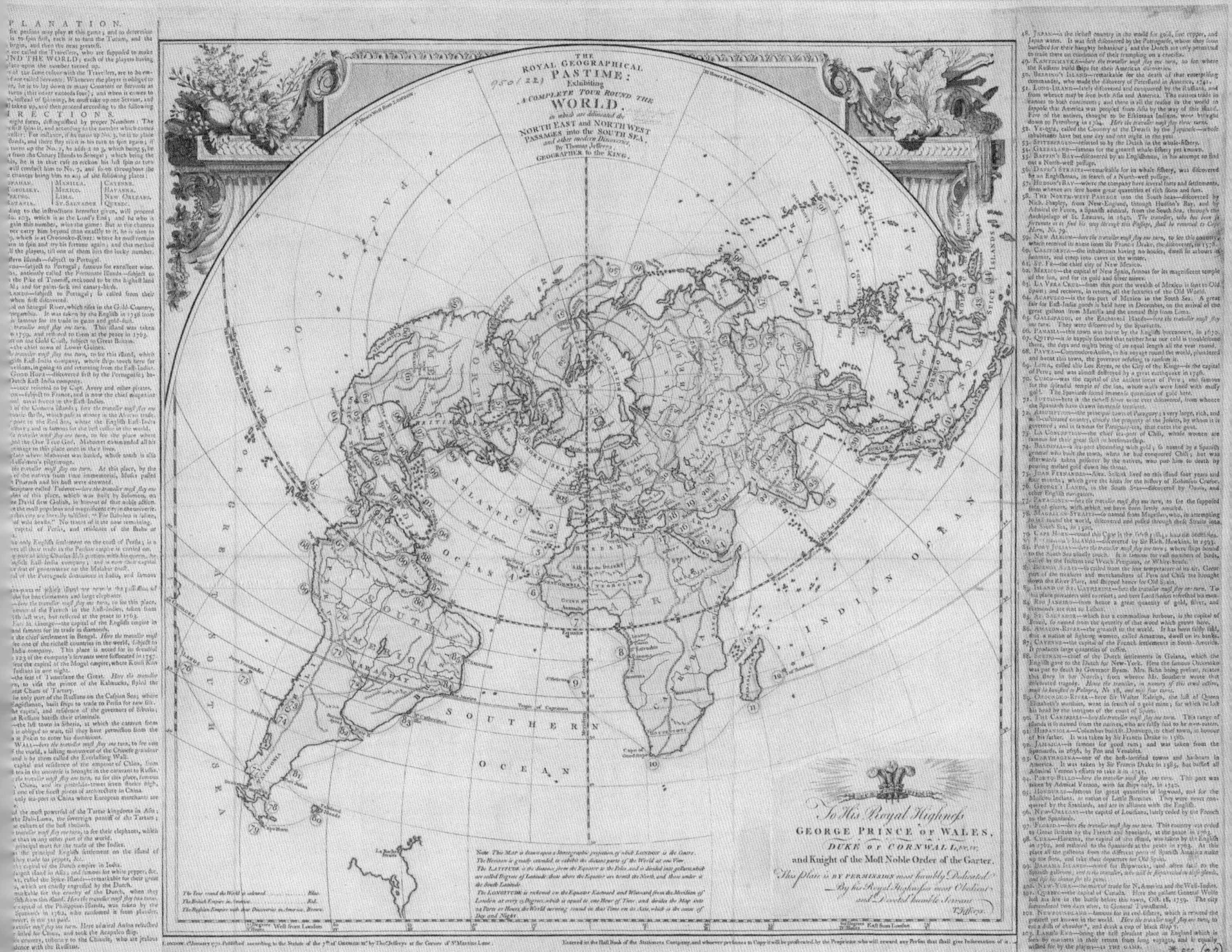
THE ROYAL GEOGRAPHICAL PASTIME: Exhibiting A COMPLETE TOUR ROUND THE WORLD, in which are delineated the NORTH EAST and NORTH WEST PASSAGES into the SOUTH SEA, and other modern Discoveries, By Thomas Jefferys, GEOGRAPHER to the KING.
ATLANTIC OCEAN
SOUTHERN OCEAN
INDIAN OCEAN
To His Royal Highness GEORGE PRINCE OF WALES, DUKE OF CORNWALL, &c. &c. and Knight of the Most Noble Order of the Garter, This plate is BY PERMISSION most humbly Dedicated By his Royal Highness's most Obedient and Devoted humble Servant T. Jefferys.

'The Royal Geographical Pastime Exhibiting a Complete Tour Thro' England and Wales', 1770

Thomas Jefferys Sr, London

Copper engraving; at widest: 509 × 440 mm (20 × 17 ⅜ in); with text: 512 × 661 mm (20 ⅛ × 26 in)

THIS IS A 'Game of the Goose' race map round England and Wales published by Thomas Jefferys on 1 January 1770. It is a companion to his earlier map of Europe (p. 90), and the map of the world (p. 92) published on the same day, with the same basic set of rules, as set out in the side panels of letterpress text. The players spin an eight-sided teetotum to progress around the map.

As usual in these Jefferys games, 'The players attending to the instructions hereafter given, will proceed regularly towards No. 111, which is at London; and he who is fortunate enough to gain this number, wins the game: But as the chances of the play will oftener carry him beyond than exactly to it, in that case he is to return back to No. 85, which is at Land's End: but in consideration of the misfortune which has befallen him, he is to spin in his turn, and try his fortune again; and this method is to be pursued by all the players, till one of them hits the lucky number.'

The 'bonus' squares are those with episcopal seats: the archbishoprics and bishoprics, as listed in the text: 'Archbishops Sees. Canterbury. York. Bishops Sees. St. Asaph. Bangor. Bath. Bristol. Carlisle. Chester. Chichester. St. David's Durham. Exeter. Ely. Glocester. Hereford. Landaff. Litchfield. Lincoln. London. Norwich. Oxford. Peterborough. Rochester. Salisbury. Winchester. Worcester.' The player landing on such a square would effectively double his throw, moving on as many squares as he had thrown on the teetotum.

The principal penalty square is '63. Coventry – famous for a yearly procession of a naked woman, in memory of Lady Godiva, who, to relieve the citizens from the oppression of her husband, consented to ride naked through this city, ordering all the inhabitants to shut their doors and windows, on pain of death. Here is to be seen the figure of a poor tailor, said to be struck blind, called Peeping Tom. Hence the traveller, in memory of this indiscretion, must be banished to Berwick, No. 11, and miss four turns.' The 'death' square is '87. Edystone Rock – famous for its light-house, so essentially necessary to the safety of the navigation of this part of the Channel. The traveller will be shipwrecked on this rock, and lose his chance for the game'.

An interesting feature of the engraving of the map is that the dedication to the Prince of Wales is printed from a second printing plate onto a label, which is pasted in position just below the title; this same printing plate was used to produce the dedication on the world map-game (p. 92). This may have been a product of haste: the dedication was later engraved onto the printing plate for the world map.

In some ways, this map is characteristic of the whole genre of map-games of England, but it has an interesting story associated with it. On both this map, and the companion map of the world, Jefferys warned '[this map is] Entered [for copyright] in the Hall Book of the Stationers Company and whoever presumes to Copy it will be prosecuted by the proprietor who will reward any person that shall give information of it'. Seeking such copyright protection was common; it was not much protection against unscrupulous rivals, but Jefferys meant what he said. The Court of Chancery papers, now deposited in The National Archives, reveal that on 15 February 1770, Jefferys filed a 'Bill of Complaint' against a rival publisher, Carington Bowles (see p. 102), subsequently obtaining an injunction against Bowles to prevent him selling rival versions of 'The Royal Geographical Pastimes'. Bowles filed his defence a month later, and the matter (with copies of the two maps) was referred to the Master of Rolls for adjudication. Unfortunately, after that, the record is silent on the outcome, although it may be assumed that Jefferys won the day.

THE
ROYAL GEOGRAPHICAL
PASTIME
Exhibiting
A COMPLETE TOUR THRO
ENGLAND and WALES
By Thomas Jefferys
GEOGRAPHER to the KING.
To His Royal Highness
GEORGE PRINCE OF WALES,
and Knight of the Most Noble Order of the Garter.
SCOTLAND
IRISH SEA
BRISTOL CHANNEL
ENGLISH CHANNEL
COAST OF FRANCE
St GEORGES CHANNEL
NORTHUMBERLAND
CUMBERLAND
DURHAM
YORK SHIRE
LANCASHIRE
LINCOLN
NORFOLK
SUFFOLK
NORTH WALES
SOUTH WALES
DEVON SHIRE
CORNWALL
DORSET SHIRE
SOMERSET SHIRE
SUSSEX
EXPLANATION.
DIRECTIONS.

'Picture of Europe for July 1772', 1772

Anonymous, London

Copper engraving; 95 × 161 mm (3 ¾ × 6 ⅜ in)

IN VIEW of its relatively small size, this unusual political cartoon map was probably engraved for a British periodical in 1772; the source has not been identified. It would appear that the title of this particular map was trimmed off by a careless binder, although it can be seen on other exemplars.

The immediate background is the decision by Prussia, Russia and Austria to dismember Poland and Lithuania by dividing their territories among themselves. The Russo-Turkish War rose out of conflict between Russia and Turkey along the border between Poland (then under Russian influence) and the Turkish possessions, a war in which the Russians achieved a decisive advantage in 1772. The specific moment in Polish history referred to in the map title is the first partition of Poland, marked by a Russo-Prussian treaty signed on 17 February 1772, and enacted on 5 May 1772.

Around the 'Map of the Kingdom of Poland' are assembled the principal sovereigns of Europe. At left, bound and with head bowed, his crown symbolically broken, is Stanisław II (1732–1798), King of Poland. Holding the map, and negotiating their share of the spoils, are Catherine II (1729–1796), Empress of Russia, Frederick William II (1744–1797), King of Prussia and Leopold II (1747–1792), Holy Roman Emperor.

Behind them, as if to keep an eye on what the three are doing, are Louis XV (1710–1774) of France and Charles III (1716–1788) of Spain; at the back and, portrayed with his eye clearly not on the events taking place before him, is George III (1738–1820) of the United Kingdom, who is sleeping through the negotiations.

The partition was a triumph for Prussia and, to a lesser extent, for Austria, and a defeat for Russia who, at the time, was more concerned with the Turkish threat along her southern borders. Mustafa III (1717–1774), the Turkish Emperor, is depicted in the background at left, in chains, representing the series of defeats his forces had suffered in the wars with Russia.

In the background, 'The Ballance of Power' is depicted as a set of scales, Great Britain's interests evidently greatly outweighed. After the recent series of wars in Europe, notably the War of the Austrian Succession and the Seven Years' War, none of the European powers wanted a continuation of fighting, but each sought a favoured outcome: while Russia wanted to defeat the Turks outright, Prussia, Austria and Britain wanted the Ottomans to remain in control of Turkey as a bulwark against Russian expansion in southern Europe, the eastern Mediterranean and in Central Asia. Austria, however, had its own designs on Ottoman possessions in the Balkans and parts of Poland, while Prussia wanted to expand her influence in central Europe. The satirist evidently believed that George III was being negligent in letting Prussia build up her strength and influence within Germany, and that the emerging situation was to the detriment of the British Crown's Hanoverian possessions, and Britain's wider interests; he seems to be portraying George, almost certainly unfairly, as almost 'sleep-walking' into a situation where the balance of power is tilting heavily against Great Britain.

The BALLANCE of POWER
GREAT BRITAIN
MAP of the KINGDOM of POLAND

'A Map or Chart of the Road of Love, and Harbour of Marriage' [*c.*1772]

'T.P.', London

'A Map or Chart of the Road of Love, and Harbour of Marriage. Laid down from the latest and best Authorities & regulated by my own Observations; The whole adjusted to the Latitude 51° 30 N. by T. P. Hydrographer, to his Majesty Hymen, and Prince Cupid.'

Copper engraving; at widest: 181 × 292 mm (7 1/8 × 11 1/2 in)

THIS MAP of the path of true love is attributed to one 'T. P. Hydrographer, to his Majesty Hymen, and Prince Cupid'. The traveller has to pass from the 'Sea of Common Life' (at the left of this engraving) to 'Felicity Harbour' and the 'Land of Promise' (at right), navigating the many false turns and hazards marked on the map, such as the 'Rocks of Jealousy', 'Henpeckd sand' and the 'Whirlpool of Adultery'. 'Felicity Harbour' could be attained only after the 'Harbour of Marriage' had been successfully traversed.

The would-be lover is given advice: 'Directions for Sailing into Felicity Harbour. Your Virtue must your Pilot be; Your Compass, Prudence, Peace your Sea; Your Anchor, Hope; your Stoage [*sic*], Love; (To your true Course still constant prove) Your Ballast, Sense, and Reason pure, Must ever be your Cynosure.'

A lengthier account is engraved along the lower border describing the route:

> Explanation. From the Sea of Common Life, we enter the Road of Love thro' Blindmans Straits, between two noted Capes or Headlands; steering first for Money, Lust, and sometimes Virtue, but many Vessels endeavouring to make the latter are lost in the Whirlpool of Beauty; from this Road are many outlets, yet some Mariners neither steer through these, nor continue their Voyage but come to their Moorings at Fastasleep Creek. Those who proceed reach Cape Ceremony, pass into the Harbour of Marriage through Fruition Straits and touch at Cape Extasy; care must be taken to keep still to the Starboard, lest we run upon sunken Rocks which lye about Cape Repentance; a good Pilot will also keep clear of the Rocks of Jealousy & Cuckoldom Bay and at least get into that of Content, some have past pleasant Straits and have arrived safe at Felicity Harbour, a Monsoon constantly blows from Fruition Straits quite up the Road, which renders a Passage back impracticable; a Tornado also arises sooner or later in those Parts and drives all Shipping tho moor'd at Content & even Felicity itself, thro the Gulf of Death, the only Outlet, terminating in the Lake of Rest.

This engraving, although independent in production, is closely linked to a contemporary satire, composed by Henry Carey, first issued in 1745 and reprinted in 1772, lengthily entitled: *Cupid and Hymen; a voyage to the isles of love and matrimony. Containing a most diverting account of the inhabitants of those two vast and populous countries, their laws, customs, and government. Interspersed with many useful directions and cautions how to avoid the dangerous precipices and quicksands that these islands abound with, and wherein so many thousands, who have undertaken the voyage, have miserably perished. By the facetious H. C. and T. B …*

While the book's title refers to a *map* of the island of marriage, this was in fact simply a sixty-eight-page textual description. The map itself was separately published by Robert Sayer (1725?–1794) in an accessible, charming and unusual image. Sayer was one of the leading and almost certainly the most prosperous of the London map and printsellers and publishers. A fine portrait by Zoffany of the Sayer family in the garden of their splendid house at Richmond is the subject of a recent monograph by David Wilson, *Johan Zoffany RA and the Sayer Family of Richmond: A Masterpiece of Conversation*, 2014.

London, Printed for ROB.T SAYER, N.o 53, Fleet Street.
Directions for Sailing into Felicity Harbour.
Fair Virtue must your Pilot be;
Your Compass, Prudence, Peace your Sea;
Your Anchor, Hope your Stoage, Love;
(So your true Course still constant prove)
Your Ballast, Sense, and Reason pure,
Must ever be your Cynosure.
Vanity Lake
KNAVES LAND
Meridian of TEENS
SEA OF COMMON LIFE
Vanity Outlet
Outlet of Reason
Cape of Goodhope
Fastasleep Creek
Fools Cape
Blindmans Straits
Lust I.
Whirlpool of Beauty
Virtue I.
Money I.
Cape Fairpromise
THE ROAD OF LOVE
COAST OF AMBITION
Outlet of Pride
Outlet of Interest
Covetous Lake
Cape Repentance
Cape Ceremony
Fruition Straits
Cape Extasy
CUCKOLDOM BAY
Cuckold's Point
Noisy Cape
CUCKOLD'S SHIRE
Henpeckd Sand
OUTLET OF DEATH
Lake of Rest.
Rocks of Jealousy
Whirlpool of Jealousy
Extravagance Bank
Cape Content
THE HARBOUR OF MARRIAGE
CONTENT BAY
Cape Goodsense
Pleasant Straits
Cape Fidelity
FELICITY HARBOUR
LAND OF PROMISE
LAND OF DESIRE
A MAP or CHART of the ROAD of LOVE, and HARBOUR of MARRIAGE.
Laid down from the latest and best Authorities & regulated by my own Observations;
The whole adjusted to the Latitude 51° 30 N.
by T. P. Hydrographer, to his Majesty Hymen, and Prince Cupid.
NB The Long.de is reckon'd from y.e Meridian of Teens.
EXPLANATION
From the Sea of common Life, we enter the Road of Love thro Blindman's Straits, between two noted ...s or Headlands; steering first for Money, Lust, and sometimes Virtue, but many Vessels endeavouring ...ake the latter are lost in the Whirlpool of Beauty; from this Road are many Outlets, yet some Mari... neither steer through these, nor continue their Voyage but come to their Moorings at Fastasleep Creek. ...se who proceed reach Cape Ceremony, pass into the Harbour of Marriage through Fruition Straits, and ... at Cape Extasy; care must be taken to keep still to the Starboard, lest we run upon sunken Rocks which lye about Cape Repentance: a good Pilot will also keep clear of the Rocks of Jealousy & Cuckoldom Bay and at least get into that of Content; some have past pleasant Straits and arrived safe at Felicity Harbour. a Monsoon constantly blows from Fruition Straits quite up the Road, which renders a Passage back impracticable; a Tornado also arises sooner or later in these Parts and drives all Shipping tho moor'd at Content & even Felicity itself, thro the Gulf of Death, the only Outlet, terminating in the Lake of Rest.
On Life's vast Sea, where Storms prevail,
...d Fortune shifts the driving Gale,
...r Realms of Bliss, a distant Shore!
... spread the Canvas, ply the Oar;
...e Road to Realms of Bliss below
... Loves alluring Name we know,
To various Islands scatter'd here
With various Views we rashly steer;
By rapid Currents born away
To diff'rent Ports a Thousand stray,
To Marriage some, a daring few!
With flatt'ring Hopes their Course pursue;
A Thousand Dangers here they run
What Points, What Rocks, what Sands to shun!
Lock'd in the winding Coast, how vain!
Are wishes to return again!
One Passage hence alone is found,
Death's dreary Gulf of depth profound!
Through this, each Bark one Tempest drives,
Where'er she sails, where'er arrives;
Happy! if previous Dangers o'er,
She coast awhile the Promis'd Shore,
Touch at Content, from Shipwreck free
And harbour at Felicity.

Map of the World, Jigsaw, 1787

Thomas Kitchin, London

Copper engraving; widest surviving: 375 × 736 mm (14 ¾ × 29 in)

THIS MAP of the world, dissected as a jigsaw, is known only in this single example. Three pieces have been lost: central Europe, Madagascar and the northern Great Lakes. It lacks the first half of the map title and everything outside the map border, but this is probably as issued. The cost of a jigsaw reflected not merely the cost of the map and the wood, but also the workmanship involved in cutting the pieces; in some cases, and probably here, the original buyer opted to buy a 'cut down' version, where everything extraneous to the puzzle itself was discarded. Geographically blank areas added to the cost of the puzzle, while not contributing particularly to the educational elements of the game.

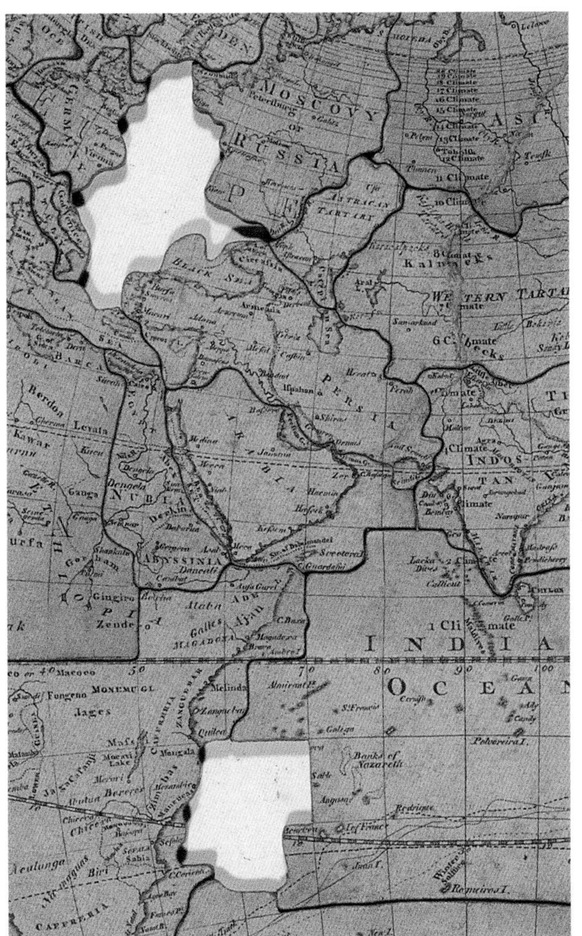

In the title the map is attributed to 'Thomas Kitchin', presumably the engraver and cartographer Thomas Kitchin Sr (1719–1784), or his son, Thomas Bowen Kitchin (fl. 1765–c.1784), who held the title of Hydrographer to the King jointly with his father. A third possibility, and perhaps the most likely, is that John Wallis, the publisher, simply applied the Kitchin name to his map, to give it additional status and prestige, without either Kitchin having a direct involvement.

After struggling early in his career – he was declared bankrupt in 1778 – John Wallis Sr established himself as one of the earliest retailers to specialise in cartographic games and, with his sons John Jr and Edward following him into the trade, built up one of the largest and most important firms in this field, responsible for producing a number of later entries in this book.

Such was Wallis' status in the field that he later claimed (despite John Spilsbury's plainly earlier examples – see p. 86) to be the first manufacturer of map puzzles in his advertising; on a generic box label, he stated:

> J. Wallis the original Manufacturer of Dissected Maps and Puzzles, (having devoted full 30 years to that particular line of business) requests the Public to Observe that all his dissected Articles are superior both in correctness & workmanship to any in London, & that none are genuine but what are signed on the label.

While it might be that his memory was a little faulty after thirty years, it seems plausible that this was simply a marketing claim to bolster his already considerable reputation, and one that he felt few could successfully challenge fifty years after jigsaws first emerged.

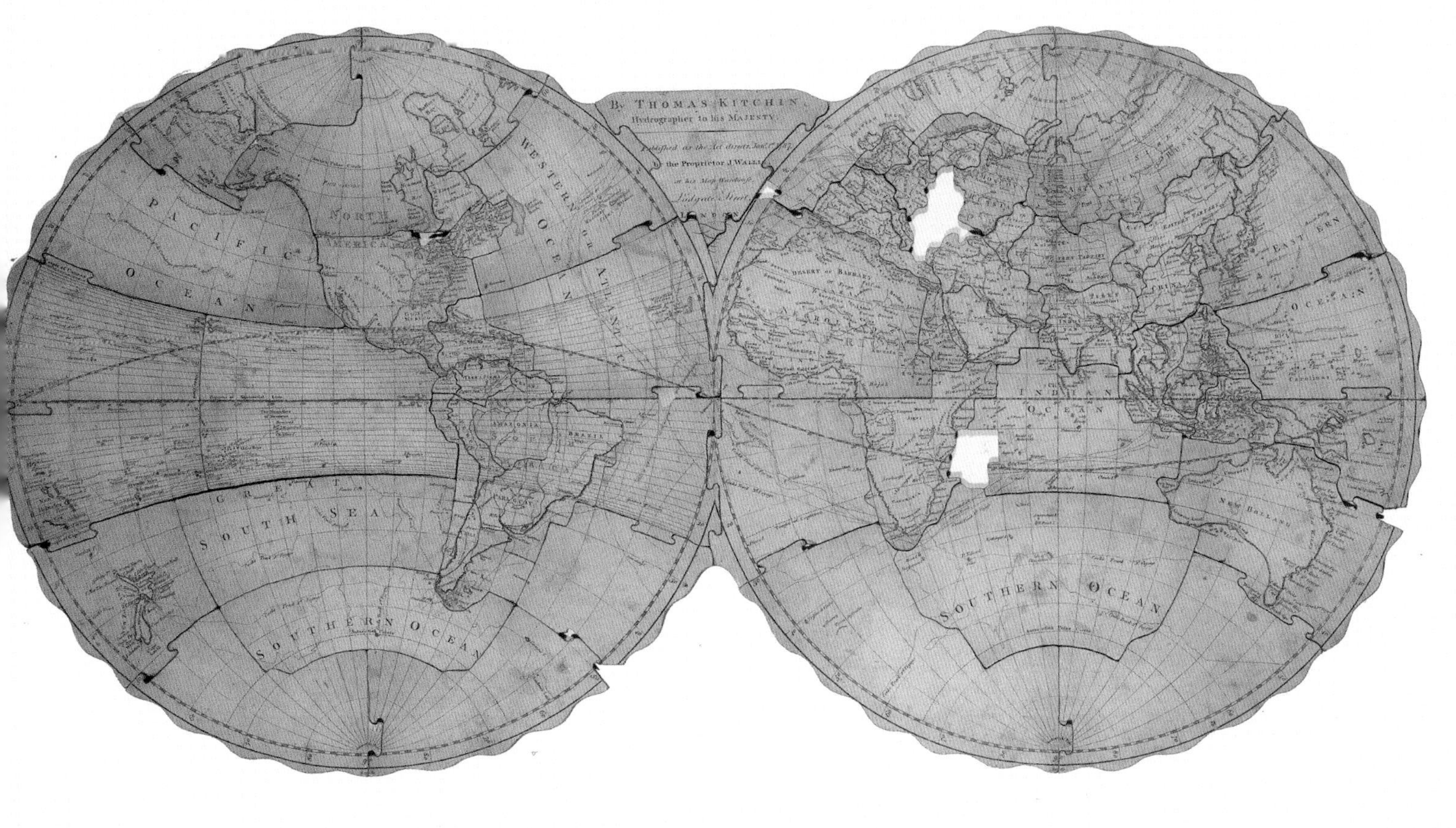
By THOMAS KITCHIN
Hydrographer to his MAJESTY
PACIFIC OCEAN
NORTH AMERICA
WESTERN OCEAN
ATLANTIC
GREAT SOUTH SEA
SOUTHERN OCEAN
DESERT of BARBARY
INDIAN OCEAN
EASTERN OCEAN
NEW HOLLAND
SOUTHERN OCEAN

'Bowles's Geographical Game of the World', 1790

Carington Bowles, London

'Bowles's Geographical Game of the World, in a New Complete and Elegant Tour through the Known Parts thereof, laid down on Mercator's Projection.'

Copper engraving; border: 376 × 630 mm (14 ¾ × 24 ¾ in);
with text: 470 × 638 mm (18 ½ × 25 ⅛ in)

THE BOWLES family were among the leading London map and print publishers of the eighteenth century, running two separate, but parallel, firms, sometimes working independently, sometimes in partnership and perhaps occasionally in competition. Thomas Bowles (I) (fl. 1683–1714) established the firm, which passed to his son Thomas (II) (1688–1767) and then to his grandson Thomas (III) (1712?–1762), although Thomas (II) came out of retirement briefly when his son died.

In his will, Thomas (I) made provision for his second son, John Bowles (1701–1779), to allow him to start his own business; John was subsequently joined in partnership by his son Carington Bowles (1724–1793) but, probably on the death of his cousin in 1762, Carington left his father's business to take over that of his uncle, Thomas (II). He continued until his own death, when he was succeeded by his son Henry Carington Bowles, trading as Bowles & Carver up to 1830.

Carington Bowles was caught up in a legal battle with Thomas Jefferys in 1770, for plagiarising Jefferys' game maps of the world and England and Wales (see p. 94) but subsequently developed his own line of games and from about 1780, he is found advertising them:

> Winter Evening's Entertainments. The night at this inclement season of the year may be rendered at once more pleasant and profitable, and the fire-side become the scene of amusement and instruction to Ladies and Gentlemen of all ages, by enlivening the gloomy hours with the following aids to improvement and pleasure ...

This world map, a simplified reduction of Bowles' great four-sheet wall-map, is unusual for the genre as being on the Mercator projection, most contemporary game-makers favouring a double-hemisphere projection.

The 'Directions for Playing' are given in the letterpress panel pasted outside the lower border of the map. It is a standard 'Game of the Goose' type of race game, a game of chance rather than skill, albeit with rather less stages than comparable maps, played by upwards of two players who set out from the Azores, journey round the world, both by land and sea, and end up in London: '76. London, (the Game) capital of Great Britain and greatest Commercial City in the World'. Winners 'consequently become entitled to the applause of the company and honor of being esteemed the best instructed and speediest traveller in the World'.

As usual, an exact throw was required to win the game. En route, the travellers were required to read out the description of the place that they landed on. Several squares have hazards attached, for example '28. Botany Bay, first European colony in New South Wales, established by the British in 1788; to prevent any idea of the traveller being brought here as a convict, he need only stay four turns to observe the manners of our new acquaintance, and then proceed' and '30. Manilla, capital of the Philippine Islands; here it is the duty of every traveller to stay three turns, to find out, if possible why the ransom money is not yet paid to the English, who took this place from the Spanish, in 1762'.

The 'death' square is placed off the northern coast of Norway: '62. Maalstroom, a most perilous Whirlpool; here you deservedly lose the chance of the game, for keeping so bad a look out, and getting into so much danger'.

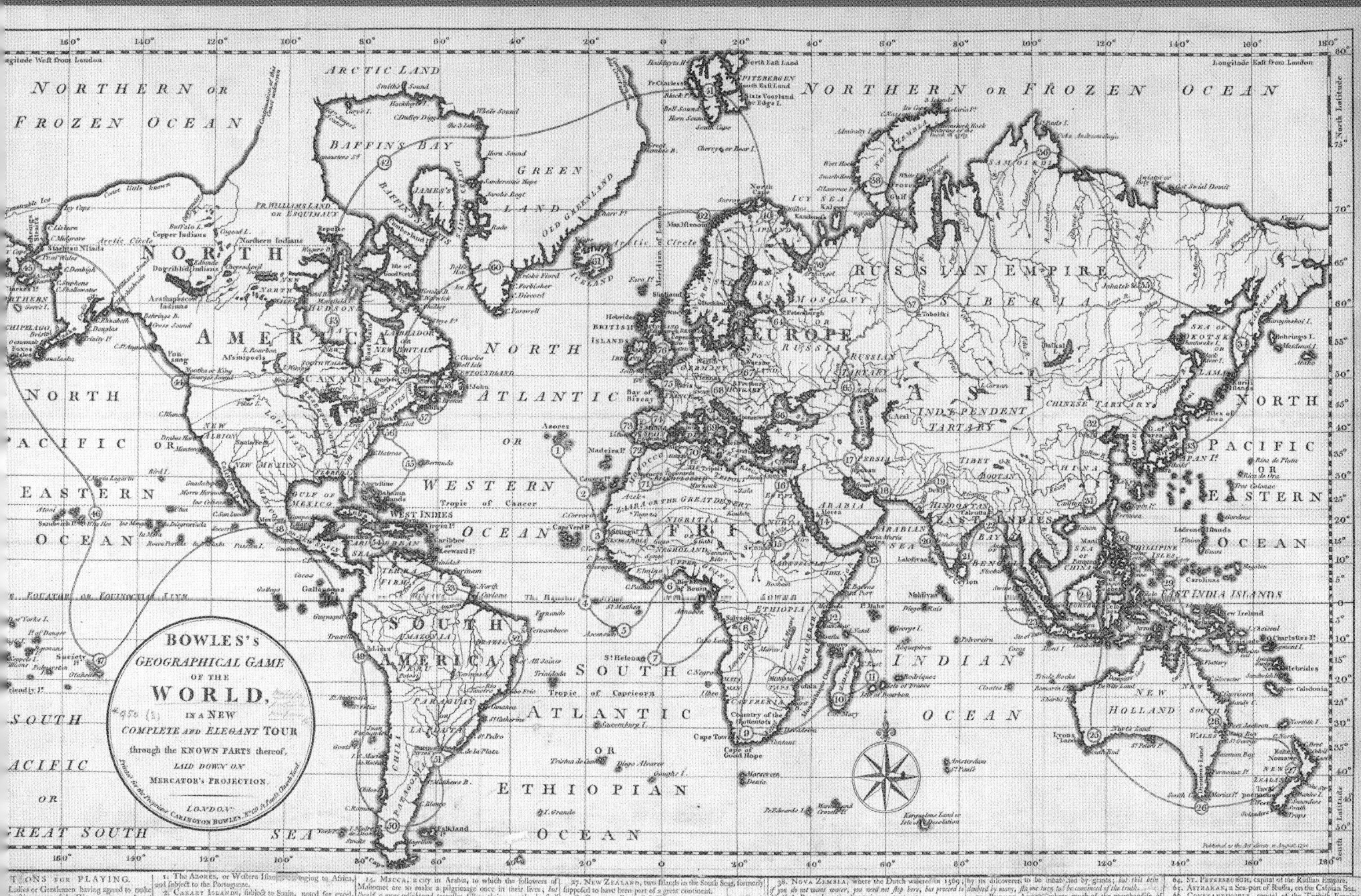
BOWLES's
GEOGRAPHICAL GAME
OF THE
WORLD,
IN A NEW
COMPLETE AND ELEGANT TOUR
through the KNOWN PARTS thereof,
LAID DOWN ON
MERCATOR's PROJECTION.
LONDON:
NORTHERN OR FROZEN OCEAN
NORTH AMERICA
NORTH ATLANTIC
WESTERN OCEAN
SOUTH AMERICA
SOUTH ATLANTIC OR ETHIOPIAN OCEAN
AFRICA
EUROPE
ASIA
RUSSIAN EMPIRE
INDIAN OCEAN
NEW HOLLAND
PACIFIC OR EASTERN OCEAN
Entered at Stationers' Hall.

Untitled Allegorical Map of *The Pilgrim's Progress*, Jigsaw, 1790

John Wallis, Sr, London

'The Pilgrim's Progress Dissected or a Complete View of Christian's Travels from the City of Destruction, to the Holy Land. Designed as a Rational Amusement, for Youth of both Sexes' (from box label)

Copper engraving; border: 277 × 177 mm (10 ⅞ × 7 in)

JOHN BUNYAN'S religious allegory *The Pilgrim's Progress from this World to that which is to Come*, first published in 1678, is one of the great works of English devotional prose, endlessly reprinted, a staple of almost every English-speaking home for centuries, never out of print and translated into over 200 languages. Bunyan (1628–1688) recounts the pilgrimage of the Everyman figure of Christian, the hero of the first part, to Heaven, via the 'Slough of Despond', 'Vanity Fair' and numerous other trials and perils which have long since become part of the common stock of English phrases.

John Wallis Sr (*c*.1745–1818) was an important English mapmaker and publisher, one of the earliest to focus on the rapidly expanding juvenile market; he became the leading maker of dissected maps, board games and so on in the late eighteenth and early nineteenth centuries. Early editions of *The Pilgrim's Progress* were not illustrated, but late-eighteenth-century publishers started to insert maps and illustrations based on the text. Wallis' map was evidently conceived as a broadsheet to be pasted up, mounted on board, framed or made into a jigsaw.

The premise is that Christian is born in the 'City of Destruction' (at the foot of the map), where he grows up. However, he realises that the life there is morally wrong and he commits to a better way, Christianity, but the knowledge of his sins revealed by the Bible creates a great burden that he has to carry with him until he can find deliverance.

The book recounts the struggles, tribulations and distractions from fellow travellers that he must endure in order to pass out of the 'City of Destruction', and its hinterland, to reach God's kingdom, lying beyond the river on the far side of 'A Pleasant Land Beulah Out of Giant Despair's reach'. Just outside the 'City of Destruction' Christian meets Evangelist, who directs him to the 'Wicket Gate', the start of the road to redemption; Goodwill, the gate-keeper, is Jesus Christ, although that is not revealed at the time. They converse. Goodwill tells him:

> ...good Christian, come a little way with me, and I will teach thee about the way thou must go. Look before thee: dost thou see this narrow way? that is the way thou must go. It was cast up by the patriarchs, prophets, Christ, and his Apostles; and it is as straight as a rule can make it: this is the way thou must go.'
>
> 'But', said Christian, 'are there no turnings nor windings, by which a stranger may lose his way?'
>
> 'Yes, there are many ways butt down upon this, and they are crooked and wide: but thus thou mayest distinguish the right from the wrong, the right only being strait and narrow.'

And so, Christian sets off. Prominent features from the book that are shown in the map include the 'Valley of the Shadow of Death' and 'Valley of Humiliation', where 'Christian fights Apolyon' (with a little vignette scene of Christian doing battle with the Devil); these are both set within the 'Dark Mountains'. Other locations are 'Apostacy', 'Country of Conceit' with a maze, 'Vain Glory', 'Carnal Policy' and 'Graceless' but also 'Morality', 'Honesty', 'Fair Speech' and 'Love-gain'. Along the road is marked 'Belzebub's Castle' with all manner of hazards, 'a bye way to Hell'.

The first book ends with Christian overcoming the hazards and being admitted into the Celestial City. The book recounts his wife's journey to join him.

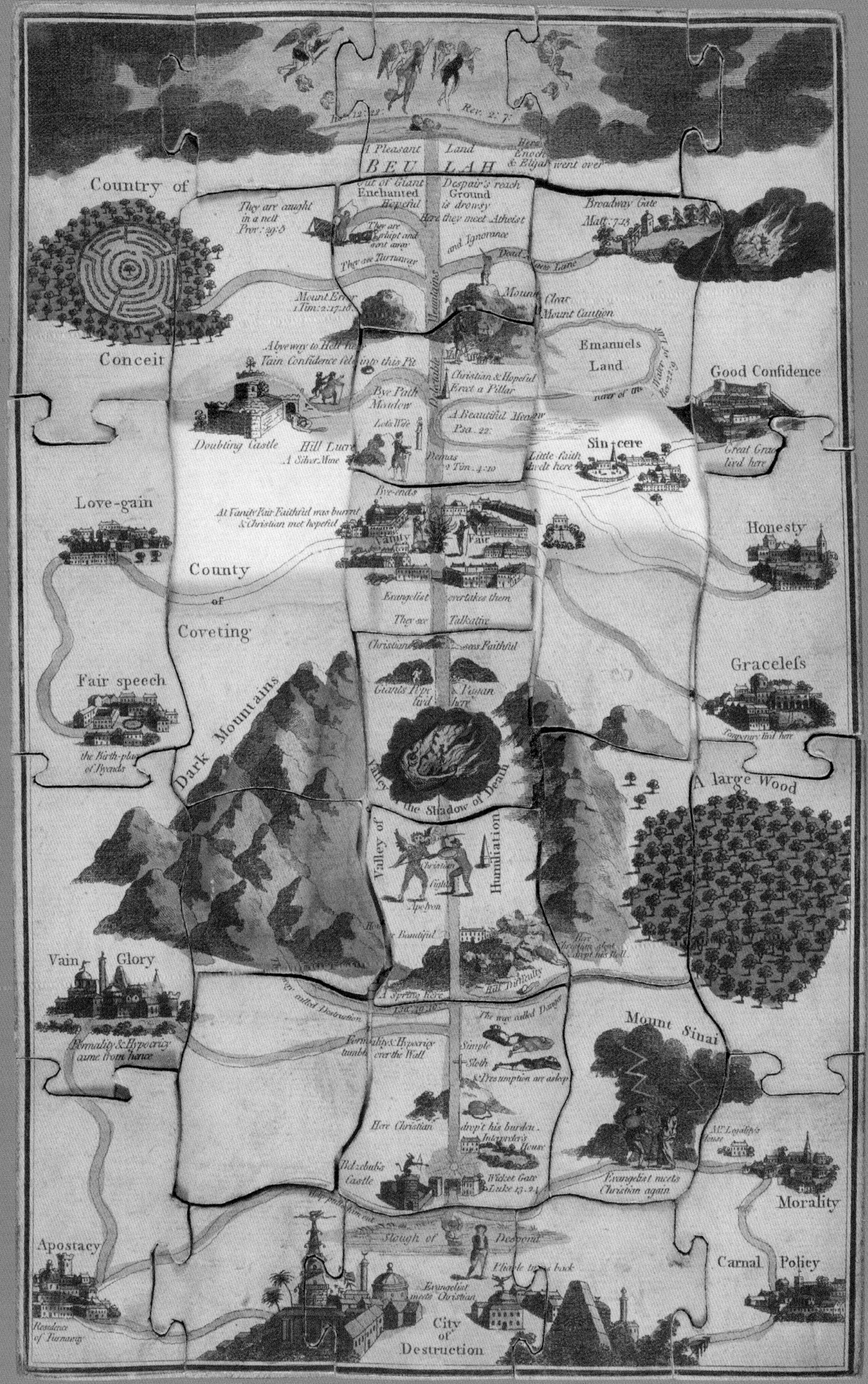

A Pleasant Land
BEULAH
Country of
Conceit
Enchanted Ground
Breadway Gate
Mount Error
Mount Caution
Emanuels Land
Good Confidence
Doubting Castle
Hill Lucre
A Silver Mine
Bye Path Meadow
A Beautiful Meadow
Sincere
Love-gain
County of Coveting
Honesty
Fair speech
Graceless
Dark Mountains
Valley of the Shadow of Death
A large Wood
Valley of Humiliation
Vain Glory
Mount Sinai
Here Christian drop't his burden.
Morality
Apostacy
Carnal Policy
Slough of Despond
City of Destruction

‘Geography Bewitched! Or, a Droll Caricature Map of England and Wales’, 1793

Robert Dighton, London

Copper engraving; image: 181 × 156 mm (7 ⅛ × 6 ⅛ in)

ROBERT DIGHTON (1751–1814) was a talented artist, illustrator and etcher, who exhibited his portraits at the Free Society of Artists and at the Royal Academy between 1769 and 1799. He enrolled in the Royal Academy Schools in 1772 and subsequently established himself as a drawing-master and painter of miniature portraits. He also had a parallel career as a singer at Sadler’s Wells and other leading places of resort.

His career came to an unhappy end; he was also a collector of old master prints, and was a well-known researcher in the British Museum, even before the Print Room opened. Dighton became firm friends with the head of the Print Room, the Reverend William Beloe. Trusted by Beloe, Dighton abused his unfettered access to steal prints from the Museum; some he kept but others he sold, and this was to prove his undoing. In 1806 a dealer who bought a print from Dighton went to the British Museum to compare his purchase with the Museum’s exemplar, and found it missing. Further searches found that other prints that Dighton had sold came from the Museum. Faced with exposure, Dighton confessed, and made a bargain: he would return all the prints he had and identify all the others he had sold, and to whom they went. The British Museum’s Trustees accepted the arrangement; Dighton avoided jail but his reputation was destroyed, and he had to leave London to live an itinerant life, struggling to make a living, while Beloe was summarily dismissed.

The great pity of it was that Dighton was a talented artist and caricaturist, as exemplified by this item, and its siblings (see pp. 108 and 110). They were published first by Carington Bowles in about 1793, and then reprinted by his successors, the partnership formed by his son Henry Carington Bowles and his former apprentice Samuel Carver, leading printsellers of the period, from their shop in St Paul’s Churchyard.

England and Wales are depicted whimsically; although the delineation may not seem familiar, this is John Bull, the popular personification of England (or Great Britain or the British Isles). The image of John Bull was popular from the eighteenth century well into the twentieth, akin to America’s ‘Uncle Sam’. Southern England is formed by a sea-monster; its head is East Anglia, its mouth the Thames Estuary, and its tail is the south-western counties of England, Cornwall and Devon. The north of England is a rather portly, jolly fellow, smoking a pipe and raising a frothing jug of ale as if to toast the viewer; Wales is formed by his cloak, blowing in the wind. The pair seem to be leaping from the sea, like a dolphin with rider, while the seascape below has many ships.

The three maps of this set are among the most famous of all curiosity maps, and spawned a host of imitations for the next thirty years. Such was their popularity that in 1806 the rival firm of Laurie & Whittle appended these lines to their ‘Whimsical Map of Europe’:

> Oft we see in the shops, a print set up for sale,
> England colour’d, an old fellow striding a whale:
> Yes! Old England’s a picture, the sea forms its frame,
> And Hibernia and Scotia they class with the same.

Dighton del.

Geography Bewitched!

or, a droll Caricature ***MAP*** *of* **ENGLAND** *and* **WALES.**

London: Printed for Bowles & Carver, No. 69 St. Paul's Church Yard.

'Geography Bewitched! Or, a Droll Caricature Map of Scotland', 1793

Robert Dighton Sr, London

Copper engraving; image: 181 × 156 mm (7 ⅛ × 6 ⅛ in)

THIS IS the second of three caricature maps that Robert Dighton drew for Carington Bowles (see pp. 106 and 110).

Here Scotland is depicted in a clown's cap with ruff collar and shirt. His cap and head represent Caithness and the Shetland Isles, while the western extent of the ruff collar takes the shape of the Isle of Skye. He is sometimes described as a hunchback, but the tartan of the 'hunchback' does not match his jacket, so perhaps it is merely a cloth bag over his shoulder. It forms the eastern bulge in Scotland, around Fraserburgh and Aberdeen, modern Aberdeenshire.

The ruff of his shirt is the area around Mull, with Fort William and Oban, while the left-hand flaps of his jacket show the Mull of Kintyre and Arran; visible at his waist are the Firths of Clyde and Forth. His left hand, under the bag, forms the county of Fife; he is shown kneeling astride a plumped-up cushion, which depicts southern Scotland, below the Edinburgh–Glasgow axis.

This delineation was quickly picked up by other publishers, and was much admired, copied and reproduced over the next thirty years.

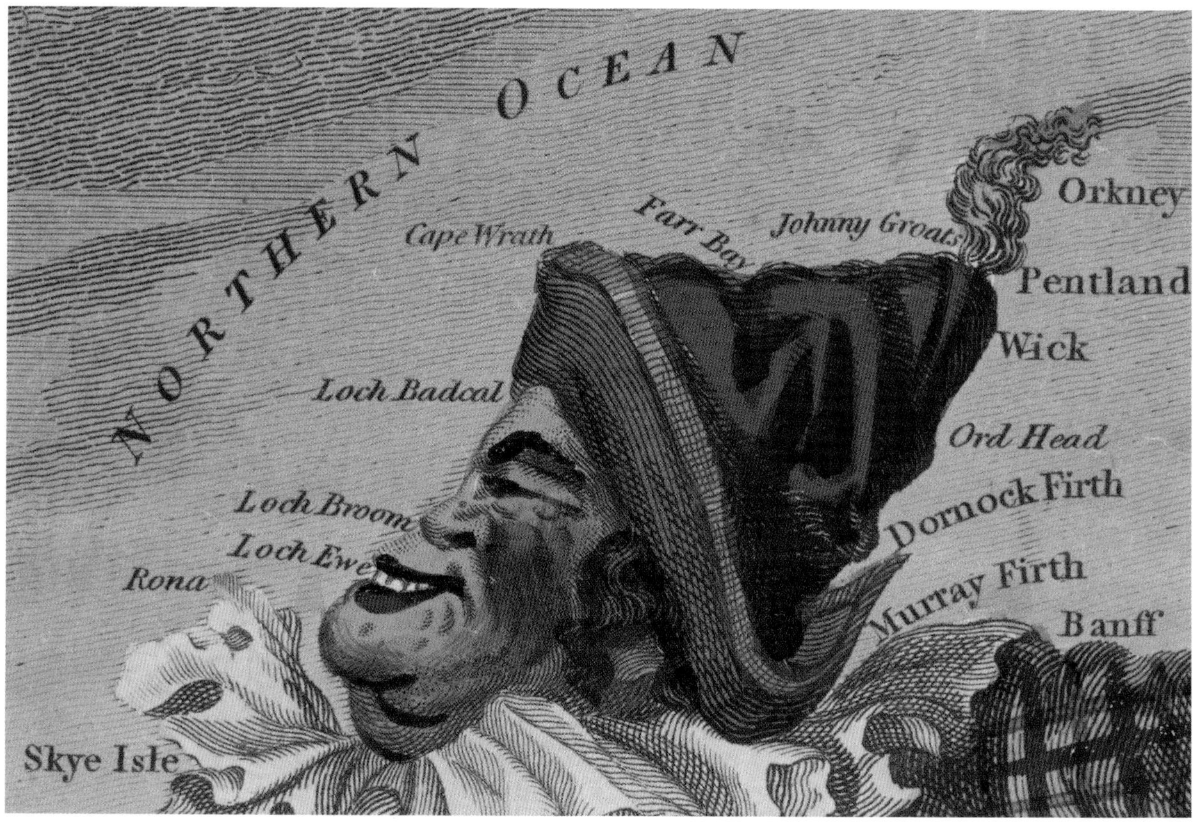

Geography Bewitched!
or, a droll Caricature MAP of SCOTLAND

'Geography Bewitched! Or, a Droll Caricature Map of Ireland', 1793

Robert Dighton, London

'Geography Bewitched! or, a droll Caricature Map of Ireland. This Portrait of Lady Hibernia Bull is humbly dedicated to her Husband the great Mr John Bull.'

Copper engraving; image: 181 × 156 mm (7 1/8 × 6 1/8 in)

THIS IS the third of the three caricature maps drawn by Robert Dighton for Carington Bowles (see pp. 106 and 108). Unlike the other two, this map bears no artist's name, but there can be little doubt that it too was made by Dighton.

Here Ireland is depicted, as described in the title, as 'Hibernia Bull', wife of John Bull, a popular personification for England or the British Isles. 'Hibernia' is shown seated, facing left, playing a harp, with a baby in a sling on her shoulder. Hibernia's head and the sling form the old province of Ulster; the harp forms the province of Connaught, and the folds of her dress Munster, while Leinster is formed by her back and the lower folds of her shawl.

The image here is a fairly benign and even amiable representation; when the Irish Question – Home Rule – became one of the most prominent and divisive issues in British politics in the 1870s and 1880s, the depictions of Ireland became rather more hostile, depicting the lady with the features of an old hag.

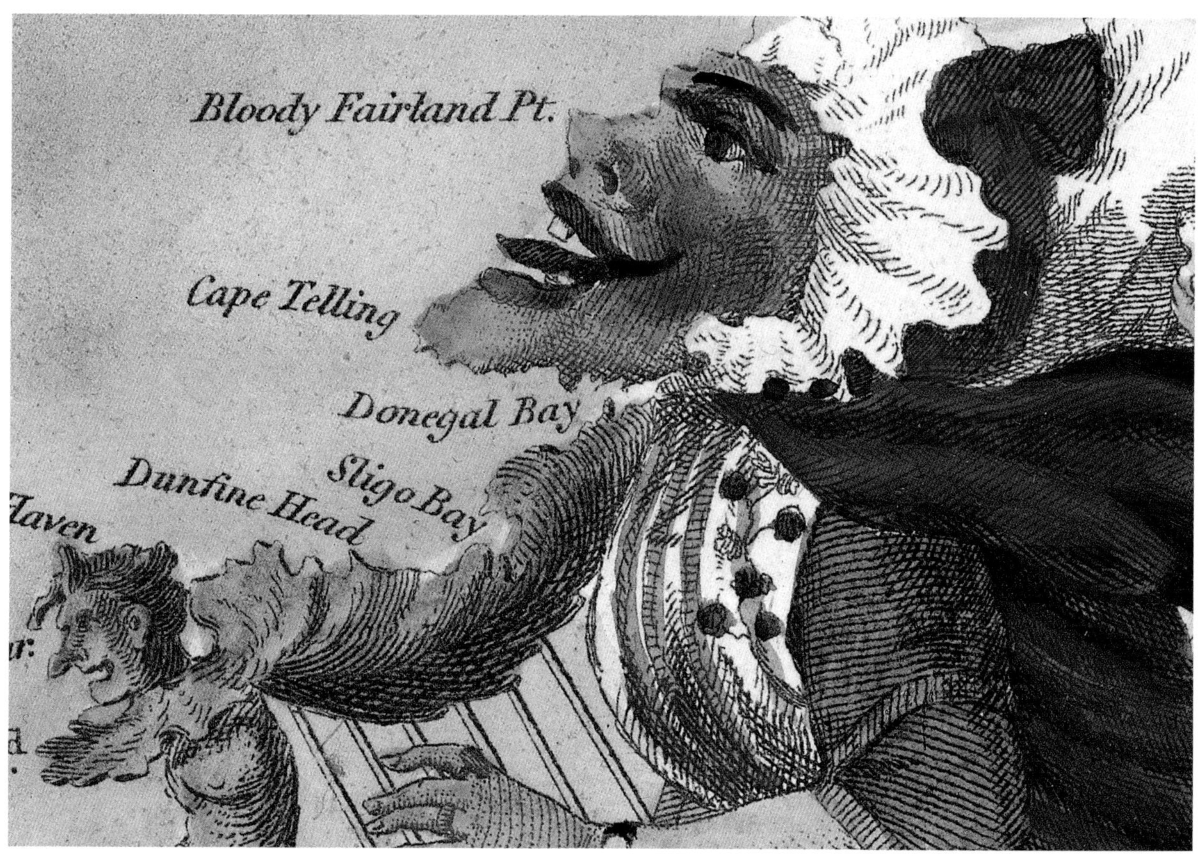

London: Printed for Bowles & Carver, No. 69, St. Paul's Church Yard.

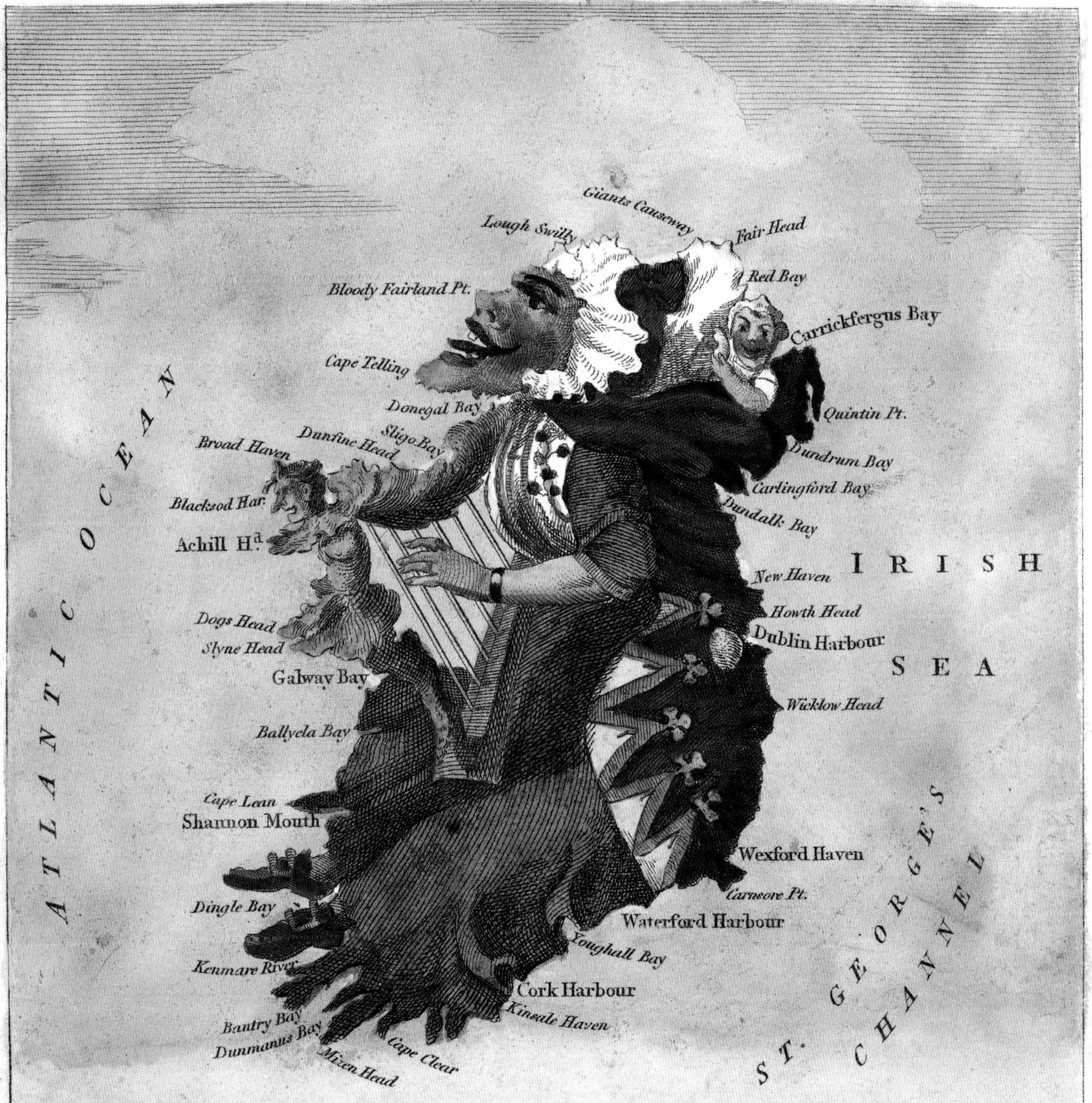

Geography Bewitched!
or, a droll Caricature ***MAP*** *of* **IRELAND.**
This Portrait of **LADY HIBERNIA BULL** *is humbly dedicated to*
her Husband the great MR. JOHN BULL.

'A New Map of England & France', 1793

James Gillray, London

'The French Invasion; – or – John Bull, bombarding the Bum-Boats'.

Copper etching; printed area 334 × 241 mm (13 ⅛ × 9 ½ in)

FOR NEARLY thirty years after the French Revolution and the storming of the Bastille in 1789, British history was dominated by conflict with France, generally termed the 'Napoleonic Wars', although this term properly refers only to one phase. Before Napoleon seized power in 1799, the period after the Revolution might more properly be called 'The Wars of the French Directorate'.

Threat of a French invasion remained a constant fear until the Battle of Trafalgar, in October 1805, when Admiral Nelson inflicted a devastating defeat on the Franco-Spanish fleet, securing British mastery of the high seas for the next century. On several occasions, the French assembled invasion barges, to be used to carry troops and artillery across the Channel but, in the event, no serious attempt at invasion was made.

This famous satirical map caricatures the situation and popular contempt for the would-be invaders in graphic style; the French barges are shown assembled along the Channel coasts of Brittany and Normandy. John Bull, here given the facial features of King George III, is shown defecating on the French, the stream originating from the great British naval base at Portsmouth, and labelled 'British Declaration', as if launching his own armada of 'boats' to wreak havoc on the invaders.

The etching is signed by John Schoebert; there seems little doubt that this was merely a pseudonym or nom-de-plume for James Gillray (1756–1815), one of the greatest of all English satirical artists and engravers. The artistry, the rather near-the-knuckle toilet humour, and the play of words on 'bum-boat' (a recognised nautical term for a small boat used to ferry supplies to larger ships) is characteristic of his style. The print was published by Hannah Humphrey, one of the very few female publishers recorded in London, with whom Gillray lodged at this period.

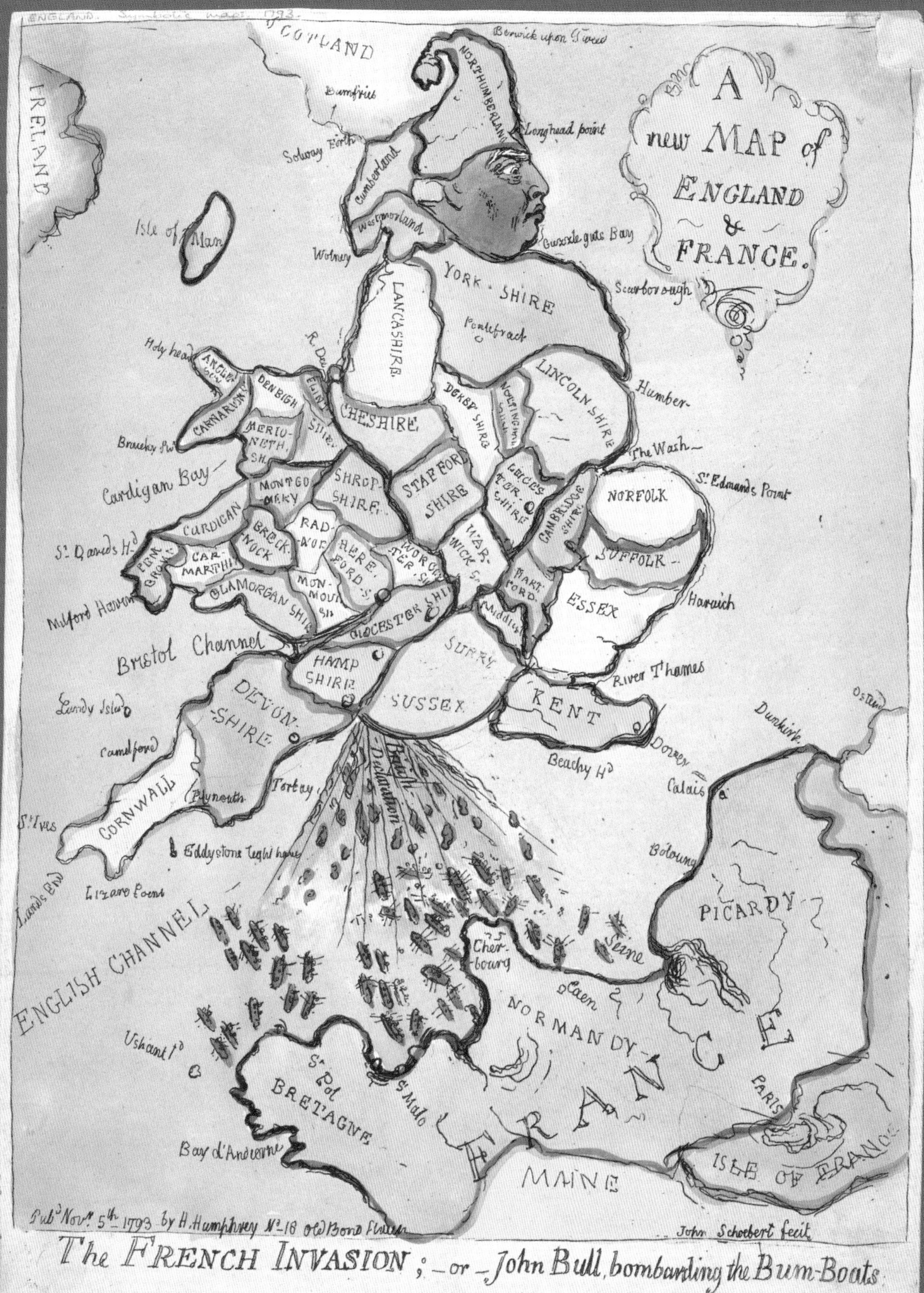

A new MAP of ENGLAND & FRANCE.
SCOTLAND
IRELAND
Berwick upon Tweed
Dumfries
NORTHUMBERLAND
Longhead point
Solway Firth
Cumberland
Westmorland
Isle of Man
Guzzle guts Bay
Wolney
YORK SHIRE
Scarborough
LANCASHIRE
Pontefract
Holy head
ANGLESEY
DENBIGH
FLINT SHIRE
CHESHIRE
DERBY SHIRE
NOTTINGHAM SHIRE
LINCOLN SHIRE
Humber
CARNARVON
MERIONETH SH.
The Wash
Cardigan Bay
MONTGOMERY
SHROPSHIRE
STAFFORD SHIRE
LEICESTER SHIRE
CAMBRIDGE SHIRE
NORFOLK
St. Edmunds Point
CARDIGAN
BRECKNOCK
RADNOR
HEREFORD S.
WORCESTER S.
WARWICK S.
SUFFOLK
St. Davids Hd
PEMBROKE
CARMARTHEN
GLAMORGAN SHIRE
MONMOUTH SH.
HARTFORD
ESSEX
Harwich
Milford Haven
GLOCESTER SHIRE
Middlesex
Bristol Channel
HAMPSHIRE
SURRY
River Thames
DEVON SHIRE
SUSSEX
KENT
Lundy Island
Dunkirk
British Declaration
Camelford
Beachy Hd
Dover
Calais
CORNWALL
Plymouth
Torbay
St. Ives
Eddystone Light house
Bolougne
Lands End
Lizard Point
PICARDY
ENGLISH CHANNEL
Cherbourg
Seine
Caen
NORMANDY
Ushant Id
St. Pol
BRETAGNE
St. Malo
FRANCE
PARIS
Bay d'Anderne
MAINE
ISLE OF FRANCE
Pub'd Novr. 5th 1793 by H. Humphrey No. 18 Old Bond Street
John Schoebert fecit.
The FRENCH INVASION; —or— John Bull, bombarding the Bum-Boats.
MAPS 187. I. 3. (3.)

'Wallis's Complete Voyage Round the World – a New Geographical Pastime', 1796

John Wallis Sr, London

Copper engraving: map at widest: 312 × 623 mm (12 ¼ × 24 ½ in); widest, with text: 482 × 623 mm (19 × 24 ½ in)

THIS POPULAR game of a voyage round the world, by John Wallis Sr (*c.*1745–1818), was first published in January 1796, and reprinted frequently thereafter, with the map unchanged, but the text with the rules reset with the imprint of various different printers. This particular example was printed by Thomas Sorrell in about 1805, though the engraved label on the slipcase retains the date 1802.

The 'New Geographical Pastime' is a game planned for two or three players (although up to six could be accommodated if extra tokens were purchased). It is played on a simple, single-sheet traditional map of the world, with the western hemisphere on the left and the eastern hemisphere on the right. The traveller's route begins at Portsmouth and ends in London – 'Game' – with one hundred numbered locations along the track, keyed to the rules, printed from letterpress and pasted along the lower border of the map, which contains the 'Directions for making a tour round the world', the rules, and descriptions of each of the locations.

Each player was denoted by a pyramid token and given four counters, called markers or servants, at the start of the game. The players would spin the eight-sided teetotum – first to see who started and then to advance around the world. As the players progressed, they would read the description of each place they landed on, finding out something interesting about their destination.

As usual in such games, players had to throw an exact number to finish: 'If the last spin does not exactly make the No. 100, but goes beyond it, he must then go back as many numbers as he exceeds it, and try his fortune again, till one of the players spin the lucky number.'

Among the stops are '43. Calcutta – the principal settlement of the East-India Company in Bengal. Stay here one turn to see the black-hole, where 123 persons was [*sic*] suffocated in 1757'.

The map is interesting for its depiction and descriptions of Australia, then still known as New Holland: '51. New Holland – the largest island in the world.' '52. Port Jackson or Botany Bay, situated on the east coast of New Holland; to this place the convicts are sent from England. Here the traveller must stay 2 turns to view this new colony.' Australia is depicted as having a dotted southern coast, demonstrating continuing uncertainty as to whether Tasmania was connected to the mainland. When the map was drawn in 1796 there was still genuine uncertainty, but by the time this particular printing appeared Tasmania had been proved to be an island: in 1798–1799 George Bass and Matthew Flinders circumnavigated Van Diemen's Land (as Tasmania was then known) sailing through 'Basses Strait' as it was to be named, shortened to Bass Strait, between the Australian mainland and Tasmania. There was, quite naturally, often a considerable time-lag between fresh geographical discovery and its adoption on commercial maps, especially those where the game was the thing.

The 'death' square is the notoriously dangerous '89. Magellan Straights – Discovered by Ferdinand Magellen [*sic*]. Here the traveller is shipwrecked, and thereby loses his chance of the game.'

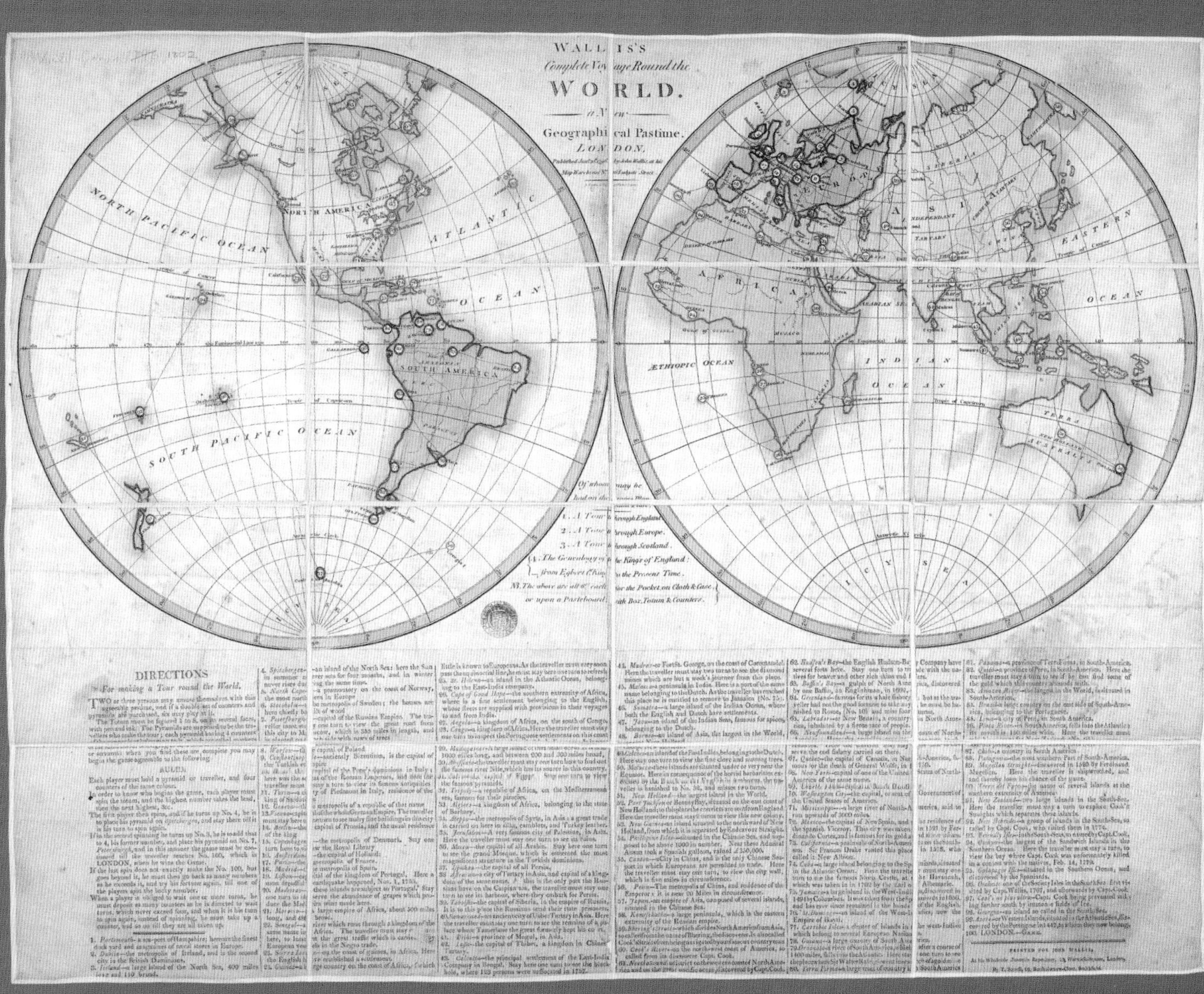
WALLIS'S
Complete Voyage Round the
WORLD.
a New
Geographical Pastime.
LONDON.
NORTH PACIFIC OCEAN
NORTH AMERICA
ATLANTIC
OCEAN
SOUTH AMERICA
SOUTH PACIFIC OCEAN
AFRICA
ÆTHIOPIC OCEAN
INDIAN
OCEAN
ASIA
EASTERN
OCEAN
ICY SEA
DIRECTIONS
For making a Tour round the World.
RULES.

'Allegorical Map of the Track of Youth, to the Land of Knowledge', 1798

Robert Gillet, London

Copper engraving; border: 137 × 185 mm (5 ⅜ × 7 ¼ in)

ROBERT GILLET (d. 1797) was a probably a French émigré who escaped to London during the early years of the French Revolution. In or before 1796 he had begun work on his book *Moral Philosophy and Logic. Adapted to the capacities of youth.* He published the accompanying map, described here, in that year, although the book itself was not to be printed for another two years, after Gillet's death.

Moral Philosophy is a lengthy essay seeking to explain and instil proper moral values in the young reader, warning of the pitfalls to those who fail to live by those values, above all by the application of reason and forethought to life's situations, all set within a Christian context. The book is illustrated with a small map, engraved by Vincent Woodthorpe, designed to complement the text, with its own key. The book is rare. The *English Short Title Catalogue*, a census of library holdings, records only three copies in British institutions, and one in the United States, testament either to poor sales or to the heavy use (and consequent wear and tear) of the book by its avid readers. This copy lacks the four-page explanation of the map, which would expand on the features in the map that are lettered as a key to the text.

The map is, at simplest, the route of youth from inexperience to maturity across 'The Ocean of Experience', again using the sea metaphor for life's experiences. This route is labelled 'the Youth's Tract', and begins at 'Dark Bay', keyed 'A', to the right of the title. Along the way are the hazards of personality normally faced by young people growing up: there are mountains of 'Idleness and Obstinacy' as well as the island of 'Dissipation'. The next landfall is the 'Land of Remorse' with 'Cape Repentance' as the young traveller realises the errors of his ways. Determination to reform takes him (or her) to 'Penance I' and the 'Archipelago of Promises' (here perhaps in the sense of unfulfilled promise and a commitment to fulfil that potential), leading on to 'Endeavour I', and 'Success I', through the 'Sands of Patience' before arriving on the mainland, which is divided by 'The Torrent of Passions' into 'Land of Knowledge' and 'Terra Firma of Happiness'.

As the traveller approaches the mainland, he is guided through the 'Rocks and whirlpools of Presumption' by the 'Light of Reason' in the 'Land of Knowledge' and the 'Light of Religion' in 'Terra Firma of Happiness', perhaps indicating that the author believed that happiness could be sought only when a career was properly established. Even here, there are possible false turns; the traveller could leave his track and sail to the 'Coasts of Hardship' and become trapped in the 'Gulph of Vanity', or be shipwrecked on the rocks around the 'Light of the Passions'.

The Ocean of Experience
Lake Tears
I
LAND of REMORSE
Cape Repentance
Penance I.
Cape Grief
Weariness
Levity I.
Trifle I.
L
J
G
Misery I.
Disappointment I.
DISSIPATION I.
F
ARCHIPELAGO of PROMISES
Docility I.
Good will I.
M
N
Endeavour I.
Cape Courage
E
Snowy and Icy Mountains
SUCCESS I.
O
Satisfaction I.
Reward I.
Q
Rocks of Idleness and Obstinacy
D
A
Dark Bay
the Youths Tract
C
R
COAST OF HARDSHIP
SANDS of PATIENCE
P
C. of Good Hope
Rocks and Whirlpools of Presumption
Light of the Passions
Light of Reason
B
V
Light of Religion
W
Diamond Mines
LAND OF KNOWLEDGE
Honey Hill
Wood of
Golden Apples
Gold Mines
The Torrent of Passions
U
S
TERRA FIRMA OF HAPPINESS
Plains of Content
Mines of Precious Stones
T
ALLEGORICAL Map of the Tract of Youth to the LAND of KNOWLEDGE
Published 1st of Feby 1796, by R. Gillet.
'd at Stationer's Hall.
V. Woodthorpe No. 27, Fetter Lane.

Battle of Trafalgar Commemorative Creamware Jug, 1805

Anonymous, [Worcester?]

'Battle off Trafalgar Gained by the British Fleet under L.d Nelson on the 21 of Oct.r 1805. Against the combined Fleet of France & Spain, in which action the intrepid Nelson fell covered with Glory and renown.'

Earthenware jug; height: 150 mm (5 ⅞ in); circumference at widest: 340 mm (13 ⅜ in)

THIS IS a commemorative creamware (a cream-coloured, refined earthenware) jug produced shortly after the Battle of Trafalgar, to commemorate that great victory and mourn the loss of Britain's greatest naval hero, Admiral Lord Nelson, who was killed during the battle.

The Battle of Trafalgar in 1805, fought between the British and a joint Franco-Spanish fleet, was a complete victory for the British. If there had ever been a possibility of French invasion, there was no longer; Britain was to enjoy the unchallenged rule of the seas for the next hundred years and more. This, coupled with the death of the conquering hero at the moment of victory, caused Trafalgar to take on an iconic aura, probably matched in public perception only by the twentieth-century Battle of Britain. It is rare that any battle can be so completely associated with a particular phrase, but Nelson's last general order before battle was joined is as well known today as in the immediate aftermath of victory: 'England expects that every man will do his duty.'

The jug is transfer-printed in black on the two faces. While there is some doubt, it is thought that the process of transfer printing images onto pottery items was developed by the engraver Robert Hancock, and introduced in Worcester in 1756. The process was complicated because of the need to create an image on a curved surface. This jug is an early example of a transfer-printed map, and shows not only the problems of working on a curved surface, but also the fragility of the medium, at risk of getting bumped, scratched, or even dropped.

On one side is a portrait of Nelson, based on an engraving by John Chapman published in December 1798. Above the portrait is inscribed Nelson's last order. On the other side is a small plan of the battle, with the note:

> The Enemy's Line consisted of 18 French and 15 Spanish Ships making 33 to our 27. The action began at Noon the Guns muzzle to muzzle and at 3 O Clock pm it terminated splendidly glorious to the British Arms. Nineteen Ships of the Line struck in which were 3 Flag Officers viz Adml. Villeneauve Commander in chief, Don Ignatio Maria d. Avila since dead Vice Admiral, and Don Baltazar Hidalgo Cisneros, Rear Admiral.

There are two known plans of the battle, but this is the better, as it clearly shows the crucial moment when the British ships, arranged in two columns line astern, met the enemy fleet at right-angles – 'crossing the T' as this classic and hazardous British naval manoeuvre is known. They passed through the enemy line and cut the opposing fleet into three parts, which could be engaged separately. The tactic was very risky as the British fleet had to sail through enemy fire to make contact, while only able to bring to bear a small number of guns to reply. Still the battle ended with the Franco-Spanish fleet losing twenty-two ships, without the loss of a single British vessel.

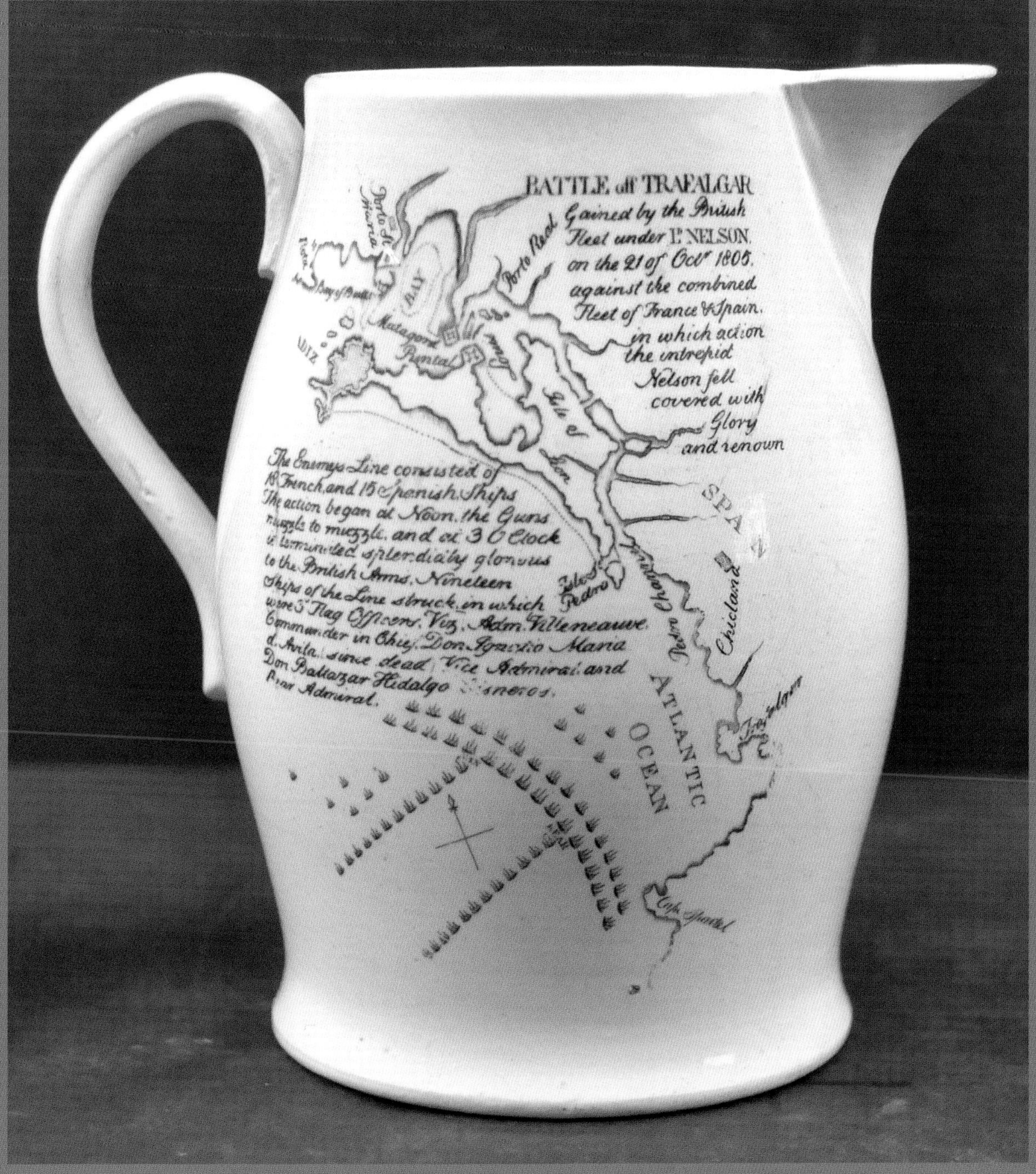
BATTLE off TRAFALGAR
Gained by the British
Fleet under Ld. NELSON,
on the 21 of Octr 1805,
against the combined
Fleet of France & Spain,
in which action
the intrepid
Nelson fell
covered with
Glory
and renown
The Enemys Line consisted of
18 French and 15 Spanish Ships
The action began at Noon, the Guns
muzzle to muzzle, and at 3 O Clock
it terminated splendidly glorious
to the British Arms, Nineteen
Ships of the Line struck, in which
were 3 Flag Officers, Viz. Adm. Villeneauve
Commander in Chief, Don Ignatio Maria
d. Avila, since dead, Vice Admiral and
Don Baltazar Hidalgo Cisneros,
Rear Admiral.
Porto Real
Puntal
SPAIN
ATLANTIC
OCEAN
Chiclana

'A Whimsical Sketch of Europe', 1806

Robert Laurie & James Whittle, London

Copper engraving; border: 181 × 251 mm (7 1/8 × 9 7/8 in);
widest with text: 347 × 251 mm (13 5/8 × 9 7/8 in)

THE PARTNERSHIP of Laurie and Whittle was formed by Robert Laurie (1755–1831) and James Whittle (1757–1818) when in 1794 they took over the business of Robert Sayer, one of the great English map and printsellers and publishers of the previous fifty years. At first they exploited the wealth of material accumulated by Sayer but gradually began to bring out their own fresh material, such as this satirical map of Europe, published in December 1806.

At the time Britain was embroiled in a war with the Emperor Napoleon's France and her allies. The decisive naval victory won by Admiral Lord Nelson at the Battle of Trafalgar (21 October 1805) had ensured that Britain was safe from the threat of invasion and Napoleon turned his attention to the east in campaigns that hardly involved the British. That image of Britain watching a European war from the sidelines is very much conveyed by this attractive map. While it is hard to gauge its later influence, this little map is the forerunner of the more elaborate allegorical maps of Europe published from the 1870s onwards, by Goggins, Rose and others.

The British Isles is depicted as John Bull astride a sea-monster (a whale according to the text), an image inspired by Robert Dighton (see p. 106), and alluded to in the first lines of the poem printed below the map:

> OFT we see in the shops, a print set up for sale,
> England colour'd, an old fellow striding a whale:
> Yes! Old England's a picture; the sea forms its frame;
> And Hibernia and Scotia they class with the same.

The allegory is constructed around the image of portraits on a landscape, with the author as the 'showman'. Initially the viewer is asked why Britain can be seen so clearly: 'Would you ask me what chrystal, so clear, 'tis before? [i.e. over Britain]; Tis the large lens of liberty plates them o'er'. By contrast, the image of France is indistinct:

> There's a picture just by these, all tatter'd and torn;
> One all blood-smear'd and red; that is France
> I'd be sworn,
> What's that covers its surface, all shining like wax?
> 'Tis the varnish of tyranny; see how it cracks!

Every verse describes a European country (or countries) in mainly derogatory terms, emphasising the malign influence of France. Each verse ends with the line 'And the sturdy old fellow's astride on his whale', contrasting that country with Britain's happy existence safe from France, ringed by the 'wooden walls of Old England' – the Royal Navy.

Holland (the Low Countries) is described as a 'drunken old sot ... how mildew'd and eat with the rot!', and Germany as a prison, with the rulers of the constituent states peering through the bars. Conquered Italy can longer 'boast of its olive retreat', while Spain is described thus:

> See that portrait! away to the left, in the rear;
> Descriptive of jealousy, anguish, and fear;
> That which tremblingly hangs by its slight golden string;
> And seems ready to drop from its great gilded ring:
> That poor portrait is Spain; and the ring is its crown:
> See! the motions of France almost shatter it down!

The Prussian soldier stands ready, while in the south-east of the map, the Turkish Sultan approaches Britannia in friendship.

A POETICAL DESCRIPTION OF THE MAP.

OFT we see, in the shops, a print set up for sale,
England colour'd, an *old fellow striding a whale:*
Yes! Old England's a picture, the sea forms its frame,
And Hibernia and Scotia they class with the same.
Would you ask me what crystal, so clear, 'tis before;
'Tis the large lens of liberty plates them all o'er.
See Hibernia! a *harp*, gilt with industry's gold;
And the *seed-spreading thistle* is Scotia we're told.
Whilst crown'd with the thistle, the harp on his side,
He hangs o'er the ocean, protector, and pride.
See! the rude hand of time has not dar'd to assail
Yet "the sturdy old fellow astride on his whale."

There's a picture just by these, all tatter'd and torn;
One all *blood-smear'd and red;* that is France, I'd be sworn:
What's that covers its surface, all shining like wax?
'Tis the varnish of tyranny; see how it cracks!
There's a portrait beside of a *drunken old sot;*
That is Holland: how mildew'd and eat with the rot!
And its station, alas! by the very first glance,
We may see does not suit; for its crowded by France.
See! these cast such a shadow throughout the whole space,
That they seem to distort every natural grace,
Save with those where their shadows don't reach to prevail,
And "the sturdy old fellow's astride on his whale."

Yon *rich landscape* behold! it has seen better days;
And illumin'd was once with bright liberty's rays:
That is Switzerland; fam'd for its picturesque views;
But France now clouds its prospects; its *vines* turns to *yews.*
View that one to the right, that to France very near,
That bore once brightest colours, transparent and clear;
A tint most harmonious, like opening dawn;
That is Italy; mark how its colours are gone!
It no longer can boast of its olive retreat,
Now its large lumb'ring neighbour so elbows its seat!
Yon *vineyard-scene's* Portugal: some tints still prevail,
Like "the sturdy old fellow's astride on his whale."

See that *portrait!* away to the left, in the rear,
Descriptive of jealousy, anguish, and fear;
That which tremblingly hangs by its slight golden string;
And seems ready to drop from its great gilded ring:
That poor portrait is Spain; and the ring is its crown:
See! the motions of France almost shatter it down!
Yon's a picture surrounded by some smaller ones,
Which seems like a *debtor that's haunted by duns;*
That is Germany, tended by all its small states,
And it looks tow'rds France like a pris'ner through grates.
They're so shadow'd by France, not a ray can prevail,
Like "the sturdy old fellow's astride on his whale."

The next portrait is Prussia, a *soldier*, afar,
Just arous'd into action, and rushing to war.
Look again, tow'rds the right; see! *three snow-scenes* appear,
See! their shadows diminish; their colours grow clear:
Russia, Sweden, and Denmark, distinguish these three,
Still there's one distant picture o'erlook'd you may see:
It is Turkey; and see it tow'rds England advance;
For its terribly lately been shadow'd by France.
Now, to tell you, as showman, it falls to our lot,
Who plac'd these around in each singular spot,
'Twas the hand of that master, which long shall prevail,
For "the sturdy old fellow's astride on his whale."

‘Geographical Recreation, or, a Voyage Round the Habitable Globe’, 1809

John Harris Sr, London

Copper engraving; border: 489 × 476 mm (19 ¼ × 18 ¾ in); at widest: 570 × 488 mm (22 ½ × 19 ¼ in); the maps each 75 × 75 mm (3 × 3 in)

JOHN HARRIS SR (1756–1846) was the foremost English publisher of children’s books in the first quarter of the nineteenth century. Originally employed by Elizabeth Newbery, herself a member of a leading family of publishers of juvenilia, he took over her business in 1801. Extremely successful, he retired in 1824 in favour of his son, also a John, having made his fortune. Many of the books that Harris published had a geographical theme, generally illustrated with maps and views; he also published numbers of board games, such as this ‘Geographical Recreation’ in 1809.

In keeping with the style of Harris’ business, the ‘Geographical Recreation’ has map elements, but is largely pictorial; it is yet another variation of the ‘Game of the Goose’, where the players have to move along a course numbered from 1 to 116. However, the board is divided into four segments, starting in the top left quadrant, with 1 on the inner circle, the player progressing along each row of the quadrant to the end, then down to the start of the next row, moving outwards to the border with the map being the final square of the quadrant. Players would then move on to the next quadrant, and go through the same sequence again, until they reached 116, the map of Europe, which is labelled ‘Game’.

Each square is numbered. The first quadrant, squares 1–28, is of scenes of Europe; the second quadrant begins with a map of Asia, 29, and then squares 37–57 depicting scenes from Asia; square 58 is the map of the Americas and 59–86 American scenes; square 87 is the map of Africa and the squares to 115 show images of Africa. Square 116 is the map of Europe.

The start, square 1, shows the traveller in the Arctic, wearing snow-shoes, with a reindeer pulling a sleigh in the background; square 3 depicts whaling, 15 ice-skating in the Low Countries, 16 the Emperor Napoleon, 19 probably Mount Vesuvius in eruption, 24 a caricature Frenchman eating a whole frog, next to 25, a caricature of a jolly Englishman with a flagon of ale and side of ham.

The images of Asia include a Chinese pagoda (33), the Emperor in a procession (34), an elephant with its turret (37) and the Colossus of Rhodes (54). For the Americas, most of the scenes depict indigenous tribes but square 59 could be Columbus, and square 74 is certainly Sir Walter Raleigh planting the royal standard in the New World. Square 77 is Niagara Falls, 81 Easter Island and 86 depicts the death of Captain Cook in Hawaii. In Africa, square 89 is a rather cruel torture scene, 93 Table Mountain, 103 a slavery scene and 111 the Sphinx in Egypt.

The central roundel is an allegorical scene with the continents in human form, imagery readily understood by someone from 200 years ago, with ‘Europe’ astride the globe, Africa with the lion and elephant, and America and Asia with the camel, all bearing gifts in tribute.

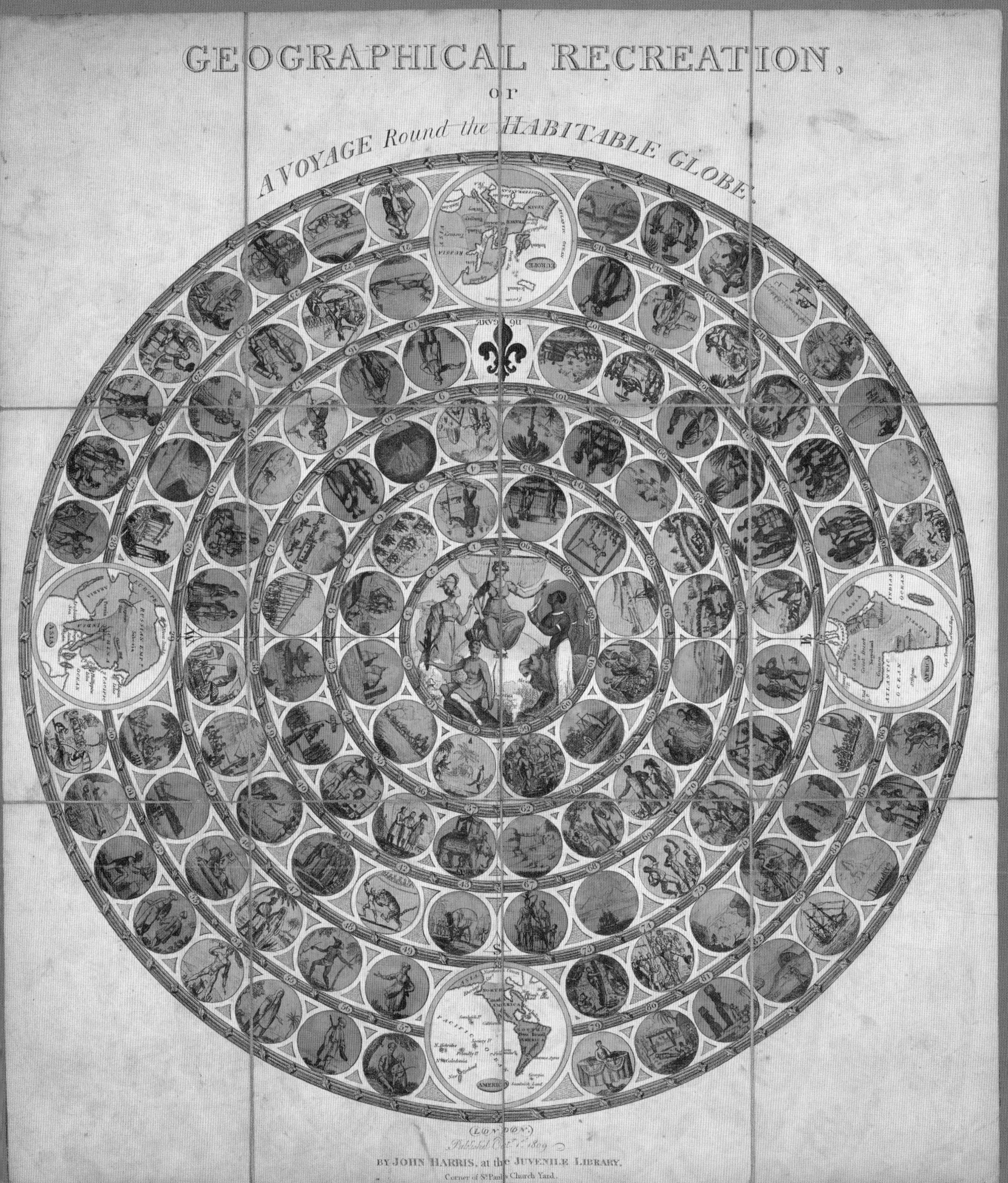
GEOGRAPHICAL RECREATION,
or
A VOYAGE Round the HABITABLE GLOBE.
(LONDON.)
Published Oct.r 1. 1809
BY JOHN HARRIS, at the JUVENILE LIBRARY,
Corner of St. Paul's Church Yard.

'Map of Green Bag Land', 1820

Joseph Onwhyn, London

Copper engraving, with letterpress text; map border: 187 × 255 mm (7 ⅜ × 10 in); with text: 247 × 397 mm (9 ¾ × 15 ⅝ in)

JOSEPH ONWHYN (1787–1870) was a bookseller, printer and publisher who dabbled as an engraver and illustrator, producing a small number of satirical illustrations, including this map. His son Thomas followed in his footsteps, producing a cartographic satire on the Crimean War.

The map is a political satire on the very messy divorce proceedings between King George IV and his estranged wife Caroline of Brunswick. The couple married in 1795 while he was Prince of Wales, but the match was never a happy one and they separated in 1796. George resolving this embarrassment by the simple expedient of paying Caroline to go away, with a settlement of £10,000 per annum to keep her in luxury.

Caroline settled in Italy, where she established her 'court'. The boot of Italy is just visible in the top right-hand corner of the map. There she employed a tall and handsome Italian, one Bartolomeo Bergami (or Pergami), successively as her courier, bodyguard, groom of the bedchamber, major-domo and then Grand Master of the Order of St Caroline – an order she founded in her own honour. Inevitably, rumours arose that the pair were lovers. In 1818 George instructed William Cooke, a barrister, J. A. Powell, a solicitor, and a Major Browne to go to Milan and gather the evidence necessary to obtain a divorce by Act of Parliament. Depositions were taken but nothing more happened until the matter became critical with George's accession to the throne on the death of his father in January 1820. The evidence collected by the 'Milan Commission' was brought home in green bags, hence the title of the map. The contents were to form the basis of the accusations in Caroline's 'trial' – in fact a bill to strip Caroline of her title and to end the marriage by Act of Parliament, brought before the House of Lords in August 1820.

Although this map is undated, publication can be established from an advertisement in the *Morning Chronicle* newspaper of 7 November 1820, the day after the House of Lords narrowly voted to approve the divorce. The Bill was never presented to the Commons, in the near certainty it would be rejected there – Caroline had become a heroine to every opponent of a deeply unpopular monarch; she was cheered in the streets on her every appearance.

The map refers to the proceedings in the House of Lords and the side panels of text clarify what is intended. The text begins, 'The Capital of the Land is the surprising City of Non Mi Recordo'. '*Non mi ricordo*' ('I don't remember') was an answer repeated over two hundred times by Theodore Majocchi, one of Caroline's servants called to give evidence. It became the archetypal catchphrase of the day, guaranteed to raise a laugh in any company for a generation. The 'Colony of Double Entendres' is self-explanatory – the proceedings at once pathetic, tragic, farcical and often hilarious. Within the map can be seen 'Gold River', which refers to George's payments to his wife, but also the bribes offered to many of the witnesses.

The caricaturists had a field day – it is said that some £2,500 was spent by the king in buying up and suppressing just the most offensive of the images. Arrayed both inside and outside the walls of Green Bag Land are various printing presses, representing the newspapers for or against the two principal characters in this Georgian-era soap opera. In 'Bull Country' (England) these include *The Times*, [Morning] *Chronicle*, *Statesman* and *Herald*; in 'Green Bag Land' are the *Guardian*, the *Post* and the *Courier*.

A fickle public soon turned against the Queen, as sinning as she was sinned against: the ditty of the day became,

> Most Gracious Queen, we thee implore
> To go away and sin no more;
> But, if that effort be too great,
> To go away, at any rate.

Caroline was not invited to George's Coronation, 19 July 1821; she turned up, nonetheless, banging on doors, unsuccessfully seeking admission. Seemingly broken by this further rejection, she died three weeks later.

MAP OF GREEN BAG LAND.

Description.

oundaries.]—This newly discovered is bounded on three sides by the ntry of the Bulls, from which it is se- ted by a *Great Wall*, well planted with on to protect it from incursions, and e fourth side it is bounded by the Sea.

---

ties.—The Capital of the Land is the sing City of *Non mi Ricordo*, which justly be termed the eighth wonder of orld; it is divided into the upper and , and it is here that the Great Fair rket for sale of *Non mi Ricordos* is d on.

the north of the City is an extensive ce, formerly used to grow Cotton ow appropriated to a *Crop of Green* which yield an extraordinary Fruit *Non mi Ricordo*, from whence both ty and the Colonists take their name. uous to this is the Colony of *Double dres*, who also drive a fine trade by ommodities, and almost rival the *Non ordos* by the afficacy of their *Double dres*.

---

rs.]—A great variety of fine Rivers this wonderful Country, the princi- which are the following:—the *Waters ivion* take their rise in the Palace of eat Hum, and after watering the nd fertilizing the respective Colonial ces, empty themselves in the great g Tub of the *Non mi Ricordos*. aters of this River are so bitter that d be death to those who drink it, t its noxious qualities dissipated by er of Gold, which forms a junction a little before entering the City.

River of Gold takes its rise in the of the Bulls, in the City of In- and flows into the Green Bag hrough one of the *Cannons* in t Wall. After forming an agree- ke in the Gardens of the Great it joins the aforesaid River,

d and Published by J. Onwhyn,

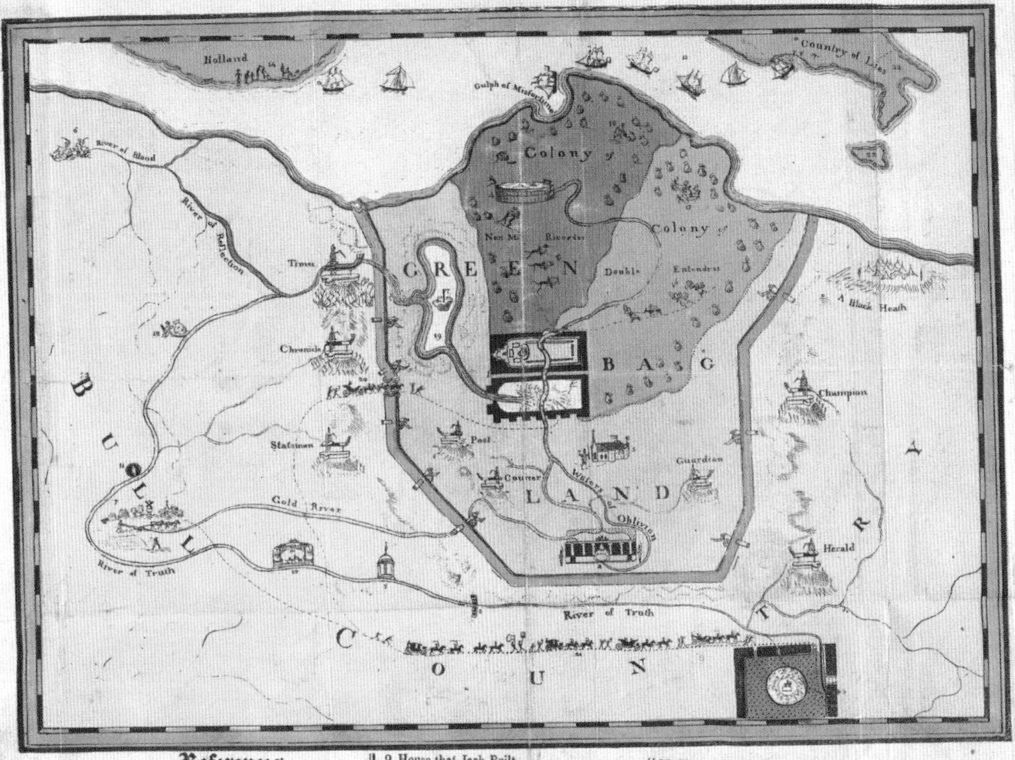

References.

1 City of *Non mi Recordo*.
2 Palace of the Great Hum.
3 Palace of the Sultana Hum.
4 Great Washing Tub of the *Non mi Recordos*.
5 The High Priest's House.
6 City of Peterloo.
7 City of Industry.
8 City of the Matrimonial Ladder.
9 House that Jack Built.
10 City of the Farce of the Green Bag.
11 City of the Man in the Moon.
12 City of the Dainty Dish to set before a King.
13 Vessels with *Non mi Recordos*, &c. on their Voyage to the Land of Green Bags.
14 *Non mi Recordos* metamorphised into Gentlemen, previous to landing in the Colony.
15 *Non mi Recordos* going to Market.
16 *Double Entendres* going to Market.
17 *Non mi Recordos* and *Double Entendres*, enjoying the Reward of Industry.
18 One of the High Priests watering the Green Bags.
19 The Lake of Despair.
20 Sultana Hum going to the Great City.
21 An Address of the Bulls going up to Sultana Hum.
22 *Precedents* for the Bulls.

Description.

and pursuing similar meanderings sweetens its oblivious Waters.

The *River of Truth* takes its rise in the small Palace of the Sultana Hum, which is situated in the Bull Country, and after passing through the various Cities which appear by the Map to be constructed on its Banks, it flows towards the Great Wall where it miraculously ascends a steep Mountain, and by means of a *Steam Engine* is precipitated through a Printing Press over the Great Wall presenting a beautiful cascade in defiance of the Cannons; it immediately forms a great Lake in the Valley of Despair, and rushing from thence in a torrent into the *lower* division of the City, it entirely disperses the before-mentioned River of Gold and Waters of Oblivion.

Mountains.]—The most celebrated Mountains are those in the Bull Country. An immense chain surrounds the Land of Green Bags, and are indeed the only security to the Inhabitants of the Bull Country; they are covered with noble forests of waving Goose Quills which give them a beautiful appearance, and the strata of the Mountains are composed of small particles of metallic ore, which, by a particular configuration in the Machines upon their summits keep the whole Land of Green Bags in a state of anxiety, in as much as they are wholly inaccessible to the range of their Cannon, as well as protect the River of Truth which the *Non mi Ricordos* and other Inhabitants of the Green Bag Land have been endeavouring to turn into another Channel.

Mountains are also observed in the Green Bag Land with similar Machines on their summits, but they are wholly artificial, and as they owe their support *solely* to the *River of Gold* and are *far* from the *River of Truth*, their influence over the Bull people is trifling unless conducted through Cannons or transmitted by Sabres as at the *Field of Peterloo*.

Catherine Street, Strand.—Price 1s.

'Labyrinthus Londoninensis, or The Equestrian Perplexed', [1830]

Charles Ingrey & Frederick Waller, London

'A Puzzle Suggested by the Stoppages occasioned by repairing the Streets. The object is to find a way from the Strand to St. Paul's, without crossing any of the Bars in the Streets supposed to be under repair.'

Lithograph; map: 115 × 215 mm (4 ½ × 8 ½ in); at widest: 195 × 268 mm (7 ⅝ × 10 ½ in)

OF THE many maps described in this volume, this one will perhaps resonate most with the modern commuter. This is a game – or perhaps a subtle complaint – born out of the constant disruption of roadworks in central London. The frustrated traveller is compelled to undertake interminable detours – 'going round the houses', as it is put – and prevented from making the simple walk from the Strand to St Paul's Cathedral directly. At its simplest, the traveller has to find a route from the Strand, at the bottom left of the map, to St Paul's Churchyard, denoted by Sir Christopher Wren's magnificent cathedral placed at the centre of the map, without crossing any of the streets 'barred' by black lines, signifying roadworks:

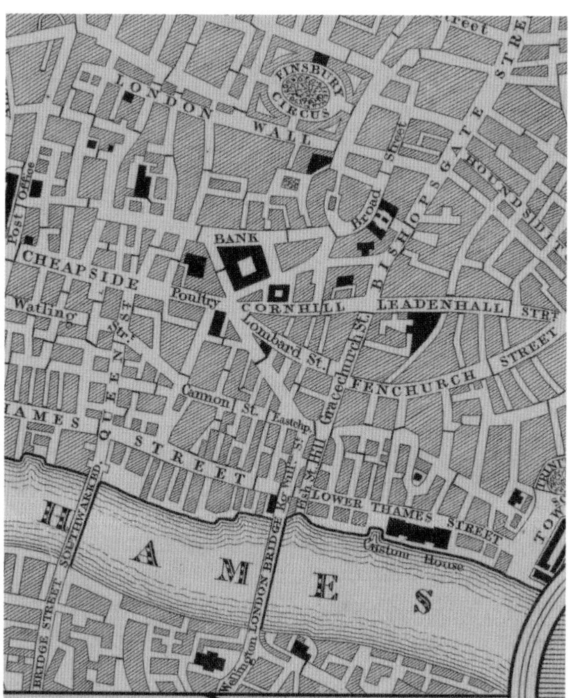

Mending our Ways, our ways doth oft-times mar,
So thinks the Traveller by Horse or Car,
But he who scans with calm and patient skill
This 'Labyrinthine Chart of London', will
One Track discover, open and unbarred,
That leads at length to famed St. Pauls Church Yard.

The publisher Charles Ingrey (1800?–1877) had a Lithographic Printing Office right in the middle of the disruption at No. 310 Strand, his exasperation perhaps the worse in that much of his recorded output was of plans relating to road improvement schemes and maps to show 'the practicability, utility and benefit of railroads'. He later became an accountant. His neighbour and co-publisher, Frederick Waller, a stationer at 49 Fleet Street, was no doubt similarly affected.

The map is bounded in the north by the Charterhouse, by Southwark and the river to the south, Covent Garden in the west and Aldgate in the east. In the corners of the map are four coats of arms: the Royal Arms, top left; the City of London arms, top right; the City of Westminster arms, bottom left; and those of Bridge House at bottom right.

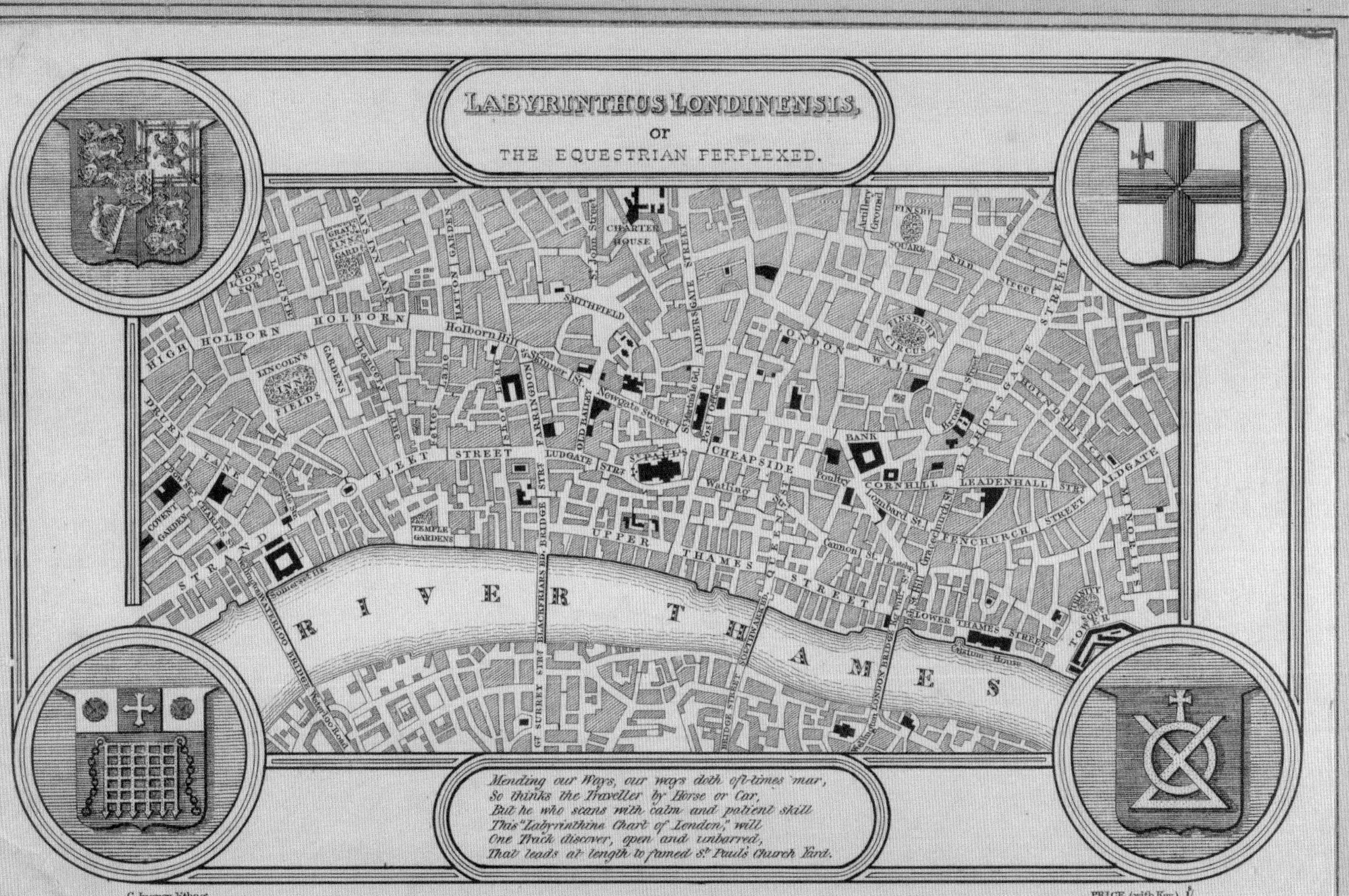
A Puzzle Map of London &c when the Streets are stopp^d up by C Ingrey.
LABYRINTHUS LONDINENSIS, or THE EQUESTRIAN PERPLEXED.
Mending our Ways, our ways doth oft-times mar,
So thinks the Traveller by Horse or Car,
But he who scans with calm and patient skill
This "Labyrinthine Chart of London," will
One Track discover, open and unbarred,
That leads at length to famed St. Paul's Church Yard.
C. Ingrey lithog.
PRICE (with Key) 1/.
A PUZZLE
Suggested by the Stoppages occasioned by repairing the Streets. The object is to find a way from the Strand to St. Paul's, without crossing any of the Bars in the Streets supposed to be under repair.
Published by C. Ingrey, 310, Strand, and F. Waller, 49, Fleet Street, London.
RIVER THAMES
STRAND
FLEET STREET
CHEAPSIDE
CORNHILL
LEADENHALL STREET
FENCHURCH STREET
UPPER THAMES STREET
LOWER THAMES STREET
HOLBORN
HIGH HOLBORN
LONDON WALL
SMITHFIELD
CHARTER HOUSE
LINCOLN'S INN FIELDS
BANK
ALDGATE

'An Illustrative Map of Human Life Deduced from Passages in Sacred Writ', 1833

John Ping, London

Copper engraving; border: 535 × 388 mm (21 × 15 ¼ in)

JOHN PING (fl. 1833–1854) and James Nisbet (fl. 1821–1854) are two minor figures from the London book and map trade. Ping was a draughtsman and engraver and Nisbet a bookseller and publisher; they co-operated in the production of this elaborate allegorical map of the path of human life.

The map is a religious allegory of the human condition from infancy to the grave, dominated by the large cross on 'Mount Calvary'. The 'straight and narrow' path of righteousness follows 'The River of the Love of God' up the left-hand side of the map; the roads running to the right represent the various perils. The two paths commence at the city of 'Infancy', where there appears the engraved label: 'Enter ye in at the strait Gate, for wide is the Gate and broad is the way that leadeth to Destruction & many there be which go in thereat; Because strait is the Gate & narrow is the way which leadeth into Life, and few there be that find it.'

The two routes (the righteous path and all the others) lead either to Heaven, labelled 'Glory' beyond the River Jordan, or Hell, labelled the 'Gulf of Death' with the motto 'Time was time is no more'. This is contrasted with the eternal life offered in the Kingdom of Heaven, 'Out of which no friend departs'.

The worthy Christian makes his or her way along the road passing all manner of suitable landmarks – 'Park of reading the Scriptures', 'Faith in Christ', 'Garden of Holiness' and finally 'Victory over the World the Flesh and the Devil', the final step before making the passage across the River Jordan. Most of the map is given over to the many ways in which a person could be led astray. The quickest way to Hell was along a road leading through 'Fields of disobedience to Parents and Masters', the 'Forest of Bad Company', through 'Desire of being a wit', 'Desire of being fashionable', over 'Mount Arrogance' and the 'Burning Mountain of Atheism' and then through 'Satan's Golden Garden of Unbelief'.

The map is dedicated to Rowland Hill (1744–1833), an evangelical preacher of strongly independent beliefs, part Calvinist, part Methodist and part Congregationalist. He founded the unique Surrey Chapel, attracting preachers from a wide range of denominations, in Blackfriars Road, only a short distance from where the mapmaker John Ping lived. Surely the two men knew each other – it may be, indeed, that Ping worshipped at the chapel – and the map reflects Hill's religious tenets.

Another prominent text counsels: 'Watch ye therefore; for ye know not when the Master of the House cometh, at even, or at midnight, or at the Cock crowing, or in the morning, lest coming suddenly he find you sleeping. And what I say unto you I say unto All, Watch.'

An
Illustrative Map
OF HUMAN LIFE,
Deduced from passages in
SACRED WRIT.
Dedicated to the Revd. Rowland Hill A.M.
LONDON,
Published Jan. 1st 1833 by JAMES NISBET, 21 Berners St. Oxford St.
MOUNT CALVARY
IN THIS WE GLORY
RIVER JORDAN
Presumption Hills
Mountains of Sin
Rocks of Unbelief
Garden of Holiness
Sands of Consideration
Worldly Sorrow
Hill of Repentance
THE ROADS TO REPENTANCE
Hill of Self Examination
Palace of
Tower of Wilful Blindness
Rocks of Self-denial
Wood of Self examination
Fields of Idleness
INFANCY
Mount Arrogance
Enter ye in at the strait Gate;
for wide is the Gate and broad is the
way that leadeth to Destruction, & many
there be which go in thereat; Because strait
is the Gate & narrow is the way which leadeth
into Life, and few there be that find it.
Watch ye therefore for ye know not when
the Master of the House cometh, at even,
or at midnight, or at the Cock crowing, or in
the morning, lest coming suddenly he find
you sleeping. And what I say unto
you I say unto All, Watch.
Ent. Sta. Hall.

'Wallis's New Railway Game, or Tour Through England and Wales', 1835

Edward Wallis Jr, London

Steel engraving; 505 × 447 mm (19 ⅞ × 17 ⅝ in); widest, with panels: 510 × 667 mm (20 ⅛ × 26 ¼ in)

THE WALLIS family was one of the leading publishers of children's games and puzzles. John Wallis Sr was the founder of the family firm; he had two sons, John Jr (*c*.1779–1830) and Edward (*c*.1787–1868). John Jr left in 1806 to set up his own business, and so Edward inherited the family business from his father, with John receiving a cash bequest. Edward continued the business with success, exploiting the shop stock that he inherited and adding a number of new cartographic games, including this game, based on the flowering railway network of England. It may be the very first printed map-game of the railways.

In later life, Edward seems to have moved away from publishing games towards selling 'toys', not perhaps toys in the modern sense, but trinkets, small items of jewellery and the like, from his shop in Islington, although, no doubt, he also continued to retail existing items from stock. In about 1847, it seems that he sold his publishing business to John Passmore, the publisher of this example of the map, and in about 1851, he retired from business, his future needs assured.

The 'New Railway Game' is primarily intended for two or three players (although up to six could play if sufficient tokens were available). It is played on a single-sheet map of England, the race beginning at Rochester and ending in London – '117. London. The Game. The principal Railway Depôt' – with 117 numbered locations along the track, which are keyed to the rules. The 'Directions for playing the new railway game of the tour of England' are printed from letterpress and pasted as side panels. They contain the rules and descriptions of each of the numbered locations on the map.

At the time the map was made, the railway network was still relatively limited (fewer than 100 miles of railway existed in Britain in 1830), so the route does not necessarily follow actual railways; however, landing on most of the main railway towns allowed the lucky player to advance rapidly along the route.

Each player was given a pyramid token to mark his or her position and four counters (used to count missing turns), the movement governed by spinning a teetotum, with the eight sides numbered from one to eight. The players would read out the text describing the place they landed, thereby learning about their destination and receiving additional instructions or penalties. The most serious of these penalties was this: '89. Isle of Man. This island is situated in the Irish Sea, from some part of which the three Kingdoms of England, Scotland, and Ireland, may be seen at once. The Traveller will be shipwrecked on this island, and thereby lose the chance of the game.'

The map gives an early indication of one of the side consequences of the developing railway system: many historic towns, by-passed by the network, were condemned to obscurity: '108. Ely. A place of little consequence, though a city with 5,000 inhabitants; and only remarkable for its fine cathedral.' By contrast, Bath had 60,000 inhabitants, Bristol 110,000, and Newcastle 60,000, but the most populous regional centre was Liverpool with a population of 250,000.

DIRECTIONS
THE NEW RAILWAY GAME
THE TOUR OF ENGLAND
RULES.
Wallis's
NEW RAILWAY GAME,
or Tour through
ENGLAND AND WALES.
LONDON:
PART OF
SCOTLAND
NORTHUMBERLAND
CUMBERLAND
DURHAM
YORKSHIRE
ISLE OF MAN
IRISH SEA
IRELAND
LINCOLN
NORFOLK
SUFFOLK
CARDIGAN BAY
ST. GEORGE'S CHANNEL
BRISTOL CHANNEL
ENGLISH CHANNEL
ISLE OF WIGHT
PART OF FRANCE

'A Voyage of Discovery; or, the Five Navigators. An Entirely New Game', 1836

William Matthias Spooner, London

Lithograph; at widest: 556 × 436 mm (21 ⅞ × 17 ⅛ in)

WILLIAM MATTHIAS SPOONER (1796–1882) was a London-based printseller and bookseller, famed for his output of map-based games from about 1836 onwards. All were drawn in a uniform pictorial style, and they were evidently extremely popular.

This one is an interesting variation on the standard race game, in which each player has their own assigned route, drawn by lot. Each then has to follow that track round the map, starting to the left of 'Mermaid's Rock' and ending to the right. Every route contains its share of dangers and rewards; on entering one circle, marked with a heavy black outer rim, the player 'loses his Ship, and is thrown out of the game'.

The tracks are overlaid on a highly decorative map of imaginary expanses of sea, filled with islands, navigation hazards and all manner of pictorial scenes to bring the 'voyage' to life. Each track contains sharp bends and circles; each of these counts as one step, so each player advances his ship as many steps as dictated. A novelty is that the 'die' is a compass teetotum, termed the 'Navigating Compass', an arrow on a spindle, with a coloured base, with five divisions, reflecting five players, the maximum number of ships available.The rim of each division is divided into five sub-divisions, each coloured to correspond to a ship, and containing a number. One player would spin the arrow; when it came to rest over one of the five main divisions, the players would look at the division in their colour, see the number and move their ship that many steps, thus moving simultaneously, rather than in turn. If that move ended on a circle, the player would read the instruction and comply. Finally, another novelty, the winner was the *last* player to return to 'Out', or pass over it, not the first.

Although the rules are quiet on this, it would seem that a store of counters would be divided among the players. Each would then start by putting three into a central pool. Each circle, in addition to the text, would require the player either to pay in or collect counters from the pool – paying in three for landing on one of the fatal circles.

As an example, red denotes the 'Track of the Red Rover Ship of the Line', fortunately with a complement of Royal Marines. The first circle notes 'Officer and Crew invited on shore, by a Native Prince'; the second 'Treacherously attacked by the Indians', with the marines depicted fighting off the attackers, before proceeding to the third circle 'Seize Canoes and Proceed down a River to the Sea'. The homeward leg is less eventful, until 'Encounter a Severe Engagement with two Pirates'. Should the red player land here and then 'throw' a one, his game would be over; defeated by the pirates, the 'Ship sinks and Crew take to the Boats'.

Each of the tracks has a different range of hazards and rewards, so the players could experience a different face of the game each time they played. The 'Prince of Orange Frigate' chiefly had to contend with poor weather and hazards in the water, the fatal circle being 'A Stiff gale, the Ship among breakers', while the 'True Blue Frigate', should the fates dictate, meets her end in a whirlpool.

The map is also interesting in demonstrating a common publishing practice, whereby publications produced at the end of one year would actually be dated for the following year (to avoid their looking out of date only weeks after publication). Although dated 1836 the map was actually published late in 1835, as shown by a glowing review from the *Court Journal* of 5 December 1835:

> A new Christmas game entitled 'A Voyage of Discovery, or the Five Navigators', has just been published by Mr Spooner of Regent street. It is a novel and ingenious game, and will afford much amusement and instruction to our youthful friends, who feel inclined to follow the tracks of the five navigators. The elegant manner in which it is got up, the ingenuity of the design, and its extreme cheapness, will no doubt procure for it an extensive circulation.

A VOYAGE OF DISCOVERY; OR, THE FIVE NAVIGATORS.

AN ENTIRELY NEW GAME.

MERMAID'S ROCK.

PRINCE OF ORANGE. Frigate. GREEN EMERALD. Frigate.

ROYAL PURPLE. Sloop. RED ROVER. Ship of the Line. TRUE BLUE. Frigate.

OUT OUT OUT OUT OUT

LONDON: PUBLISHED BY WILLIAM SPOONER, 259 REGENT STREET, OXFORD STREET. 1836.

'The Journey or, Cross Roads to Conqueror's Castle. A New and Interesting Game', [1837]

William Matthias Spooner, London

Lithograph; at widest: 549 × 409 mm (21 ⁵⁄₈ × 16 ¹⁄₈ in)

WILLIAM SPOONER was a prolific publisher of children's games, many based around maps (see pp. 132 and 136). This particular example is based on an imaginary region, in which the travellers have to reach a castle at the top of the sheet, passing various hazards invented by the mapmaker. The players begin at the 'Starting Place', in the centre of the lower border. Unusually, this is not a dice game; numbers are only important in selecting the sequence of play. Thereafter the 'Circle of Chance', a freely rotating arrow mounted on a spindle, points the direction each player must take: 'Left', 'Right', 'Forward' and 'Backward', moving from one circle to the next only (a single step), to await the next spin of the wheel, and subject to the conditions laid down along the way.

The map might almost be subtitled 'A city-dweller's view of the perils of the countryside' because that is very much the nature of the challenges to be surmounted on the road – a road surrounded by all manner of amusing caricatures. Some are realistic, but some are drawn from the realms of imagination – the toad on 'Short Cut' lane, with pipe and tankard of ale, being one example. All are rather humorous, though the traveller on 'Hoki-Poki' lane being kicked by his donkey might not find it so amusing. He seems to have taken both hind hooves full in the chest.

Similarly, the penalties on the route range from the everyday to the out-of-the-ordinary. A lucky traveller might travel down 'Fillpot Walk' and be distracted by the appeal of the Anchor public house, where weary legs could be rested for two turns. Otherwise the country folk depicted all seem to have their hands out for payment: an old lady wants payment for not sweeping dust over the traveller; a man with two peglegs is guarding a crossing over a ditch; the physician and the toll-booth keeper both require money for their services.

More sinister are characters such as the masked highwayman with loaded pistol who demands the traveller's wallet on 'Crackskull Common', and the 'Black Forest' guide who offers, for a fee, to guide the traveller through 'Witches Glen' and past the three witches who are tending their cauldron with their cat. Even more scary is 'Giant Grumbo', who demands a ransom from all passers-by.

Then there are the natural hazards of the countryside: one family have taken a short cut across 'Hobble Heath' and ended up in mud up to the elder son's knees; two others have tried to climb 'Breakneck Hill' and ended up not at the top, but in a pile at the bottom, the fall of one cushioned by his rotund companion (with an appropriate fee payable to a local for helping them up); while an older gentleman rues his shortcut through 'Mad Bull Lane'.

A NEW AND AMUSING GAME.

THE JOURNEY;

OR,

CROSS ROADS TO CONQUEROR'S CASTLE,

DIRECTED BY THE CIRCLE OF CHANCE.

DIRECTIONS TO FIX THE CIRCLE OF CHANCE.

Place the Circle on the top of the wooden pedestal, and fix it by passing the peg through the centre of the circle into the aperture in the pedestal, having first placed the arrow on the wooden peg close to the knob.

TO DECIDE WHO SHALL BEGIN THE GAME.

Let each Player turn the CIRCLE of CHANCE, and the Player who obtains the highest number when the circle stops, plays first—the next highest second, and so on. Two Players turning to the same number, to turn the Circle again: the one who gets the highest number plays before the other. When the arrow does not distinctly point out the number or direction, but lies on a separating line, the Circle is to be turned again.

RULES OF THE GAME.

I. Each Player has a numbered mark with which he moves in his turn from Circle to Circle on the roads to the Right or Left, Forward or Backward, as he may be directed by the "CIRCLE of CHANCE." From two to five, or even more persons, can play the Game at the same time, having first provided themselves with Counters for forfeits. On the commencement of a Game each Player is to put *three* Counters into the Pool.

II. The Game commences at the Circle, marked "*The Starting Place,*" and the Player who begins, turns the "CIRCLE of CHANCE," and when it stops, the division in which the arrow rests, specifies in which direction the Player is to move. The rest of the Players do the same in rotation, and move their marks according as they are directed.

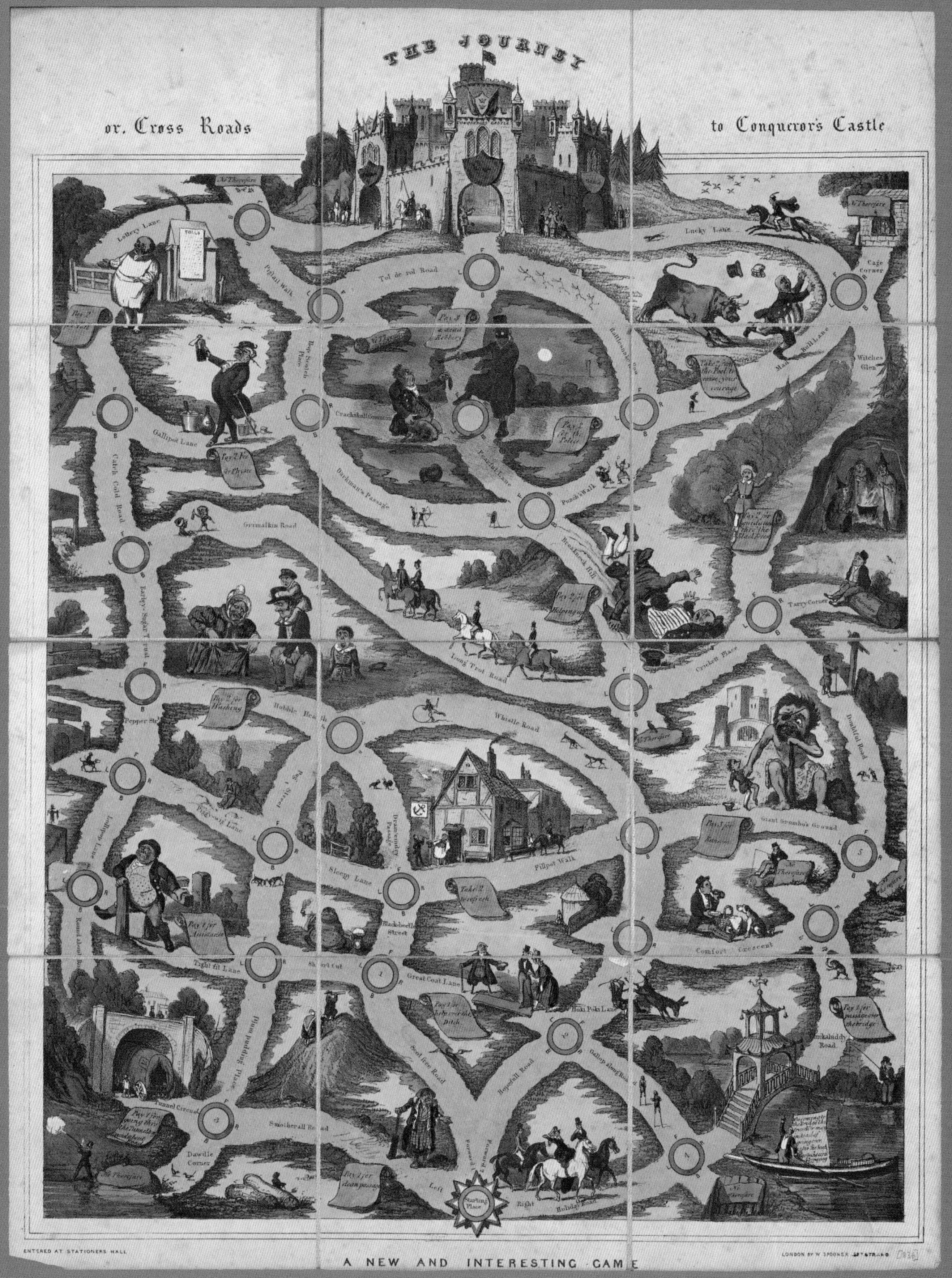
The Journey
or, Cross Roads
to Conqueror's Castle
Lottery Lane
Tol de rol Road
Lucky Lane
Cage Corner
Mad Bull Lane
Witches Glen
Gallipot Lane
Catch Cold Road
Grimalkin Road
Darkman's Passage
Punch's Walk
Long Trot Road
Tarry Corner
Whistle Road
Doubtful Road
Giant Grumbo's Ground
Sleepy Lane
Comfort Crescent
Great Coat Lane
Short Cut
Tight fit Lane
Funnel Circus
Dawdle Corner
Starting Place
Left
Right
Holiday Road
ENTERED AT STATIONERS HALL
A NEW AND INTERESTING GAME
LONDON BY W SPOONER

‘The Travellers, or, a Tour Through Europe’, 1842

William Matthias Spooner, London

Lithograph; at widest: 497 × 612 mm (19 ⅝ × 24 ⅛ in)

SPOONER’S EARLIER games (see pp. 132 and 134) were set against fictional landscapes, where the pleasure of participation in the game was paramount. In this map-game, he made the transition to using actual countries as a backdrop, employing maps in a more informative and educational way. The setting is a general map of Europe with the African littoral of the Mediterranean, extending west to the Caspian Sea. The outlines of the constituent countries are given, and a few of the principal cities.

A very striking feature is the inclusion of finely observed vignette views of the cities of the region, intended to give young (and old) armchair travellers a broad-brush impression of the different locations. The vignettes are miniature works of art. For example, in Africa and the Near East, the cities Algiers, Morocco, Tunis, Tripoli, Alexandria, Cairo, Jerusalem and Acre are depicted, along with a non-specific site captioned ‘Scenery of Asia Minor’. These views, placed in the periphery of the map, are significantly larger than those placed centrally in the game – such as London, Edinburgh, Dublin, Madrid, Paris, Venice, Rome, Vienna, Athens and Constantinople.

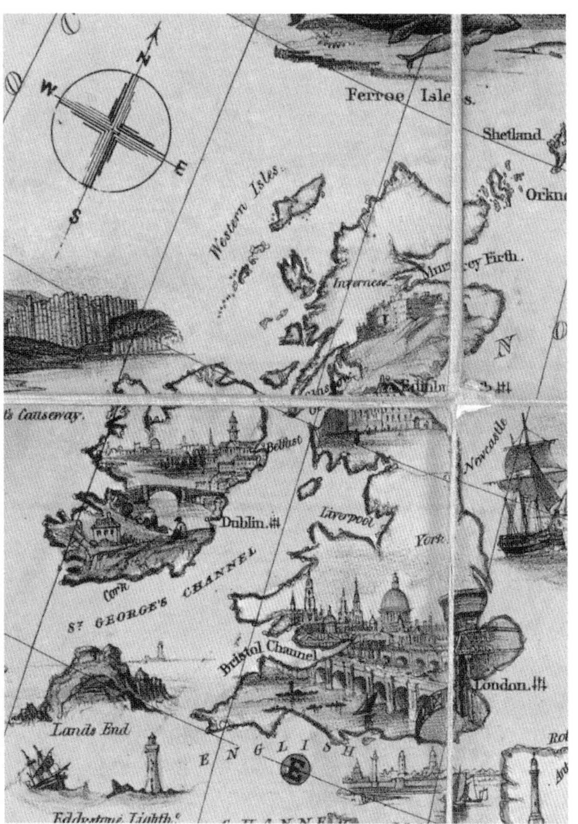

There are no indications within the game itself of how to play (the rules were printed in a separate pamphlet), but St Petersburg is lettered ‘A’, Stockholm ‘C’, Berlin ‘D’ and London ‘E’, with Warsaw or Dresden ‘B’. These were the starting or finishing points for the players. No routes are marked but lines of latitude and longitude are shown and the game was played out on their intersections. Movement was determined by the spin of a four-sided teetotum bearing the letters ‘N’, ‘E’, ‘S’ and ‘W’ for the points of the compass. The player’s marker was moved in the appropriate direction as far as the next intersection. Each player began from a different starting point and was required to journey to a different destination: a player starting from Jerusalem, for example, had to reach Vienna, while from Cairo the aim was to reach St Petersburg. Certain intersections required the player to pay or take from the pool – players losing all their initial twenty counters (‘expenses’ for the journey) became ‘bankrupt’ and had to leave the game. Travellers reaching any principal city (marked by three daggers) had to announce the name of the country or region of which it was the principal city or else pay a fine of two counters. The first player to reach his or her destination scooped the pool (the author is very grateful to Adrian Seville for this explanation). This game is among the most educationally challenging of all the games described in this book and also one of the most instructive. This would be a very simple way of fixing the location of cities and countries in the minds of the players, while also giving them a picture reference to associate with the most important cities.

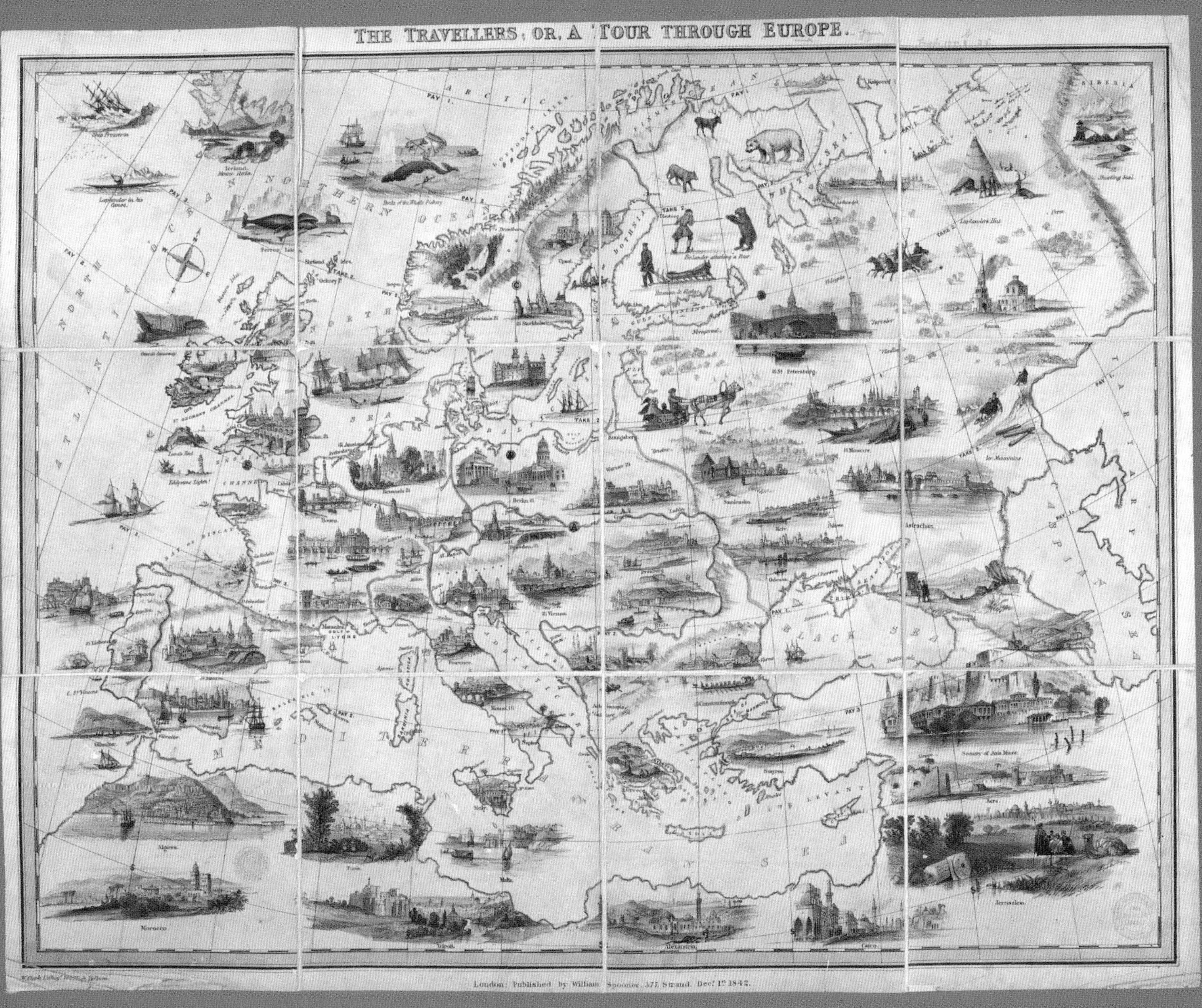
THE TRAVELLERS; OR, A TOUR THROUGH EUROPE.
London: Published by William Spooner, 377 Strand, Decr 1st 1842.

‘Game of the Star-Spangled Banner, or Emigrants to the United States’, [1842]

Edward Wallis, London

Aquatint and etching; border: 663 × 489 mm (26 ⅛ × 19 ¼ in)

THIS MAP published by Edward Wallis (1787?–1868) is a fascinating pictorial guide to the United States as it was perceived by an English publisher in the early 1840s. It was also published as a jigsaw and was soon copied by an American publisher in Philadelphia. It is frequently stated to be the first map-game of the United States. Although undated and quite a wide range of dates have been suggested, it seems likely that the map was published in 1842, based on the (admittedly contradictory) evidence provided by the game itself. In the first instance, the slipcase depicts the American flag with twenty-six stars (states), so should date between the admission of Michigan (the 26th state) on 26 January 1837 and that of Florida (the 27th state) on 3 March 1845.

As against that, the accompanying rules refer to Michigan as a ‘Territory’, its status before statehood in 1837 (108), while the text on Arkansas implies it is a state (90; acceded 15 June 1836). Vandalia is given as the capital of Illinois, which it was between 1819 and 1839 (26), but then the description of Augusta (Maine) notes ‘126. Augusta. The capital of Maine, and seat of its government. – Draw twice more, to commemorate the settlement of the boundary of this state, so long the subject of dispute between Great Britain and the United States’, which presumably refers to the Treaty of Washington, signed on 9 August 1842.

The rule book is entitled ‘Explanation to Wallis’s Star-Spangled Banner, or game of emigrants to the United States’. The booklet contains sixteen pages; two pages of ‘rules for playing the game’ and fourteen describing the 147 locations.

The game begins at sea, with the sighting of ‘The great Sea Serpent. Much astonishment was created in 1817, by accounts of this monster having been seen, 90 feet long, and rearing its head out of the water as high as the topmast of a ship!’ The travellers then make landfall on Long Island, and disembark at Brooklyn, before making a circuitous tour ending back at New York, where ‘Whoever first arrives at exactly this number, wins the Game’. The tour includes each of the state capitals, and those of the new territories, as well as Washington, with a diversion into Canada, and includes important geographical features and landmarks such as Niagara Falls, Chesapeake Bay, the Great Lakes, and the rivers Mississippi, Missouri, Tennessee and Ohio.

There is also commentary on contemporary life in the United States, notably on slavery: ‘48. Slaves. The slave-holders of the southern states are extensively supplied from the markets of Virginia, where negroes are reared for the purpose of sale and traffic. While here, however, they are maintained in a state of comparative comfort. As many as four thousand have been sold in one year, and the least taint of negro blood subjects an individual to this degraded condition.’ Frontier justice is also described: ‘90. Lynch Law. (Arkansas) An odious practice, too frequently indulged in, in the states which are at a great distance from the general government. It is no other than a mockery of justice, by which persons who offend against the popular opinion, are tried and executed by illegal and self-constituted judges.’

If this is all rather grim for the young traveller, there are other ‘stops’ that range from the gross to the increasingly alarming. The Turkey Buzzard ‘feeds on carrion, and if attempted to be taken, vomits the contents of its stomach in the face of its pursuer, emitting the most intolerable stench’ (10). ‘A Skunk. Here is a pretty little creature. But stop! do not hunt him or you will repent; for when attacked he emits such an insufferable and suffocating stench, that dogs, wolves, – aye, and even men, are glad to run away from it’ (38). The Glutton ‘drops from the trees on the backs of deer and other prey which pass beneath, tearing their flesh and drinking their blood, till life is extinct’ (130) and Alligators ‘abound in the Mississippi, and destroy numbers of pigs, calves, and sometimes even children’ (72). Surely enough to cause even the keenest migrant to have second thoughts?

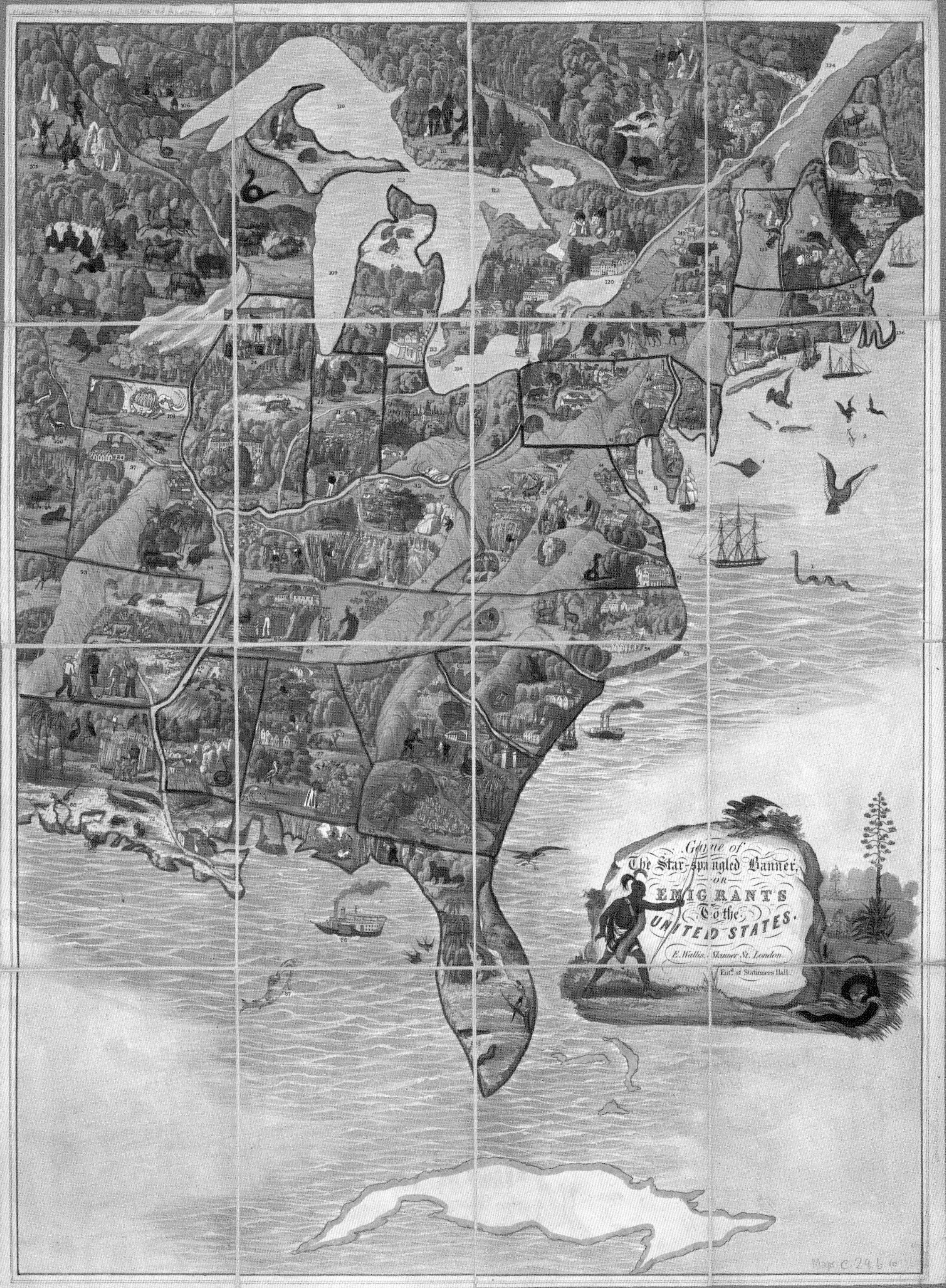
Game of
The Star-spangled Banner,
OR
EMIGRANTS
To the
UNITED STATES.
E. Wallis, Skinner St. London.
Entd. at Stationers Hall.
Maps C.29.b.10

'Wallis's New Game of Wanderers in the Wilderness', [1844]

Edward Wallis, London

Aquatint and etching, border; 654 × 491 mm (25 ¾ × 19 ⅜ in)

EDWARD WALLIS' map-game of the 'Wanderers in the Wilderness' is based in South America. It was produced as a companion to his 'Game of the Star-Spangled Banner, or emigrants to the United States' (p. 13=). The present map is signed by the artist and engraver John Henry Banks (1816–1879), who also produced *A Panoramic View of London* for Wallis in 1845 and is best known for his much-reproduced *A Balloon View of London (As seen from Hampstead)*, which first appeared in 1851. The similarity in style between this and the previous map suggests that he was responsible for both.

The game comprises eighty-four stops around Latin America; 85, 'Game' is not pictured. The accompanying twelve-page booklet, gives the rules of play and then a description of each of the stops that the traveller had to negotiate to win. The game starts on a plantation in the British colony of Demerara, where the writer sets the scene for the players: 'Welcome, gallant friends, to the shores of South America. I shall soon be ready to attend to you on your journey to explore this land of wonders, for such you will indeed find it to be: but we shall have many a weary mile to travel across plains and swamps of vast extent, and through forests as old as the creation.'

Then, perhaps to create a deliberate distinction with the United States, our host continues, 'While my servants are preparing some refreshment, I will just shew you over my estate. I employ here about two hundred negroes, who were formerly slaves, but I now pay them regular wages; and find I am the gainer by the abolition of the old system. The soil of all the plantations on the river, is the richest in the whole world, and we raise on it abundance of coffee, cotton, and sugar.'

He goes on to advise his guests of their equipment needs: 'You will want neither shoes nor stockings, for we shall find no burning sands as in Africa, and rocks in this part of the continent are rare. A hat, a shirt, and a pair of light trowsers, will be all the clothing you will require. We will take a couple of native Indians for our guides, who will provide us with game and fish, for our support; they will also carry the painted cloth, which ... will serve us for a tent, at night, or during heavy rains.'

A final piece of advice, and they are off: 'Let me advise you to be cautious where you step, lest you tread on the poisonous Libarri snake, the scorpion, or centipede; though, unless provoked, they will not attack us. The fatal Rattle Snake will, by the sound of his tail, give us timely notice to get out of his way before he makes a spring, and our numbers are enough to scare away the malicious and cruel alligator called the Cayman. Mind, too, how you wash your hands in the pools, lest you touch a Gymnotus, or Electrical Eel, which would give you a shock you would not easily forget. Now, my friends, are you all prepared for the bush?'

Unlike the United States tour, there are few cities to visit. Rio de Janeiro, Buenos Ayres, Valparaiso and Lima, devastated by an earthquake in 1755, are the principal ones mentioned. The majority of stops and their related images refer to the peoples, plants and animals of the regions traversed, notably various different types of boa constrictor. These include the 'Camoudi Snake, or Bull-killer, above thirty-six feet in length' (no. 3); the 'Coulacanara. It is of the boa constrictor kind, and has dined off a stag'; or 'Hark at the horrid sounds which proceed from the forest close by! It is the death roar of a Jaguar, which an immense Boa constrictor is in the act of crushing to jelly. How fearfully the scaly monster hisses as his foe strives to tear him with his teeth and claws; but resistance is in vain; the serpent will break every bone in his skin, and then devour him whole' (22).

But any player would quickly come to know and recognise the many monkeys, birds (such as eagles, condors and humming birds) and other animals (such as bears, horses and pumas), as well as more unfamiliar creatures (such as tapirs, ant-eaters and armadillos).

Wallis's
New GAME of
WANDERERS
in the
WILDERNESS

'Picturesque Round Game of the Produce and Manufactures of the Counties of England and Wales', [1844]

Edward Wallis, London

Copper engraving; border: 643 × 488 mm (25 ¼ × 19 ¼ in)

THIS MAP-GAME by Edward Wallis could justly be described as a pictorial map of the Industrial Revolution, at a time when the face of the industrial heartlands in the Midlands and north of England were changing for ever. Although it is undated, and earlier editions have been cited, the text of the present example refers to a fire in the Royal Dockyard at Plymouth in 1840 and so was presumably published shortly after that.

It is another 'Game of the Goose' derivative – a circular race map round England and Wales, with the rules and descriptions of the many stages set out in the accompanying booklet. However, there are a number of variations from more standard format versions. The game is played without dice; the accompanying sheet of symbols served for all purposes. Each player would get one of the letters to use as a marker; the other symbols (the nine numbers, the two crosses and the blank) would be put into a bag (the rules suggest a lady's reticule) and then be drawn out by each player. The numbers served for the dice; the blank acted as zero, so the player would remain where he was. The crosses carried a penalty, as specified in the rules: 'But if he draw a cross, he is to draw again, till he obtain a number, which number is to be deducted from, instead of added to, his former station, and his letter moved back accordingly.'

The players started on the first square – the River Thames at London, number 1, and London was also number 151, the finishing post. As usual, players were required to throw the exact number to finish; any overthrow had to be counted backwards, and they would have to wait to throw again. There are the expected penalty squares, where the traveller is either forced to remain for a number of throws or sent back to an earlier stage. There is also a 'death' square: '143. Ramsgate (Kent). A Royal harbour, and fashionable place of resort for sea-bathing. Off here are the dangerous Goodwin Sands. – The player is shipwrecked here, and loses his chance of the game.'

The rules are written in a rather high-minded high-Victorian tone: '150. Dulwich (Surrey). Here is a famous college, founded by All[e]yn, an actor, who is said to have endowed it in consequence of the Devil having appeared to him, while personating that enemy of souls. Certain it is, that if all the profligate men who follow that profession were to build hospitals, the land would be full of them.' Several of the penalty squares are well-known gambling resorts, such as '149. Epsom (Surrey.) Celebrated for its races, and its saline waters. – Go back to the Eddystone rock, No. 118, and do penance for frequenting gambling places.'

While vast swathes of England were by-passed by the Industrial Revolution, many regions and cities were transformed, particularly in Lancashire, Yorkshire and Warwickshire. Mines (for example, square 41), canals (51 River Mersey), and bridges (66 the Menai and 108 the Clifton Suspension Bridge), feats of engineering in their own right, were built to ensure that the furnaces had the fuel and the factories the raw products they required. Ports expanded to enable these goods to be carried all round the world (see 119 Plymouth Breakwater). The description for Manchester reads '57. Manchester (Lancashire). The largest manufacturing town in the world, with 300,000 inhabitants, and adorned with several noble buildings. – Stop while the others draw once; examine the vast magazines of goods produced by the woollen, silk, and cotton factories, and admire the powers of steam, from which this place derives its importance.' Sheffield gets a rather less flattering note: '28. Sheffield (Yorkshire). This town makes cutlery and plated goods for the world. It is a dirty, black, smoky place, but has many fine buildings. Iron and coal abound in the immediate neighbourhood. – Go on to No. 40.'

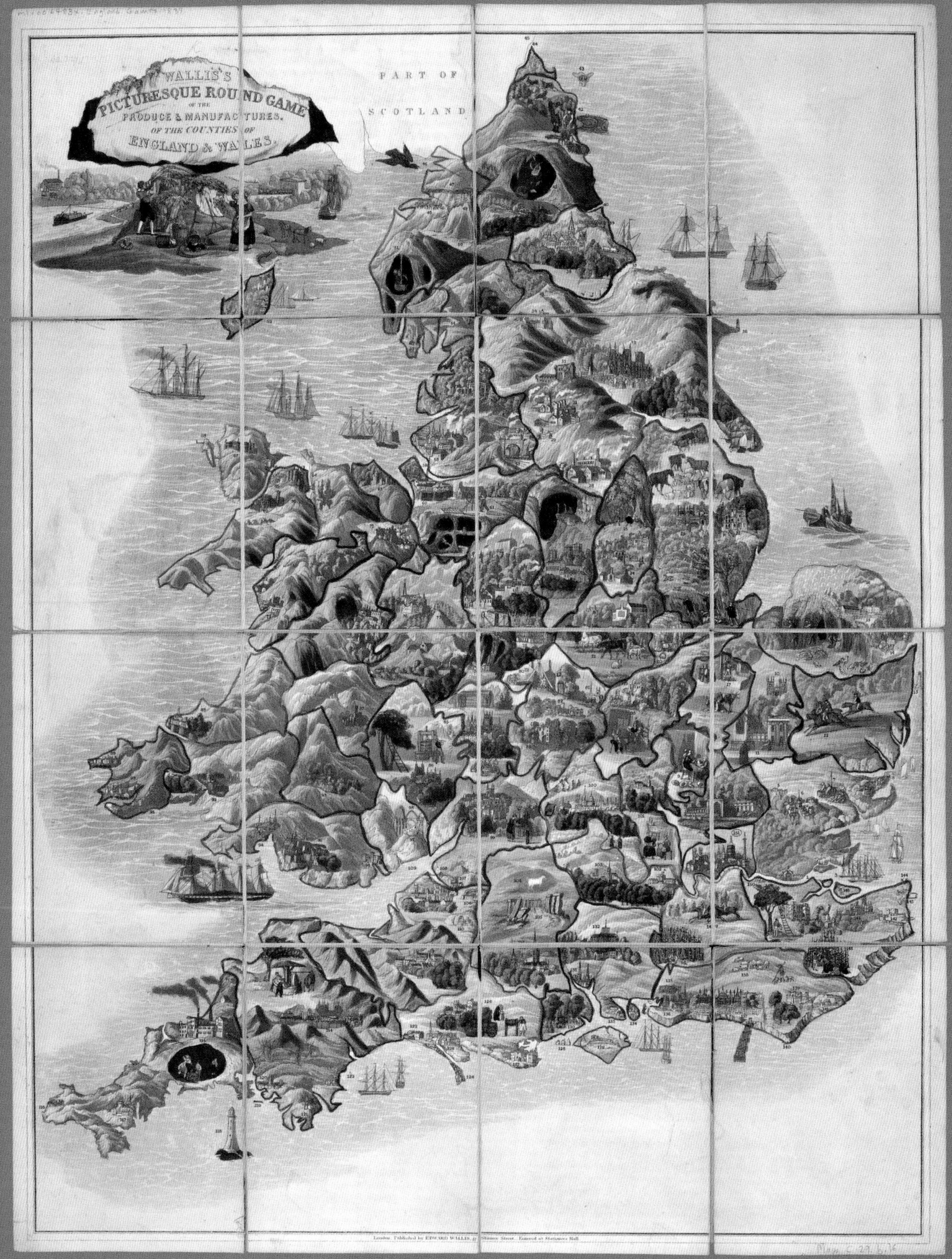
WALLIS'S
PICTURESQUE ROUND GAME
OF THE
PRODUCE & MANUFACTURES,
OF THE COUNTIES OF
ENGLAND & WALES.
PART OF
SCOTLAND
London. Published by EDWARD WALLIS,
Skinner Street. Entered at Stationers Hall

‘Spooner’s Pictorial Map of England & Wales Arranged as an Amusing and Instructive Game for Youth’, 1844

William Matthias Spooner, London

Lithograph; at widest: 609 × 483 mm (24 × 19 in)

LIKE OTHER geographical games by William Spooner (see pp. 132, 134 and 136) this map is a mix of the cartographic and the pictorial. Here, superimposed on the map of England, are a series of oval views, with other views in the blank areas of the sea, depicting cities, towns and famous landmarks. These represent the successive stages (squares) the player would advance along.

The game caters for up to five players. Marked in the engraved surround of the map are the starting points for each player, lettered A to E, and their final destination, labelled ‘Game’. Thus, player A starts from the northern border of the map and has to make his or her way to the final destination in the border next to Kent.

The game was originally accompanied by a thirty-six-page pamphlet giving descriptions of each of the counties, which the players probably had to recite from, as well as directions for play. It is believed that it was played in much the same way as Spooner’s *The Travellers, or, a Tour Through Europe* (p. 136), with a four-sided spinner indicating the points of the compass, although variants are known in which the track is numbered in the conventional way. In either case, it was a test of knowledge at every turn.

As Spooner noted of the Europe game: ‘Little, it is presumed, need be said in favour of a mode of information at once so pleasing and so exact; or of an amusement which will leave on the youthful mind many of the solid advantages of a more regular instruction. The publisher, therefore, presents this little game with equal confidence both to parents and to children, in the full assurance that to the former it will afford all the satisfaction, and to the latter all the amusement, which either he or they can desire. To those little wayfarers who, by so many various routes, seek the object of the heart’s constant affections – Home – he recommends not only the patience and perseverance of more matured travellers, but that as they meet or cross each other’s path on this extended stage of travel, they should cultivate that kindness and good humour which sweeten the path of life – to the child as well as to the man – in the city and in the desert, as well as by their own cheerful and contented firesides.’

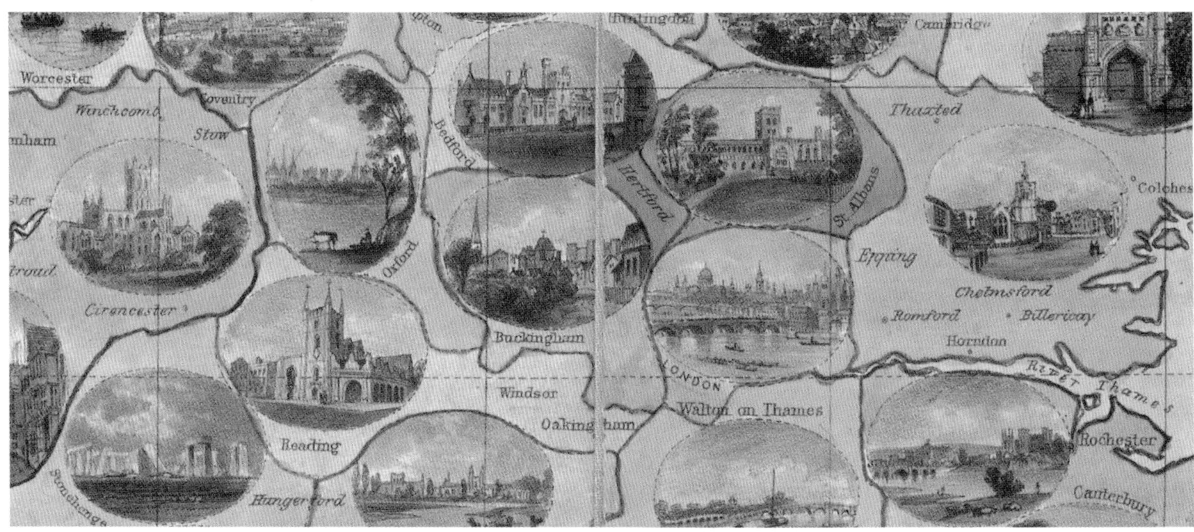

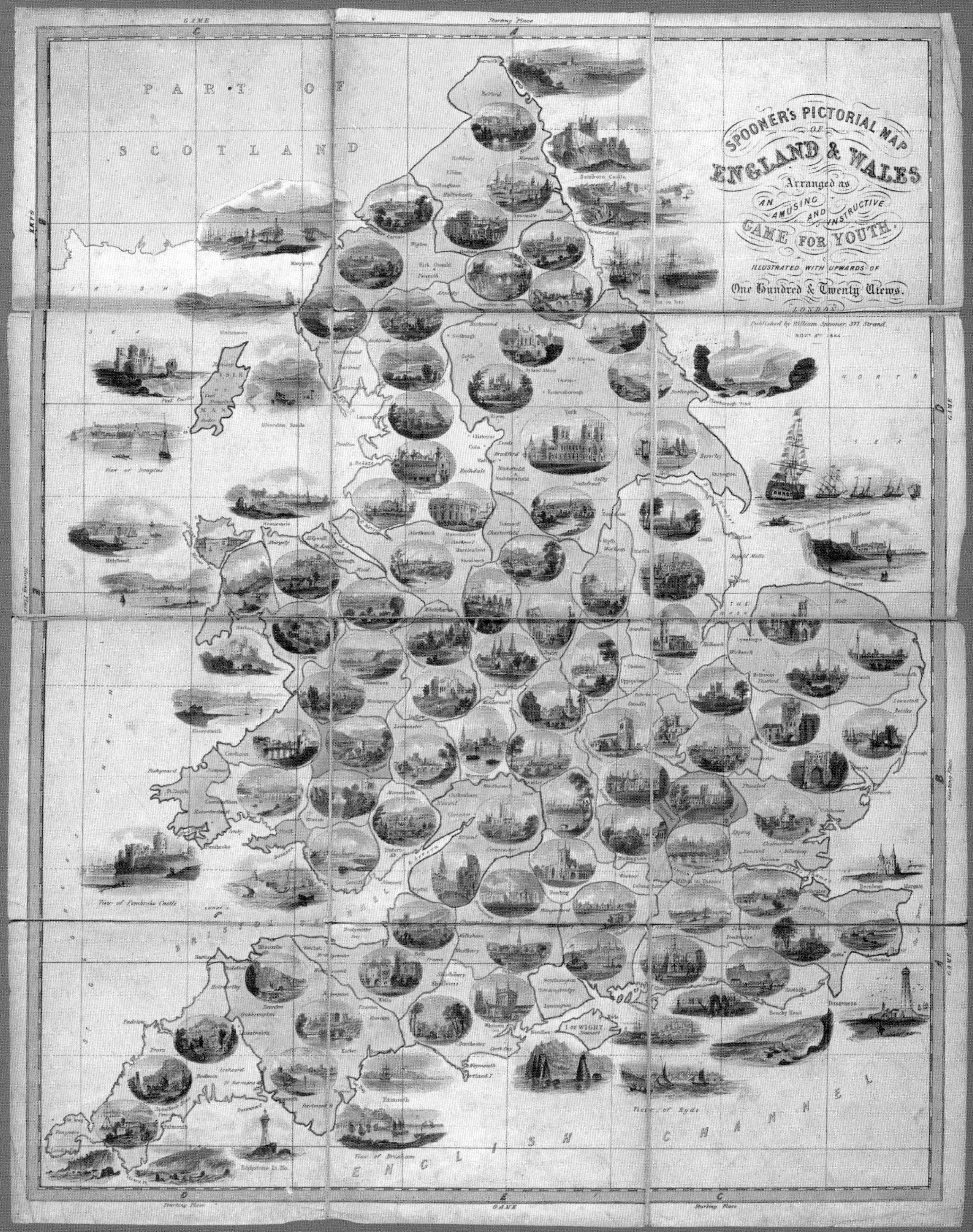
SPOONER's PICTORIAL MAP
OF
ENGLAND & WALES
Arranged as
AN AMUSING AND INSTRUCTIVE
GAME FOR YOUTH.
ILLUSTRATED WITH UPWARDS OF
One Hundred & Twenty Views.
LONDON
Published by William Spooner, 377, Strand.
NOVr 8th 1844.
PART OF SCOTLAND
NORTH SEA
ENGLISH CHANNEL
View of Pembroke Castle
View of Ryde
I. OF WIGHT
Starting Place
GAME

'The Cottage of Content or Right Roads and Wrong Ways. A Humorous Game', 1848

William Matthias Spooner, London

Lithograph; at widest: 545 × 390 mm (21 ½ × 15 ⅜ in)

THIS IS a race game of a rather different kind from William Spooner (see pp. 132–136 and 144). It has a strong underlying moral theme, with the player rewarded for acts of kindness, and punished for moral lapses. The game is set against an imaginary pictorial map, in a countryside setting, with humorous vignette illustrations depicting scenes visible from the road, or relating to the benefits and forfeits that are an integral part of the game. Notable scenes are the Maypole towards the bottom right, an early form of trampolining, four men participating in a sack race – although one is face down in the grass! – two children playing badminton and a game of cricket in full flow.

The game is played with a teetotum that directs each player in a particular direction ('F' Forward; 'R' Right; 'L' Left; and 'B' Back), where they proceed one step. On many of the roads there are red notices, which the player has to read and obey. One of the first hazards encountered is the armed highwayman, on 'Rifleman Road', who demands three counters to let the traveller carry on; to the right, along 'Punishment Path', a careless or indiscreet traveller has provoked a swarm of bees which pursues him down the road, while another traveller has tried to take a short cut across a field – 'Mad Bull Field' – and been forced to take refuge up a tree.

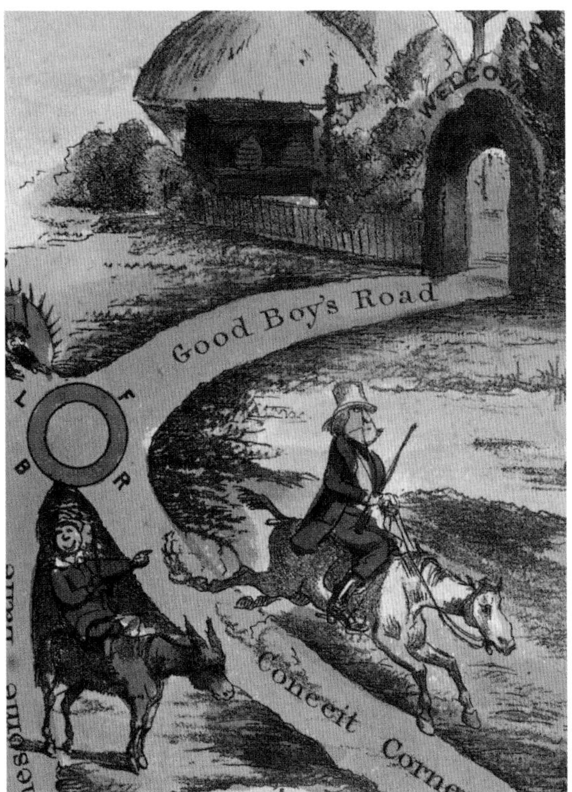

Many of the scenes relate to morality and honesty: a figure at bottom right has stolen a piglet and is put to terrified flight as its mother and siblings pursue him; a young boy, on 'Bad Boy's Road', has grabbed a goose by the neck and is chased by the rest of the flock; in the centre of the left-hand border, a young lady skinny-dipping watches in horror as a thief runs off with her clothes; while three boys laugh at a man confined in the stocks – in 'Laughing Stocks Lane' – for which the traveller is fined two counters. A man is depicted carrying a child through a ford, and for his kindness receiving three counters from the traveller.

Signs of the developments taking place in Victorian England are everywhere: two balloons can be seen, as well as a stagecoach and a train, both of the latter pursued by latecomers. The train is chased by a couple coming down 'Too late row', who are fined two counters for loitering, as a pig and a goat laugh at them from the train.

Sights that are more unexpected in the English countryside include the two elegantly dressed frogs, dancing to a frog fiddler (lower left), or the crocodile waiting for an unsuspecting traveller in a pond along a cul-de-sac. Another traveller seems shocked to be face to face with a stork. Rather incongruously in the upper centre of the map, a smuggler is shown crawling to a cliff edge, while three of his companions celebrate a successful night with a mug of ale, in their cave hideout. For the traveller directed that way, there is even a chance to meet Robin Hood and Friar Tuck, along 'Robin Hood's Walk'.

THE COTTAGE OF CONTENT, OR, RIGHT ROADS AND WRONG WAYS.
THE COTTAGE OF CONTENT.
Good Boy's Road
Gate Road
Help Row
Long sought Lane
Lonesome Lane
Conceit Corner
Patience Pond
Tattle Corner
Turnabout Lane
Smugglers' Lane
Robin Hood's Walk
Blanket Row
Prudence Passage
Rattle away Road
Bridge Toll
Stand off Lane
Lucky Lane
Holiday Road
Shuttlecock Walk
Boatman's Passage
Cooks Corner
Laughing Stock Lane
Bad Boys Road
Chance Road
Odds Bobs Street
Refuge Row
Recruit Road
Ruination Row
Welcomers' Walk
Mad Bull Field
Poverty
Industry Row
Probity Pass
Punishment Path
Consequence Corner
Path
Frog Island
Shooter's Hill
Forethought Road
Misfortune Lane
Rifleman Road
Common
Gazer's Road
No 1
Pay 1 for the Bridge
Pay 1 for a Sight
A HUMOROUS GAME.
LONDON, PUBLISHED BY
WM SPOONER, 379, STRAND, NOVR 1ST 1848.

BRITISH ISLES
SCOTS
IRISH SEA
NORTH SEA
EASTERN QUESTION
ENGLISH CHANNEL
HOLLAND
DENMARK
LIBERTY EQUALITY FRATERNITY
B. OF BISCAY
CORSICA
SARDINIA
Gibraltar

THE VICTORIAN ERA AND GROWTH OF THE MASS MARKET

(1850–)

'The New Game of The Royal Mail or London to Edinburgh by L. & N.W. Railway', 1850

John Jaques, London

Wood engraving; inner border: 227 × 1203 mm (9 × 47 ⅜ in)

THE FAMILY firm of John Jaques & Son (nowadays simply Jaques of London) has every claim to be the oldest surviving manufacturer of games and sporting equipment in the world. The family is credited with the invention of Croquet, Ping Pong, Snakes and Ladders, Ludo, Tiddledy Winks, Happy Families, Snap and many other perennial favourites. Founded by Thomas Jacques (1765–1844) in 1795, the firm had passed to his son, partner and successor, John Jaques I (1795–1877) by the time this game was produced.

In contrast to Edward Wallis' railway game (p. 130) which shows the whole of England, this Jacques game is based on the route of the London and North-Western Railway line from Euston to Edinburgh, composed of 104 steps, with Edinburgh numbered 104. For this purpose, the track is given a meandering path along the game board, representational rather than geographically accurate.

The London and North-Western Railway Company was founded in 1846, an amalgamation of three existing railway companies; it operated what is now called the 'West Coast Line' from London to Birmingham, and from Manchester and Liverpool on to Glasgow and Edinburgh.

Twenty-two stations along the route are named, and each is illustrated with a small vignette view of a principal landmark or other scene (in the case of Preston a picnic hamper). Euston Station is represented by its famous Doric Arch, with the London skyline behind; Lichfield by its cathedral; Crewe by the engine works; Wigan by its coal mines; and Lancaster by its castle; while Edinburgh is represented by the Castle and the Sir Walter Scott Memorial (completed in 1844). In view of the route, it is noteworthy that there is no representation at all of Birmingham, Manchester or Liverpool.

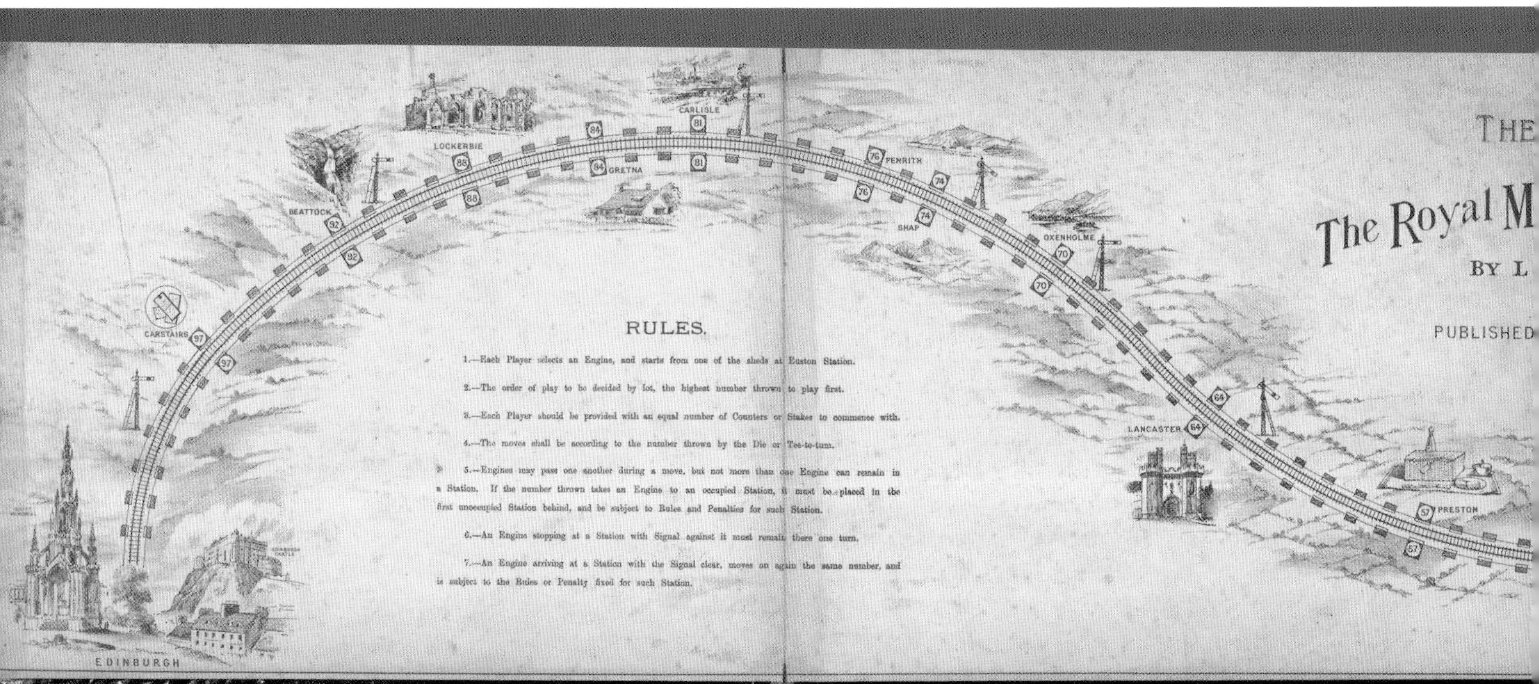

The simple rules are printed on the face of the map. There were four engines (not present here) – one for each shed at Euston. The players would throw a die or dice to choose the order of play, share out the tokens among the players and then start the game. The track is controlled by a series of signals; if a player lands on a station and the signal is set to 'stop' (with the arm horizontal) he or she has to miss a turn, but if the signal is set to 'clear' (the arm angled downwards) the player proceeds on as many steps as he or she threw on the dice. The other rule was that no two engines could be on the same step; if a player would otherwise land on an occupied square, he or she had to stop on the square behind, and would be subject to the penalties for that square.

The penalties, listed on the right-hand side of the board, relate to the stations en route. Most require the player to pay a forfeit, normally a 'stake' or counter, to go sight-seeing – for example (9) 'Berkhamstead – Pay 2 Stakes to see the Castle'. Other squares have harsher penalties: (40) 'Crewe – Remain here until every Player has passed, and visit the N. W. Railway Engine Works', (24) 'Nuneaton – Remain here while other Players throw twice, and make a visit to Burton-on-Trent.' The square especially to be avoided is (84) 'Gretna – Visit the Smithy, and go back to London and begin again'. It is not quite a 'death' square, but it is a severe punishment, nonetheless.

There is only one station with a 'reward' – (33) 'Stafford – Exchange places with the Engine furthest in advance', while the player who lands on the game square, Edinburgh, 'Wins the game, and takes all the contents of the Pool'. There is no mention of what happens if the player does not throw the exact number to arrive at Edinburgh, so presumably any throw that reached the station was a winning throw.

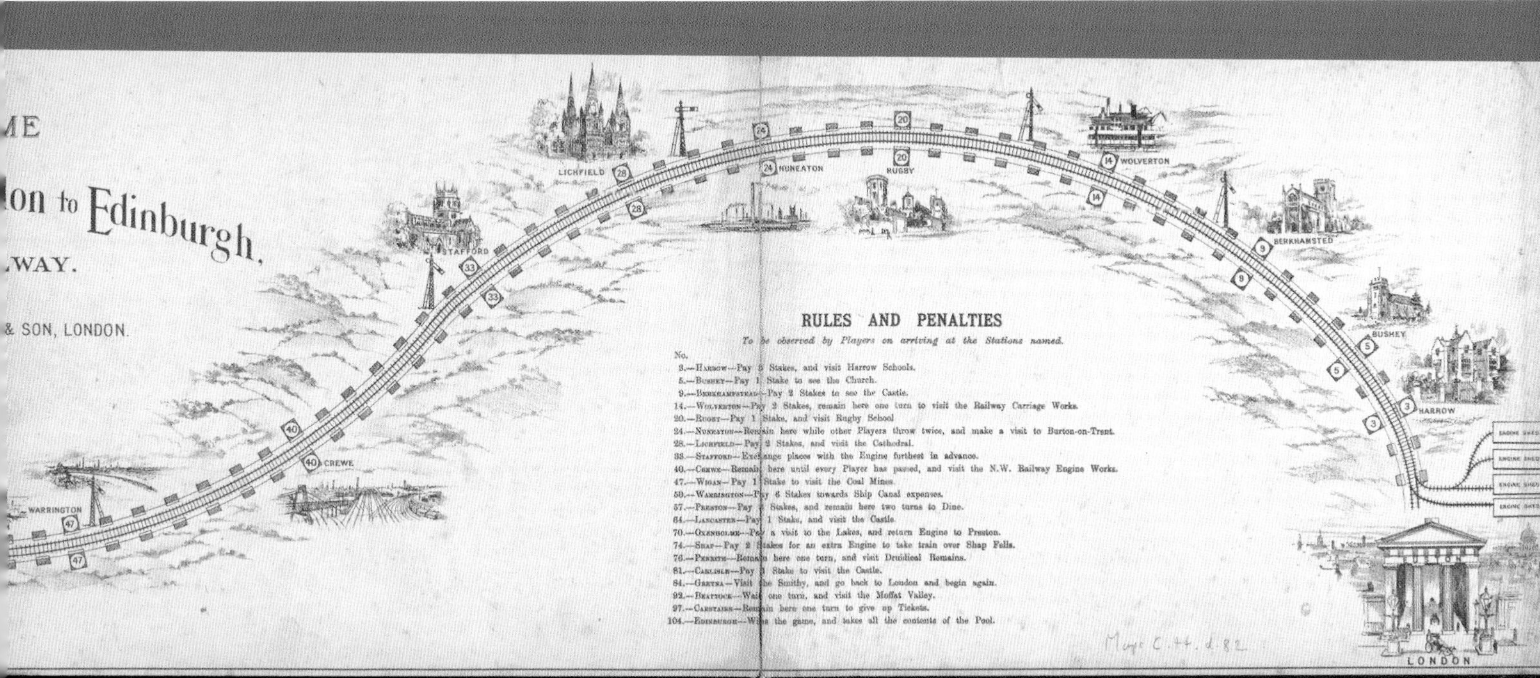

‘Dame Venodotia, Alias Modryb Gwen; A Map of North Wales’, 1851

Hugh Hughes, Caernarfon

Lithograph; border: 219 × 234 mm (8 5/8 × 9 1/4 in);
at widest: 305 × 234 mm (12 × 9 1/4 in)

HUGH HUGHES (1790–1863) was an engraver, caricaturist, artist and portrait-painter, born at Pwll-y-Gwichiad, near Llandudno, in North Wales in 1790, and baptised there on 20 February 1790. He worked for a while in Liverpool and London. In about 1835, he published a cartoon map of North Wales, ‘Dame Venodotia, alias Modryb Gwen: a lady incog’. The original version is very rare, but later copies were published by the London and Glasgow publishing house of Maclure, Macdonald & Macgregor, and this version also appeared under the imprint of Hugh Humphreys. Humphreys’ version was lithographed by Joseph Josiah Dodd (1809–1894), a designer and lithographer originally from Liverpool, who settled in Caernarfon with his wife in about 1850.

The map depicts North Wales as an old woman, named ‘Dame Venodotia’ (Venodotia was the name given by the Romans to the area); she is shown bare-footed, with a bag slung over her shoulder, walking to the left. Her head is formed by the island of Anglesey, and her outstretched arms the Llyn peninsula; her torso is north-western Wales (roughly modern Gwynnedd), the bag the counties of north-east Wales and her skirt the central band (modern Ceredigion and Powys).

The principal geographical features are listed in the key below the map: lakes; towns and locations divided by the traditional counties (Anglesey, Carnarvonshire, Merionethshire, Flintshire, Cardiganshire, Denbighshire and Montgomeryshire); English towns in the marches; principal light-houses; rivers and mountains.

The image is quite subtle, and repays close examination. The shoulder bag is actually a lady in a voluminous ball-gown; within her dress can be seen a billy goat (below ‘11’), perhaps a fox’s face (below ‘64’), a human and rabbit’s face (to the right of ‘63’), and perhaps even a ram’s head (below ‘37’). Within the lower folds of the jacket seem to be a fox, several fox cub faces and a flock of sheep running downhill. Other faces are visible in the skirt, but are difficult to identify because of the colour overlaid on the image. Below ‘73’ appears to be the outline of a human face; below ‘72’ seems to be a tri-horned animal head, and below ‘79’ an indistinct animal’s face; there are at least another three possible animal faces, but one would really need an uncoloured example to give a better sense. Most surviving examples, as here, were printed in colour and finished by hand, particularly the stripes of the dress.

Designed by H. Hughes, and Drawn on Stone by J. J. Dodd.
DAME VENODOTIA, ALIAS MODRYB GWEN;
A Map of North Wales.
6096
LAKES
1 Llanberis
2 Nantlle
3 Cywellyn
4 Dinas
5 Capel Curig
6 Ogwen
7 Crafnant
8 Geirionydd
9 Conway
10 Aled
11 Alwen
12 Bala
13 Talyllyn
ANGLESEY
8 Amlwch
9 Llanerchymedd
10 Holyhead
11 Gwalchmai
12 Mona Inn
13 Llangefni
14 Beaumaris
15 Pentraeth
16 Menai Bridge
16 Britannia Bridge
17 Aberffraw
18 Newborough
19 Malltraeth
CARNARVONSHIRE
20 Conway
21 Bangor
22 Aber
23 Carnarvon
24 Clynog
25 Rivals
26 Bethgelert
27 Bettws y coed
28 Nevin
29 Porthdinllaen
30 Tremadoc
31 Port Madoc
32 Pwllheli
33 Criccieth
MERIONETHSHIRE
34 Festiniog
35 Maentwrog
36 Harlech
37 Corwen
38 Trawsfynydd
39 Bala
40 Barmouth
41 Dolgelley
42 Dinas Mowddwy
43 Mallwyd
44 Llwyngwril
45 Towyn
46 Aberdovey
FLINTSHIRE
53 Rhyl
54 Rhuddlan
55 St Asaph
56 Flint
57 Holywell
58 Caerwys
59 Hawarden
60 Mold
61 Bangor
CARDIGANSHIRE
47 Machynlleth
48 Borth
49 Aberystwith
DENBIGHSHIRE
62 Abergele
63 Denbigh
64 Ruthin
65 Wrexham
66 Llanrwst
67 Chirk
68 Llangollen
69 Oswestry
ENGLISH TOWNS
50 Liverpool
51 Parkgate
52 Chester
71 Shrewsbury
MONTGOMERYSHIRE
70 Llanfyllin
72 Llanfair
73 Welshpool
74 Machynlleth
75 Montgomery
76 Newtown
77 (Ch. Stretton)
78 (Bishop's Castle)
LIGHT HOUSES
1 Bardsey
2 South Stack
3 Skerries
4 Penmon
5 Point of Air
6 Leasowe
7 Black Rock
RIVERS
A Mersey
B Dee
C Conwy
D Elwy
E Mawddach
F Dyfi
MOUNTAINS
G Gt Ormshead
H Penmaen mawr
I Snowdon
L Moel Famau
M Mignant
N Berwyn
O Cader Idris
P Plinlimon
PUBLISHED BY H. HUMPHREYS, CASTLE SQUARE, CARNARVON

'The Crystal Palace Game, Voyage Round the World, an Entertaining Excursion in Search of Knowledge, Whereby Geography Is Made Easy', *c.*1854

Henry Smith Evans, London

Lithograph; at widest: 483 × 666 mm (19 × 26 ¼ in)

HENRY SMITH EVANS was a member of the Royal Geographical Society and a noted mapmaker; he is best known for his 'Map of the World on Mercator's Projection shewing the British Possessions, with the date of their accession, population, &c., all the existing Steam Navigation, the Overland Route to India, with the proposed extension to Australia, also the route to Australia via Panama ...', which was published under a variety of slightly different titles between 1847 and 1852 and possibly later.

The knowledge he acquired in creating that made him ideally placed to produce this map-game, lithographed by the highly skilled John Anthony L'Enfant (1825?–1880) and based on a voyage round the world. 'The Crystal Palace Game' was almost certainly produced to coincide with the removal to Sydenham of the great 'Crystal Palace' originally built for the Great Exhibition. The exhibition is properly called 'The Great Exhibition of the Works of Industry of All Nations'. The brainchild of Prince Albert, consort of Queen Victoria, the exhibition was held in Hyde Park, in London, between May and October 1851. Ostensibly designed to showcase the innovations of all countries, the exhibition was really intended to highlight the United Kingdom's pre-eminence as world leader in arts, science and technology and her global dominance. As the motto at bottom centre attests – 'Britain upon whose empire the sun never sets'.

Built to a revolutionary pre-fabricated design involving over 293,000 panes of plate-glass, the main exhibition building in Hyde Park immediately became known as the Crystal Palace. It was designed by Sir Joseph Paxton and constructed in rather less than nine months. The building was relocated from Hyde Park to Sydenham, south-east London (the area is now called Crystal Palace) in 1854, but destroyed by fire in 1936. The large vignette on the right side of the sheet is the opening ceremony of the Palace in its new home, performed by Queen Victoria and Prince Albert. Around the border of the map are fourteen further vignettes depicting scenes from around the world – including slavers, tiger-hunting in India and Australian aborigines. There are smaller vignettes within the map proper, including the famous scene of the death of James Cook in Hawaii, Columbus making his first landfall in the West Indies and Alexander Selkirk on the Island of Juan Fernandez, the inspiration for Daniel Defoe's novel *Robinson Crusoe*. Also noticeable is the track of a voyage by a genuine Captain Flint in the *Alfred*, perhaps a latent source of inspiration to Robert Louis Stevenson, who would have been just the right age to have enjoyed this kind of game as a boy.

The track of the players round the game is marked with numbered steps, starting from the Azores (number 1) and continuing round the world, coasting Africa, through Arabia, round India, through the East Indies, along the Pacific rim, back to Australasia then round the coast of South America, past Cape Horn and then up round Brazil through the West Indies, along the eastern seaboard of the United States to Newfoundland and from there making the transatlantic crossing to the British Isles.

The index booklet, which would explain the various images, is not present. The large images are all numbered, but the numbers do not tally with the numeration of the map. Without the rules it is impossible to gain a complete understanding of the intricacies of the game, although it may well be that squares 11, a ship threatened by a large sea-monster, and 46 and 58, mariners being killed by hostile natives (58 being Cook) are 'death' or at least penalty squares.

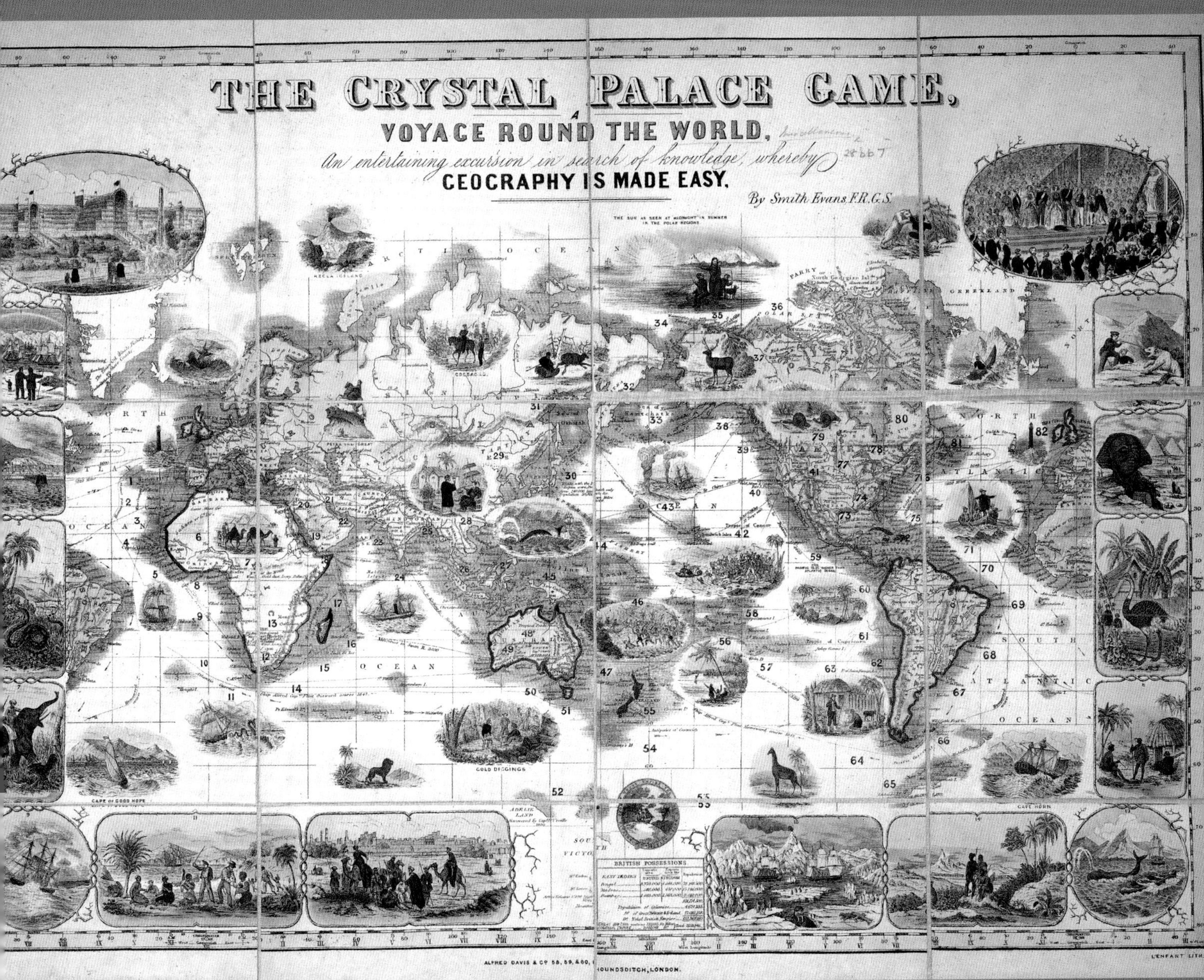
THE CRYSTAL PALACE GAME,
A VOYAGE ROUND THE WORLD,
An entertaining excursion in search of knowledge, whereby
GEOGRAPHY IS MADE EASY.
By Smith Evans F.R.G.S.
THE SUN AS SEEN AT MIDNIGHT IN SUMMER IN THE POLAR REGIONS
CAPE OF GOOD HOPE
GOLD DIGGINGS
CAPE HORN
BRITISH POSSESSIONS
ALFRED DAVIS & Co 58, 59, & 60, HOUNDSDITCH, LONDON.

'Betts's New Portable Terrestrial Globe', 1852

John Betts, London

'By the Queen's Royal Letters Patent, Betts's New Portable Terrestrial Globe Compiled from the Latest and Best Authorities'

Lithograph printed in colour, mounted on a collapsible metal frame; 400 mm (15 ¾ in) diameter when opened out, the frame 710 mm (28 in) high

JOHN BETTS (1803?–1889) was a famous maker and retailer of games in the Victorian period. He was apprenticed to William Darton Jr on 11 August 1819 – the Dartons being at that time among the leading retailers and publishers of books, maps, prints, games and puzzles for the juvenile market.

By the late 1820s Betts had established himself in business. In 1827 he published a *Key to the interrogatory geographical game of the world*, illustrated with a map of the world by the engraver and mapmaker Sidney Hall (d. 1831), with a second edition in 1831. He published two atlases, *Betts's Juvenile Atlas* and *Betts's School Atlas*, both of which gave him a ready supply of maps to make jigsaw puzzles. He also produced a stream of other maps. As a retailer with a shop in a prime location on the Strand in London, he aimed not only at the educational market but also at a more general passing trade.

Perhaps the most innovative of his many educational products was his patent portable globe, apparently inspired by the design of an umbrella; the globe has a central metal strut, with two rings at either end to which the metal spokes are attached. When closed, the spokes are straight. As the rings are pushed inwards, the spokes curve outwards forming the inner ribs of a ball, and pull the loose, baggy fabric of the globe taut, so it takes shape as a globe. Fixed to the top of the central strut is a ring, so that the globe could be hung from the ceiling at home or in the classroom, and there is a knob at the bottom which serves as a handle.

Although the globe and its box are undated, on geographical grounds the globe appears to date from the mid-1850s, and is one of the earliest Betts versions. The United States, for example, shows Texas (1845), New Mexico (1850) and California (1850) as having become part of the part of the Union, but seems to predate the Civil War (1861–1865).

Betts also issued a *Companion to Betts' portable globe and diagrams*, undated on the title page but probably dating from the second half of 1851 – the accompanying *Catalogue of maps, atlases, dissected maps, amusing, instructive and educational games, etc.*, bound with it includes a publisher's letter bearing the date 17 June 1851 and referring to a number of items expected to be available in February 1852.

This globe was very popular, if rather fragile, and Betts went on to make a number of later, corrected, versions; the successor company, George Philip, continued to manufacture such globes on into the 1920s.

BETTS
GLOBE

'Comic Map of the Seat of War with Entirely New Features', 1854

Thomas Onwhyn, London

Lithograph; border: 479 × 681 mm (18 ⅞ × 26 ¾ in)

THIS IS one of the most famous nineteenth-century caricature maps. It was the inspiration for, and progenitor of, many later 'serio-comic' war maps of Europe. The map was previously attributed solely to its publishers, the partnership of Rock Brothers & Payne, but a recently discovered early printing of the map clearly shows the initials of the actual artist responsible – 'Done by T. O.' – on the southern coast of Asiatic Turkey, which is partially concealed on this example. This attribution allows us to elevate the illustrator Thomas Onwhyn (1813–1886), son of Joseph Onwhyn to a position alongside Fred W. Rose and other famous artists in this field, and accord him the honour of being the originator of this genre.

The immediate stimulus for the map was the commencement of the Crimean War (1853–1856), in which Britain, France and their allies waged war against Russia. The four principal protagonists are depicted as animals: the British lion, who appears to be on sentry duty; the French imperial eagle, denoting Napoleon III's Second Empire; Turkey in Europe (inevitably) as a turkey; and Russia as a rather harmless-looking bear, albeit wielding a flail of many strands, with the knots in each shown as skulls. The bear's body is labelled 'despotism', 'bigotry', cruelty', 'slavery', 'ignorance', 'oppression' and other unflattering terms. Russia's neighbour, Poland, is shown as enslaved, with even her name spelled out in bones.

In the Baltic, the allied fleet has penned the Russian fleet in its harbour. The Anglo-French fleet is being blown eastwards by the Danish bellows towards Cronstadt, with the Swedish exhortation 'Go it Charley', and the diminutive figure of Admiral Charles Napier crying, 'I'll give him a flea in his ear'.

The main theatre of war was the Crimea; the allied fleet is shown encircling the peninsula, optimistically trimming the toenails of the Russian bear. Below the map title, the 'Balance of Power' is shown shifting decisively in favour of the allies, represented by a French cockerel, two turkeys (Asiatic Turkey and European Turkey), and the British lion. In fact, the Crimean War turned out to be a brutal affair, perhaps the first modern war, with neither side an outright winner, even if the allies achieved their original objective of capturing the Russian naval base at Sebastopol.

Despite the wartime setting of the map, many of the national representations are altogether more peaceful. Tunis is depicted as a dancing lioness, in harem trousers and curly-toed slippers, playing the banjo; Italy is a startled dog, with Sicily as a battered kettle – presumably alluding to the volcanic Mount Etna – tied to its tail. The island of Elba, where Napoleon Bonaparte was exiled, is denoted by his bicorn hat.

Throughout the map are numerous references to alcohol, coupled with some truly terrible puns: the Caucasus mountains are depicted as a series of foaming bottles, labelled 'Cork as us Mountains & bottle him'; the bottle clutched by Turkey is labelled 'The sublime Port[e]'; Malta is a tankard of malt beer; Munich is a stein; and western Germany a champagne flute.

Onwhyn's map is crowded with all manner of other symbolism relating to the *casus belli*, the struggle between the competing Russian and Turkish spheres of influence in southern Europe. Some of the references are obvious but some are now unclear; this is one of the most detailed of all 'serio-comic' maps of the type, and a rewarding subject for future study.

COMIC MAP
OF THE
SEAT OF WAR
WITH ENTIRELY NEW FEATURES
SCALES
BALANCE OF POWER
RUSSIAN EMPIRE
POLAND
FINLAND
SWEDEN
NORWAY
NORTH SEA
BRITISH ISLES
IRELAND
SCOTLAND
WALES
ENGLAND
ST GEORGE'S CHANNEL
ENGLISH CHANNEL
GERMAN OCEAN
SKAGER RACK
DENMARK
BALTIC
GULF OF BOTHNIA
GULF OF FINLAND
PRUSSIA
GERMANY
BELGIUM
FRANCE
BAY OF BISCAY
SPAIN
BOHEMIA
SWITZERLAND
AUSTRIA
HUNGARY
TRANSYLVANIA
ITALY
Gulf of Venice
BOSNIA
SERVIA
BULGARIA
TURKEY
GREECE
SLAVERY
OPPRESSION
TYRANNY
TREACHERY
BOUNDARY OF POLAND
RUSSIA PROPER
POLTAVA
R VOLGA
RIVER VOLGA
ASTRAKHANE
ASIA
COSSACKS OF THE BLACK SEA
CAUCASUS
CIRCASSIA
SEA OF AZOV
CRIMEA
BLACK SEA
GEORGIA
ASIA MINOR
ARMENIA
THE SUBLIME PORTE
ASIATIC TURKEY
SMYRNA
KARAMANIA
KURDISTAN
ALGEZIRA
CORSICA
SARDINIA
SICILY
MEDITERRANEAN SEA
AFRICA
CANDIA
May 30th 1854
PUBD BY ROCK BROTHERS & PAYNE LONDON

Game Based on the Geography of France, [*c.*1855]

Massuet, Paris

'Le clerogéographie ou loterie géographique de la France'

Lithographed maps on card; each 135 × 82 mm (5 ¼ × 3 ¼ in)

THIS IS a boxed game containing eighty-six lithographed maps, engraved by Massuet with the lettering by L. Aubert *père*, each on a card measuring 135 × 82 mm, and divided into five 'suits' – 'Bassin de la Garonne' (twenty cards), 'Bassin de la Loire' (twenty-one cards), 'Bassin de la Seine' (sixteen cards), 'Bassin du Rhône' (nineteen cards) and 'Bassin du Rhin' (ten cards), thus dividing France (unevenly) into the catchment areas of the principal rivers.

There is a spindle with a wooden base and an elaborate removable metal pointer which rotates on the spindle; each suit has its own card, listing the principal towns (on the outer edge) with the *département* in which it is found.

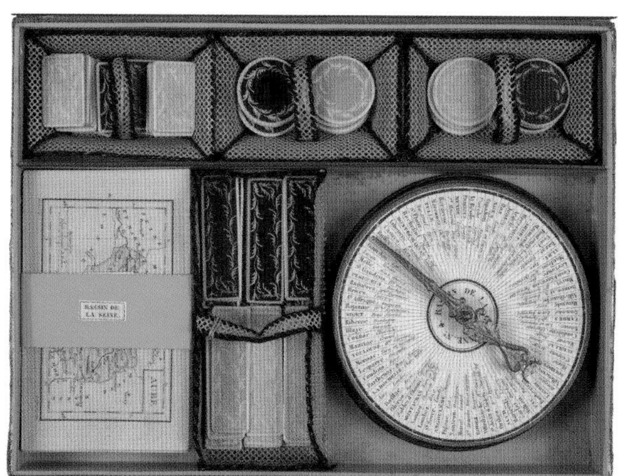

Three smaller baskets and a longer rectangular basket hold three different sizes of ivory counter: circular, short rectangular and long rectangular, each in four colours: red, green, purple and yellow. They are evidently tokens or counters of some form, presumably for four players.

This game has no rule book present, and there is no indication of how it might be played. The author has been unable to find any kind of reference to another example for comparison.

‘Map of Europe, 1859. Illustrative War Scene’, 1859

Jacobus Johannes van Brederode, Haarlem

‘Kaart van Europa 1859. Aanschouwelijk oorlogstooneel, zamengesteld naar de beste telegraphische berigten, waarop met een oogopslag de politieke toestand en ligging der grootere en kleinere Mogendheden van Europa kunnen erkend en beoordeeld worden’

Lithograph; at widest: 251 × 278 mm (9 ⅞ × 11 in)

THIS IS an unusual satirical map of the political situation in Europe at the time of the Franco-Austrian War of 1859, also known as the Second Italian War of Independence. At the time the Austro-Hungarian Empire controlled large areas of Italy, particularly in the north, including Lombardy and Venetia. Italian attempts to gain independence and to drive the Austrians out had failed. As a result, the Kingdom of Sardinia, which included Piedmont, made a secret treaty with France, in which Savoy and Nice would be ceded to the French in exchange for military aid against the Austrians. The long-term aim was the complete reunification of the Italian peninsula under an Italian ruler.

This map depicts the war in its early stages. Symbolising their position as interlopers from outside, both France (in the person of the Emperor Napoleon III) and Austria (the Emperor Franz Josef I) have one foot in their own country and one foot in Italy. Vittorio Emanuele II (Victor Emanuel) of Savoy, King of Sardinia, is shown clinging to the advanced leg of France, while Italian rebels are sticking miniature swords (pins) in Franz Josef's boot.

Denmark, Holland, Spain, Portugal and the United Kingdom (the last with his hands firmly in pockets labelled ‘Neutral’) look on in uninterested fashion, while the German Emperor has binoculars and sword in hand, to observe and perhaps intervene with the soldiers arrayed along the border with Italy. In central Italy, the Pope looks on helplessly, while southern Italy is shackled and unable to assist.

While the principal countries are focused on events in Italy, the Russian Tsar is pushing a large boulder, labelled ‘Panslavisimus’ (i.e. Russian hegemony over all the Slavic peoples) westwards through the countries of eastern Europe. The seated figure of European Turkey watches this advance, ready with sword in hand, while little Greece dances. European Turkey is also unaware that tiny Montenegro, with sword in one hand and a Turkish head affixed atop a pike in the other hand, is running up behind him. In the Crimea, the graves of the Crimean War (1853–1856) fought by the British, French, Piedmontese and Turks against Russia are still fresh.

Although France was successful in the pitched battles, the war dragged on, and Napoleon III was persuaded to make peace with Austria, not least because of concerns that Germany might actually intervene. Although peace was made, France's allies refused to abide by the terms; indeed, they took advantage of Austria's weakness and, in 1861, finally achieved their aim of a united Italy when Vittorio Emanuele II was acclaimed King of Italy, although it was to take another war in 1866 finally to bring Venetia into Italian hands.

AANSCHOUWELIJK OORLOGSTOONEEL,
zamengesteld naar de beste telegraphische berigten, waarop met een oogopslag de politieke toestand en ligging der grootere en kleinere Mogendheden van Europa kunnen erkend en beoordeeld worden.
KAART VAN
EUROPA
1859.
NEUTRAAL
STOKHOLM
LONDEN
LIBERTÉ
NATIONALITÉ
GLOIRE
HANOVER
CHERBURG
HENRI V.
L'EMPIRE C'EST LA PAIX!
PARIJS
DRESDEN
MÜNCHEN
PANSLAVISMUS
WEENEN.
MADRID
TURIJN
FLORENCE
MONTENEGRO
NAPELS
KRIM
CONSTANTINOPEL
ATHENE
UITGAVE van J.J. VAN BREDERODE.
PRIJS 15 CENTS.
Lith v. Emrik & Binger, Haarlem.

'The Evil Genius of Europe', 1859

William Nicholson, London

'On a careful examination of the Panorama the Genius will be discovered struggling hard to pull on his Boot. It will be noticed, he has just put his foot in it. Will he be able to wear it?'

Lithograph, with hand colouring; at widest: 285 × 211 mm (11 ¼ × 8 ¼ in)

BOTH WILLIAM CONEY and William Nicholson are otherwise unknown as mapmakers or publishers. Nicholson is not to be confused with the famous lithographic artist of the same name (1872–1949), active in the late Victorian era and most famous for his series of plates representing the letters of the alphabet. Nonetheless, this other Nicholson was obviously talented, and he has brought his full abilities to bear on this rare map of Italy.

From about 1854 onwards, a small number of English publishers, particularly Day & Haghe and Thomas Packer, introduced a new genre of maps, depicting regions or countries from the air as if from the point of view of a bird flying over them. This highly decorative style was pleasing to the untutored eye, and proved remarkably popular among impulse buyers, who wanted a map on which, in a general way, to follow events then current in the news. That is why the vast majority of such maps show theatres of war, as here.

For the greater part of its history, Italy was governed by a series of warring city states dominated by Spain or Austria. In 1859, inspired by the wave of revolution passing across Europe, the Italian city states, led by Piedmont and Sardinia, rose up against Austria in the Second Italian War of Independence. The French, allied to Sardinia, saw this as opportunity to weaken Austria and spread their own influence south of the Alps and sent an army to assist the Italians, commanded in person by the French Emperor, Napoleon III. This caricature, inspired by the leg-like shape of the Italian peninsula, shows Napoleon III trying to put his leg inside the Italian boot, an allusion to his attempt to gain control over Italy for France. Although the colour is deliberately a little pale, the Emperor's head can be seen in south-eastern France.

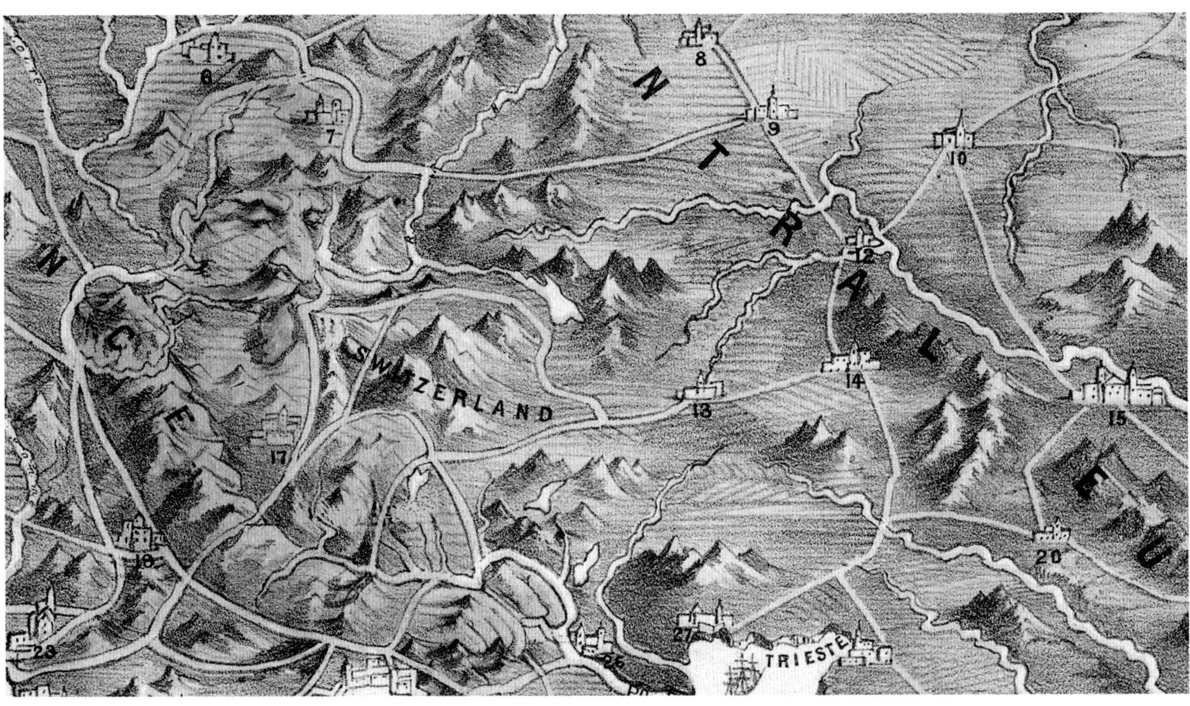

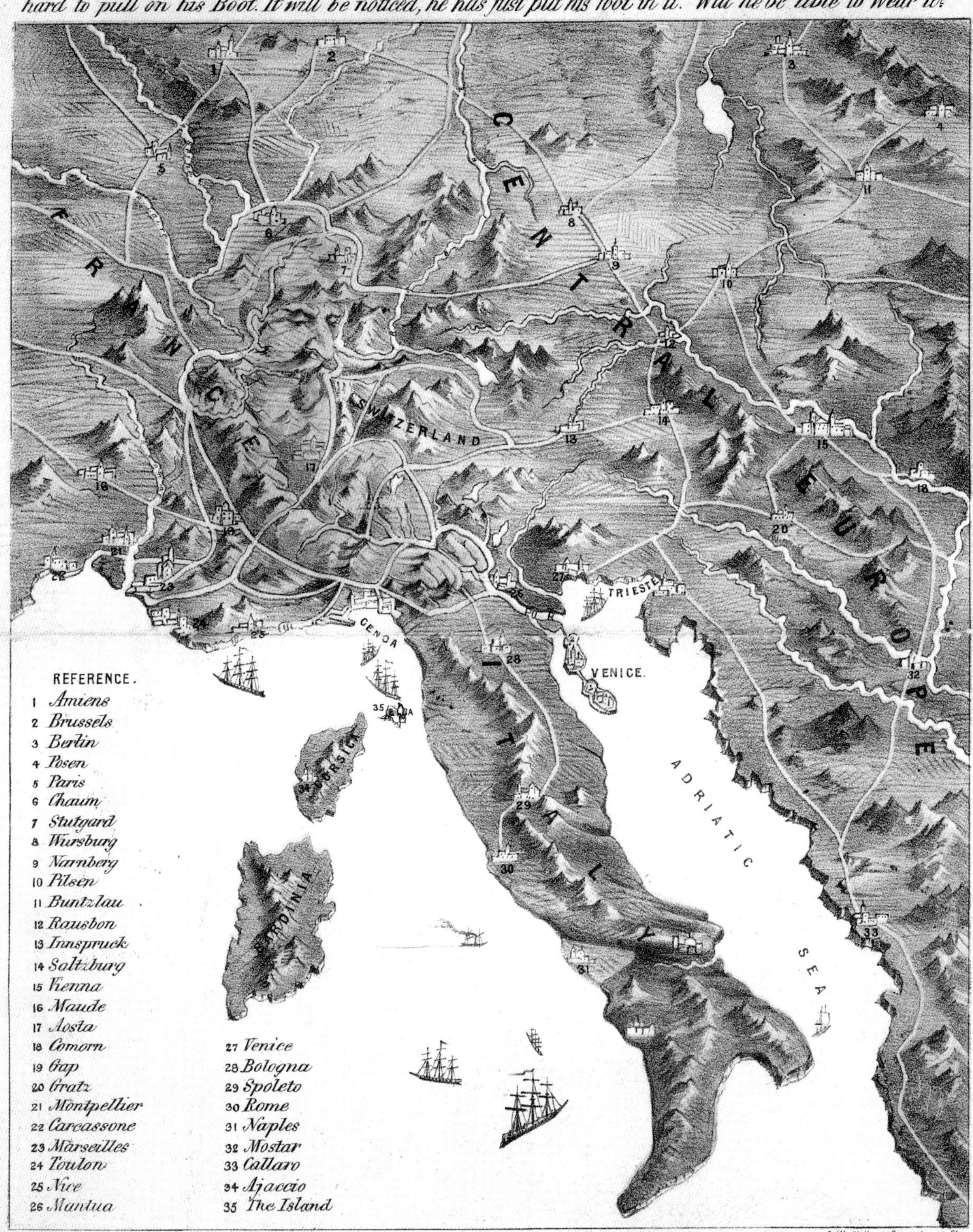
THE
EVIL GENIUS OF EUROPE
On a careful examination of this Panorama the Genius will be discovered struggling hard to pull on his Boot. It will be noticed, he has just put his foot in it. Will he be able to wear it?
FRANCE
CENTRAL EUROPE
SWITZERLAND
ITALY
GENOA
TRIESTE
VENICE
ADRIATIC SEA
CORSICA
SARDINIA
REFERENCE.
1 Amiens
2 Brussels
3 Berlin
4 Posen
5 Paris
6 Chaum
7 Stutgard
8 Wursburg
9 Narnberg
10 Pilsen
11 Buntzlau
12 Rausbon
13 Innspruck
14 Saltzburg
15 Vienna
16 Maude
17 Aosta
18 Comorn
19 Gap
20 Gratz
21 Montpellier
22 Carcassone
23 Marseilles
24 Toulon
25 Nice
26 Mantua
27 Venice
28 Bologna
29 Spoleto
30 Rome
31 Naples
32 Mostar
33 Callaro
34 Ajaccio
35 The Island
LONDON: W. CONEY, 61, WARDOUR St. OXFORD St.
3. Bell Yard, Gracechurch St.

'New Map of France', 1862

Victor Guillon, Paris

'Nouvelle carte de France'

Steel engraving; border: 456 × 461 mm (18 × 18 ⅛ in)

THIS IS a detailed map of France, with an inset of Corsica, drawn by Victor Guillon, printed by Alfred Lemercier and published by Augustine Logerot. Although undated, the companion book was published in 1862, hence the date attributed here.

France is coloured in four parts, seemingly in watersheds: northern France and the environs of Paris are green; the Loire valley catchment area reaching down to the Auvergne mountains is pink; south-western France is blue; and south-eastern and eastern France yellow. These regions are separated by mountain chains, for example the chain 'Monts Corbieres', 'M[onta]gnes Noires' and 'M[on]ts. de l'Espinouse' separate the south-western and south-eastern regions.

The map was intended to accompany a book entitled *La géographie versifiée de la France d'après un nouveau système de dénomination des localités dit système du cadran* ('Geography of France versified, based on a new system of denominating the localities, called the dial system'). *Cadran* means dial, and can be used to mean a clock face, a sundial or the rotary dial on a telephone, among other uses. A diagram on the upper board, labelled 'Seine et Oise' (a *département* comprising a half-circular ring round the outskirts of Paris containing the northern, western, and southern suburbs), depicts the dial, with two large 'hands' and a smaller dial inset, similar to the face of an old pocket-watch.

Without the accompanying book, some of the symbols are hard to fathom. Within the map, many towns are connected by straight broken (pecked) lines; a lot have arrows pointing from them, mostly in a westerly direction. It is possible that the game revolves around some sort of memory game – reciting poetry as a mnemonic device to help pupils learn the geography of France.

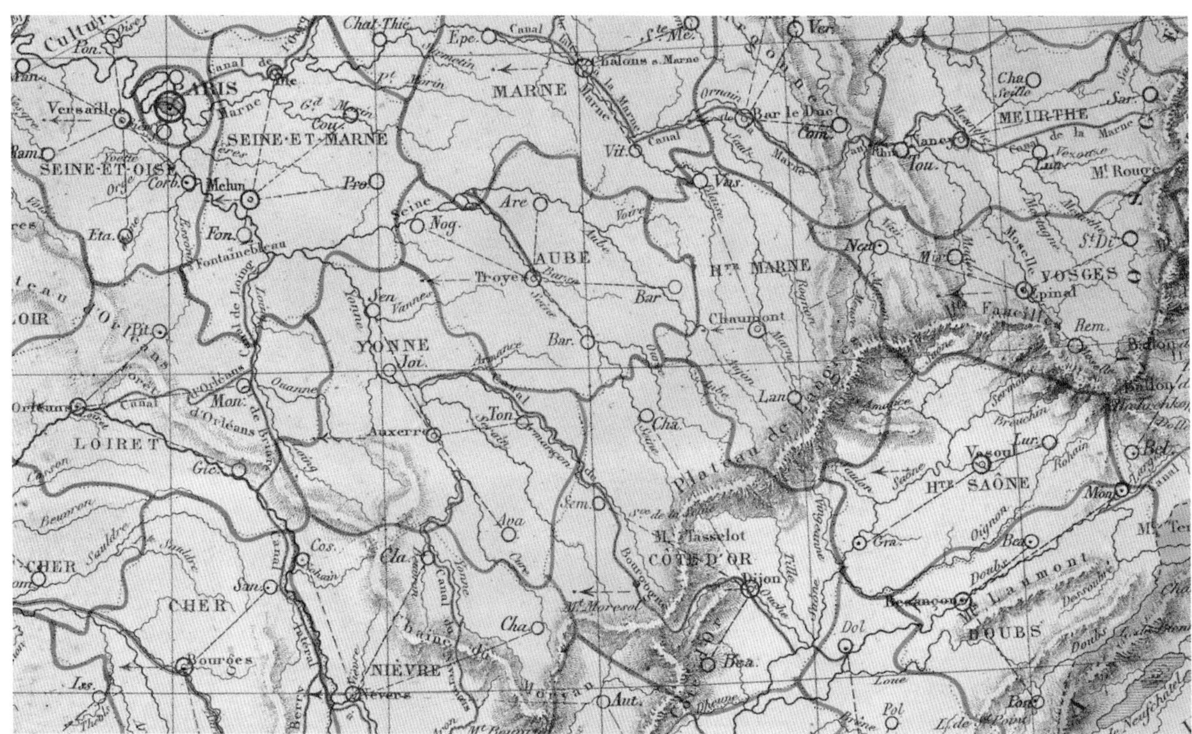

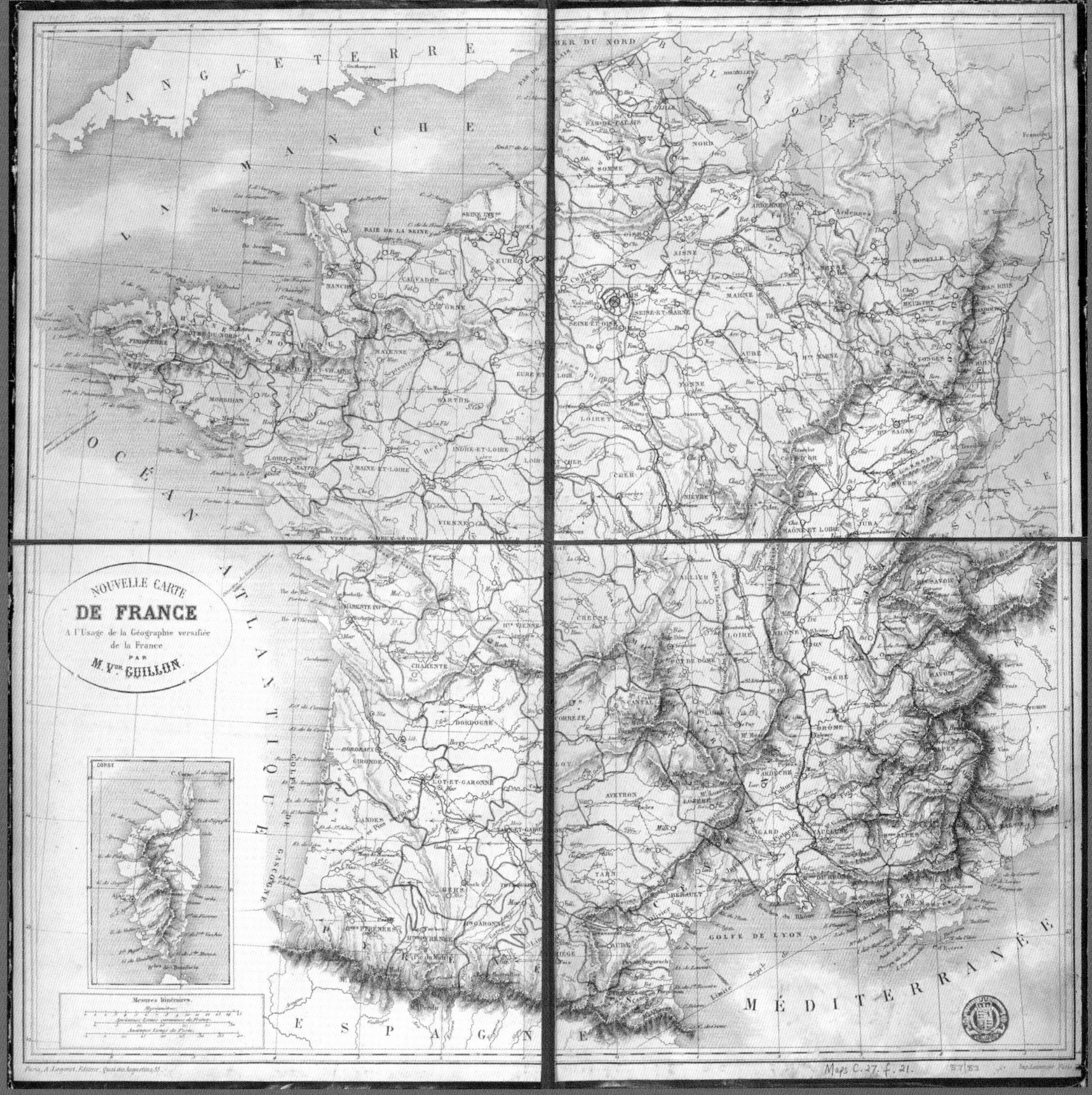
NOUVELLE CARTE
DE FRANCE
A l'Usage de la Géographie versifiée
de la France
PAR
M. Vve GUILLON
ANGLETERRE
LA MANCHE
MER DU NORD
BELGIQUE
OCÉAN ATLANTIQUE
SUISSE
GOLFE DE GASCOGNE
GOLFE DE LYON
MÉDITERRANÉE
ESPAGNE
CORSE
Mesures Itinéraires

'The Dissected Globe', [1866]

Abraham Nathan Myers, London

Lithograph; globe of 170 mm (6 ¾ in) diameter

ABRAHAM NATHAN MYERS (1804–1882) was a toymaker, toy-seller and importer of fancy goods, working in London; one of the strands to his toymaking was the manufacture of dissected globes, in both wooden and cardboard segments, as well as a range of geographical games. From 1842 to 1864 he worked in partnership with Aaron Joseph, Solomon Joseph and Maurice Solomon as A. & S. Joseph, Myers & Co. The partnership was formally dissolved on 31 December 1864, whereupon he established his own business, trading as A. N. Myers & Co.

This particular globe appears under Myers' own name, and appears to date from about 1866; it is an unusual type of globe – a globe jigsaw. It is composed of thirty-eight segments, laid out in eight layers, or cross-sections; the North and the South Pole are each one segment, with the South Pole balancing on a wooden base, with a recess for the plug in the base. The further segments, triangular but with a curved outer edge, then butt into position on the base, placed layer by layer, thus forming the completed globe. The North Polar segment has a little knob on the top, perhaps to assist in placing the final piece without overbalancing the puzzle.

What is particularly interesting about this globe is that the inner surfaces of the segments also form jigsaw images, with maps and descriptions of the 'continents' of the world, with the map on the uppermost surface, and the description on the lower face.

On the second layer (immediately below the North Polar cap) there is a map of Polynesia on the lower face and a description of the region on the upper. The layer below has a map of North America and an accompanying description of the region, illustrated with polar bears on ice floes, that gives an account of its discovery by Columbus, the contribution of Amerigo Vespucci, the confines, length and breadth and a guesstimate at the population.

N O R T H
A T L A N T I C
O C E A N
G of Mexico
VENEZUELA
GUYANA
ECUADOR

Map of Prussia, [1868]

'Aleph', London

'His Majesty of Prussia – grim and old –
Sadowa's King – by needle guns made bold;
With Bismarck of the royal conscience, keeper,
In dreams political none wiser – deeper.'

Chromolithograph; border: 240 × 200 mm (9 ½ × 7 ⅞ in);
at widest: 263 × 203 mm (10 ⅜ × 8 in)

THE NAME 'aleph' is a pseudonym for William Harvey MD (1796–1873), a journalist and antiquary who wrote particularly about London. He was the author of *London scenes and London people: Anecdotes, reminiscences and sketches of places, personages, events, customs and curiosities of London city, past and present*, a volume of personal recollections assembled from his column in the *City Press* newspaper.

Harvey also wrote one of the most famous books of cartographic caricatures, from which this map comes. His 1868 *Geographical Fun: Being humorous outlines of various countries with an introduction and descriptive lines* was illustrated with twelve colour printed maps of the countries of Europe. The artist is not named, or otherwise identified, but the 'Introduction' adds some information about the mapmaker, and the purpose of the maps:

> The young lady who is responsible for these Sketches is now in her fifteenth year, and her first idea of Map Drawing is traceable to her meeting with a small figure of Punch riding on a Dolphin, and contrived to represent England. The thought occurred to her when seeking to amuse a brother confined to his bed by illness. It is believed that these illustrations of Geography may be rendered educational, and prove of service to young scholars, who commonly think Globes and Maps but wearisome aids to knowledge, by enabling them to retain the outline of the various countries so humorously caricatured in the work, by associating them in their mind's eye with odd fancy figures. The bluffs and headlands of Scotland would be identified with the struggling Piper, and La Belle France with the grotesque looking madam dancing before a portable glass. Nearly every system of artificial memory supposes arbitrary way-marks, with which dates and events must be connected, and we have all remarked, when we are at a loss to recollect words or things, how vividly they are recalled when the mind is able to grasp objects associated with them, as a favourite book or a beautiful prospect. If these geographical puzzles excite the mirth of children; the amusement of a moment may lead to the profitable curiosity of youthful students, and embue the mind with a healthful taste for an acquaintance with foreign lands. No history, no journal can be understood without a knowledge of maps, and good service is done when we make such information more easy and agreeable. Aleph.

Recent research has established that the young artist was Eliza Jane Lancaster (1852–1939). She is better known as Lilian Lancaster, and was later to become famous as a pantomime artiste, comic actress and singer, much celebrated for her hit comic song, 'Lardy dah, Lardy dah!'

This map of Prussia depicts the Prussian king, Wilhelm I (1797–1888), and his Chancellor, Otto Eduard Leopold von Bismarck (1815–1898). At Bismarck's suggestion, Prussia established a professional standing army which overwhelmed the Austrians at the battle of Sadowa (or Königgrätz; 3 July 1866), the Danes in 1864 and in 1870–1871, France. These victories led to the unification of Germany with Wilhelm I as the first Emperor.

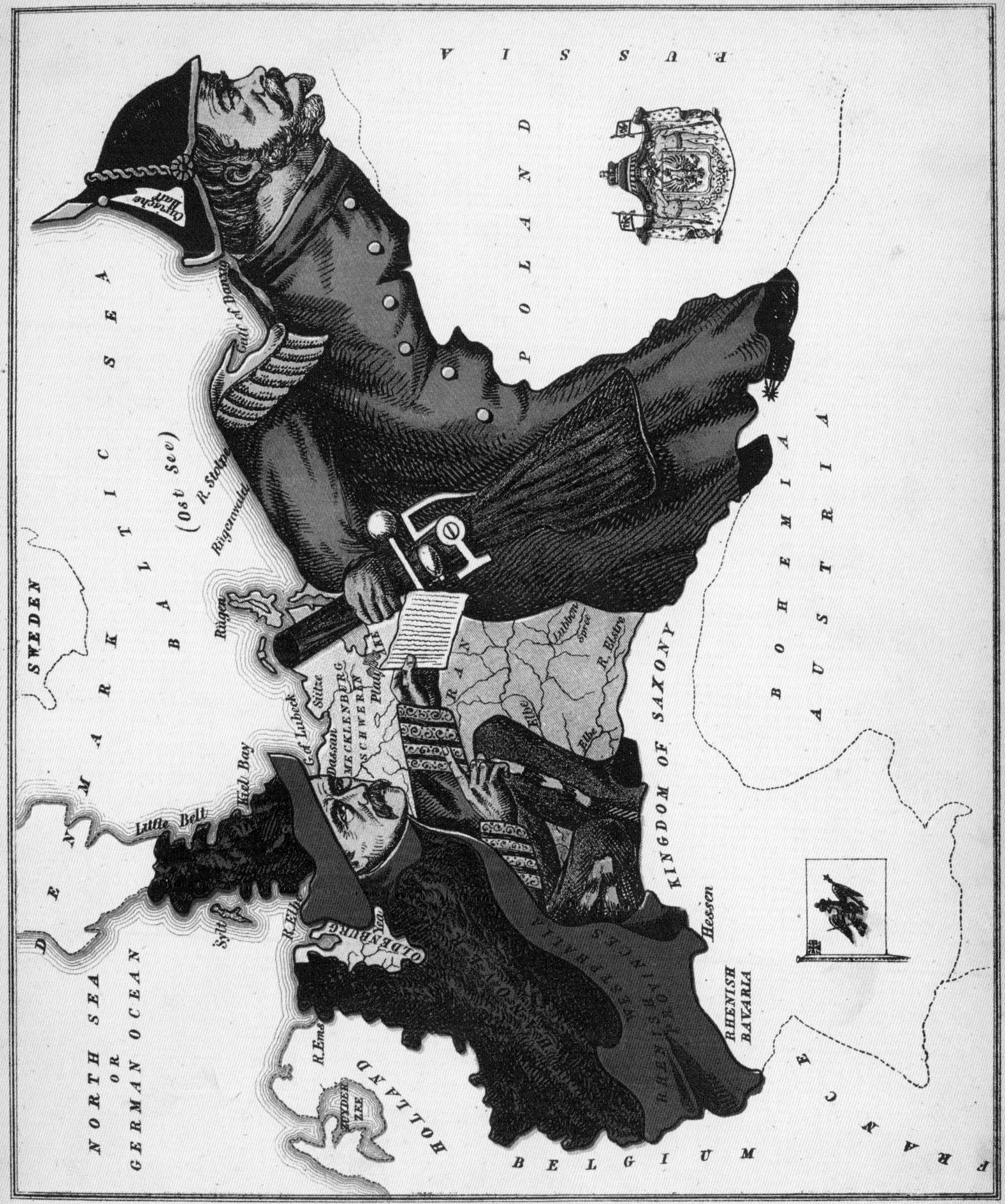

Vincent Brooks, Day & Son, Lith. London, W.C

His Majesty of Prussia—grim and old—
Sadowa's King—by needle guns made bold;
With Bismark of the royal conscience, keeper,
In dreams political none wiser—deeper.

Map of Spain and Portugal, [1868]

'Aleph', London

'These long divided nations soon may be,
By Prim's grace, joined in lasting amity.
And ladies fair – if King Ferdinando rules,
Grow grapes in peace, and fatten their pet mules.'

Chromolithograph; border: 243 × 202 mm (9 5/8 × 8 in);
at widest: 268 × 202 mm (10 1/2 × 8 in)

THIS MAP comes from *Geographical Fun: Being humorous outlines of various countries with an introduction and descriptive lines* by 'Aleph' (real name William Harvey MD; see p. 170). It depicts the Iberian peninsula, with Spain as a lady in elaborate costume, with a substantial veil. She is a representation of Isabella II (1830–1904), Queen of Spain from 1843 until September 1878, when she was deposed and sent into exile. She is offering the bear, representing Portugal, a bunch of grapes as a sign of friendship.

The 'Prim' mentioned in the verse was Don Juan Prim y Prats, Marquis of Los Castillejos (1814–1870). He was one of Isabella's supporters in the Carlist Wars that bedevilled her reign; the Carlists, supporters of Isabella's uncle Carlos, Count of Molina, wanted a male succession to the throne. Ultimately, Prim threw his lot in with Isabella's enemies and assisted in her overthrow. He is mentioned here as being a proponent of union between the crowns of Spain and Portugal. Also mentioned is Ferdinand II of Portugal (1819–1885), King Consort to Mary II of Portugal, and regent, 1853–1855, after her death, for their underage son, Pedro V. After the overthrow of Isabella, Ferdinand was offered the throne of Spain, but declined. The doubt over the succession was one of the underlying causes of the Franco-Prussian War: the French wanted a Spanish Bourbon to succeed to the throne, while the Prussians backed a Hohenzollern candidate, a relative of their king.

Geographical Fun is generally attributed to 1869: there is no date on the title page. However, the book was advertised in *The Standard* (issue 13850) for 24 December 1868:

> Hodder and Stoughton's Illustrated Christmas Gift-Books for the Young. 1. In 4to. ornamented boards, 5s. or in cloth elegant, 7s. 6d. A Novelty – Geographical Fun: Humorous Outlines of Various Countries. Printed in Colours by Vincent Brooks, Day and Son. With Descriptive Verses.

The artist was Eliza Jane Lancaster (1852–1939), better known as Lilian Lancaster, a famous Victorian actress (especially in comedy), stage performer and singer. According to the Introduction of *Geographical Fun* (quoted in full on p. 170) she drew these maps for her brother, William, when he was confined to bed through illness. If she was indeed in her fifteenth year when she drew them, she displayed a remarkable precocious talent, which she continued to demonstrate throughout her career. As part of her act, she would draw caricatures at the request of the audience, as described in the *Illustrated Sporting and Dramatic News* (issue 317) for 21 February 1880: 'she has availed herself of the opportunity of introducing her talent as an artist, which is considerable, by drawing crayon sketches of well-known faces in sight of the audience, taking only 50 to 60 seconds for each head'.

In addition to this map, and its companion, she drew the maps on pp. 184, 186, 200 and 202, the last two signed with her married name, Lilian Tennant, following her marriage to William Edward Tennant in 1884.

Vincent Brooks, Day & Son, Lith. London, W.C.

These long divided nations soon may be,
By Prims' grace, joined in lasting amity.

And ladies fair—if King Fernando rules,
Grow grapes in peace, and fatten their pet mules.

‘Novel Carte of Europe Designed for 1870’, 1870

Joseph Goggins, Dublin

‘England, isolated, filled with rage, and almost forgetting Ireland whom she holds in leash. Spain smokes, resting upon Portugal. France repulses the invasion of Prussia, who advances with one hand on Holland, the other on Austria. Italy, also, says to Bismarck, “Take thy feet from hence.” Corsica and Sardinia, being a true Gavroche, laughs at everything. Denmark, who has lost his legs in Holstein, hopes to regain them. Turkey-in-Europe yawns and awakens. Turkey-in-Asia smokes her opium. Sweden, bounding as a panther. Russia resembles an old bogy who would wish to fill his basket.’

Lithograph; at widest: 355 × 412 mm (14 × 16 ¼ in)

JOSEPH JOHN GOGGINS, manufacturing stationer, lithographic artist and designer, was born to a Roman Catholic family in Dublin in or about 1842. His sole recorded contribution to the world of maps is this attractive cartoon map of Europe from the early part of the Franco-Prussian War (1870–1871). Although the artist may be all but unknown, this map, whether originating in Dublin, or perhaps more likely in an almost identical version published in Paris, spawned a large number of derivatives, published all over Europe as late as 1914, which capture the same spirit as the original.

Goggins' view seems anti-British; Great Britain is shown as a hag, turning her back on the European conflict, and covering her ears to cut out the disturbance of arms. British governments had a long history of trying not to get embroiled in European conflicts, and there was little reason for Britain to involve herself in a dispute between France and Germany. Ireland is shown as a faithful hound, on a lead.

Prussia, portrayed as Bismarck, is a fat figure in a *pickelhaube* (a Prussian infantryman's spiked helmet), crushing Austria under his weight and reaching for Belgium, but seemingly driven backwards by France. Many military experts at the time thought that the French would win the Franco-Prussian War because they had better equipment in supposedly decisive areas, but in fact the French were comprehensively out-generalled and out-fought, and suffered a humiliating defeat. In the west, the Spanish senorita reclines at leisure, smoking a cigarette, supported by poor little Portugal. Turkey-in-Europe and Turkey-in-Asia seem similarly unconcerned, the former as if waking from deep sleep, the latter smoking opium from a hookah.

The peninsula formed by Norway and Sweden is shown as a panther. Little Denmark stands to attention on his wooden legs because the province of Holstein, the base of the peninsula, had been lost to Prussia in an earlier war. Russia, in unflattering guise in an almost animalistic human form, is portrayed with a large bag slung over his shoulder, as if looking for spoils in eastern Europe while Prussian attention was diverted westwards, and with a pack of wolves at his back.

One can imagine that, with its anti-British slant, the map's sales within the British Isles were slow. However the pictorial narrative clearly resonated with European publishers, explaining the number of derivatives.

COMIC CARTE OF EUROPE FOR 1870
BRITISH ISLES
ATLANTIC OCEAN
NORTH SEA
NORWAY
SWEDEN
BALTIC SEA
WHITE SEA
RUSSIA
DENMARK
HOLLAND
SWITZERLAND
FRANCE
AUSTRIA
CRIMEA
SEA OF AZOV
BLACK SEA
TURKEY IN ASIA
ADRIATIC SEA
ITALY
CORSICA
SARDINIA
PORTUGAL
SPAIN
TURKEY
ARCHIPELAGO
ENTERED AT STATIONERS HALL
DUBLIN

Satirical Map of Europe, 1871

Manfredo Manfredi, Bologna

'L'Europa geografico-politica veduta a vola d'oca'.

Chromolithograph; map 342 × 528 mm (13 ½ × 20 ¾ in); with text: 428 × 657 mm (16 ⅞ × 25 ⅞ in)

ONE OF the most unusual and rarest of the satirical European political maps was published by Manfredo Manfredi in Bologna, in or shortly after 1871. The images are quite unlike any of the other maps of this genre, and Manfredi has constructed the accompanying verses in an archaic poetical Italian style reminiscent of Francesco Petrarca (1304–1374), commonly known as Petrarch, the humanist philosopher and prolific writer. Much of the imagery is hard to interpret at this remove, and the Italian verses are little help in that regard.

The map seems to be set in the immediate aftermath of the Franco-Prussian War of 1870–1871; the portly German soldier seated in Alsace-Lorraine, drinking copious amounts of beer, surely represents the German acquisition of the provinces in the Treaty of Frankfurt, May 1871, which ended the conflict. In Algeria, the scene depicts the uprising of the Kabylie region which began in March 1871; an Arab figure, presumably Muhammad al-Muqrani, the leader of the revolt, is shown spanking a French soldier, reflecting his early successes. When the Franco-Prussian War ended, fresh troops were sent to Algeria and the revolt was finally crushed in 1872.

On the island of Corsica is the bust of a soldier, labelled in the key as the 'Eroe di Sedan', referring to the Battle of Sedan, 1 September 1870, in which an entire French Army under the Emperor Napoleon III surrendered to the Prussians; the bust possibly refers to Maréchal Patrice de MacMahon, commander on that day, although he was wounded early in the battle and hence played little part in the defeat.

Within France, the three-headed Hydra, apparently slain, presumably relates to the bloody suppression of the Paris Commune, a socialist alliance that took over Paris at the end of the Franco-Prussian War, from March to May 1871, before the French Army stepped in to restore government control. MacMahon was commander of the Armée de Versailles that retook Paris.

In Portugal, João Carlos de Saldanha, Duke de Saldanha, is depicted pulling the strings of a puppet king; Saldanha, a distinguished soldier, played a leading role in Portuguese politics for some fifty years, mounting some seven *coup d'états*; he was Prime Minister for a short period from 19 May 1870 to 29 August 1870, but was sent to London as Portuguese Ambassador and died there in 1876.

In the east of the map, Russia is depicted as a bloodthirsty Cossack waving a blood-stained knife, holding Poland in chains, while a German figure, perhaps Chancellor Bismarck, looks on.

Throughout the map is depicted all manner of unrest, but the Swiss lawyer shelters behind fortress walls, while England remains blissfully unaware, gnawing on the bone that is India.

L' EUROPA GEOGRAFICO-POLITICA
VEDUTA A VOLO D' OCA
OCEANO ATLANTICO
MAR BALTICO
FRANCIA
MARE DI BISCAGLIA
MARE MEDITERRANEO
MARE ADRIATICO
MARE JONIO
ARCIPELAGO
TURCHIA
RUSSIA
Bologna presso Manfredi Manfredo Editore Via Venezia N. 1749

Puzzle Game Blocks [1875]

Jean-Guillaume Barbié du Bocage, Paris

Lithographs on wood; each map approx. 235 × 310 mm (9 ¼ × 12 ¼ in), the blocks 45 mm (1 ¾ in) square.

THIS IS a set of six maps, dissected and mounted on forty-two six-sided wooden blocks, each block therefore with one section of each of the six maps. The skill in the game, therefore, is not merely correctly aligning forty-two sections to make a map, but also distinguishing which of the six faces of each block bears the correct section for the map being made.

To assist the 'players', the blocks are accompanied by the six maps in sheet form for guidance in case of difficulty. The blocks are housed in the original case with an elaborate embossed title on the lid, and paper pasted over a wooden frame.

Jean-Guillaume Barbié du Bocage (1793–1843) was the son of Jean-Denis Barbié du Bocage (1760–1825), a distinguished French geographer, trained by Jean-Baptiste Bourguignon d'Anville (1697–1782), *premier géographe du Roi*. His mother was the daughter of Guillaume-Nicolas de la Haye (1727–1802), *premier graveur du Roi*, and member of a distinguished family of map engravers.

These maps were apparently first published in Barbié du Bocage's *Géographie générale* in 1842; the plates passed into other hands after his death, and were much reprinted, here by the Parisian publishing house Bouasse Lebel. Although undated, the maps probably date to about 1875. The boundaries of France are those established after the end of the Franco-Prussian War (1870–1871), in which the provinces of Alsace and Lorraine were lost to Prussia, while the boundaries of North America have been heavily revised since the first printing. The boundaries of Mexico have been pushed to the south, with the absorption of Texas, New Mexico and California, all once part of Mexico, into the United States.

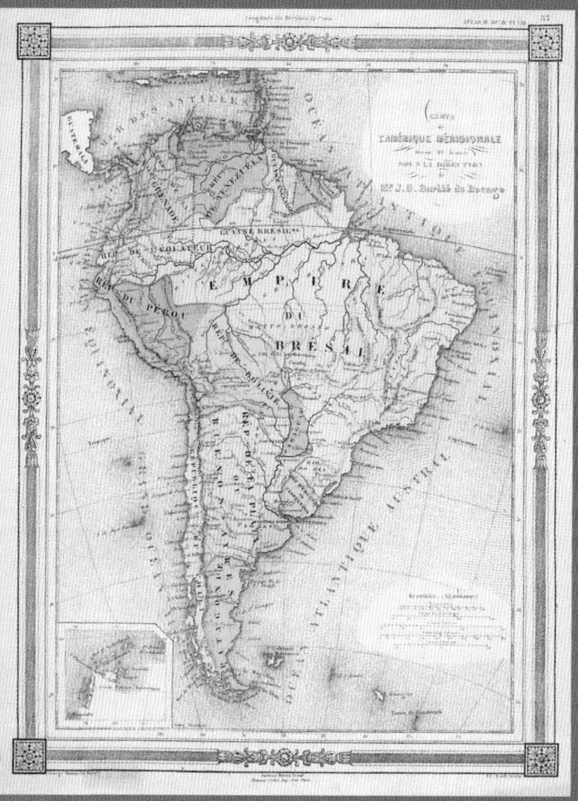
EMPIRE
DU
BRÉSIL

OCÉAN GLACIAL ARCTIQUE

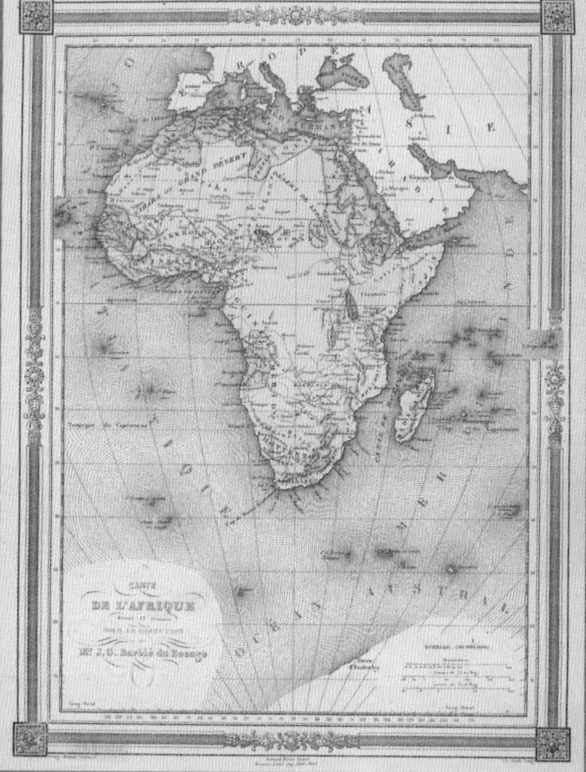
CARTE
DE L'AFRIQUE
Mr. J.G. Barbié du Bocage
OCÉAN AUSTRAL

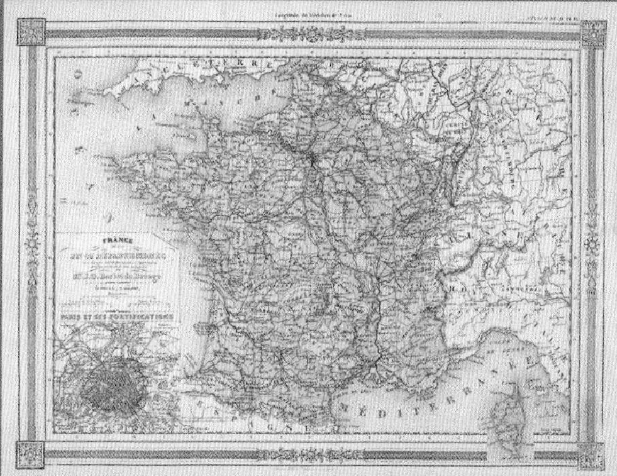
MÉDITERRANÉE

ASIE
AMÉRIQUE

‘Serio-Comic War Map for the Year 1877’, 1877

Fred W. Rose, London

‘The Octopus – Russia – forgetful of the wound it received in the Crimea, is stretching forth its arms in all directions ...’

Chromolithograph; border: 426 × 579 mm (16 ¾ × 22 ¾ in)

THIS IS the first of a sequence of fine allegorical – ‘serio-comic’ – maps of Europe, inspired by Thomas Onwhyn, but drawn by the elusive ‘F.W.R.’, identified from later map titles as Fred W. Rose. Rose was clearly a gifted artist and caricaturist, whose work spanned the period 1877 to 1900, but it was only in 2014 that he was positively identified as Frederick William Rose (1849–1915), an occasional artist whose principal employment was as a fairly senior civil servant, although he was also the author of two somewhat sensational novels.

This is a political cartoon, with explanatory text in English, referring to the situation surrounding the Russo-Turkish War of 1877. While Britain was no great friend of the Ottoman Turks, they provided a convenient counterweight against Russia in south-eastern Europe and the Middle East. Britain was concerned about Russian intentions and expansion in Central Asia extending southwards into Afghanistan, with the potential threat to British India – a fear reflected in Rose’s occasional comments on European politics in his *Notes on a Tour in Spain* (1885). Equally, Germany preferred to have a distracted neighbour to her east, rather than one intent on expansion into Poland and threatening Austria and the Balkan countries.

This is the earliest known cartoon map of Russia to portray the country as an octopus; while the image of the tentacles reaching in different directions aptly captured the regional aspirations of the Russian Empire, the prevalence of the octopus motif in later maps suggests that the octopus also spoke to humanity’s primeval fears, evoking a terrifying and mysterious creature from the depths (the dark outer places of the world) that convincingly conjured a sense of limitless evil. Here the Russian octopus has seized Turkey’s wrist and ankle, tugging him in opposite directions, and has a stranglehold on Poland, the Crimea and the Persian Shah. One tentacle is being fended off by the German Chancellor Bismarck, while the Austro-Hungarian Empire, with dagger drawn, is ready to join the war.

REVISED EDITION
SERIO-COMIC WAR MAP
FOR THE YEAR 1877.
BY F. W. ROSE
ERKLARUNGEN.
REFERENCE.
RUSSIA
NORWAY
SWEDEN
G. of BOTHNIA
G. of FINLAND
BRITISH ISLES
NORTH SEA
BALTIC SEA
ENGLISH CHANNEL
BAY OF BISCAY
PORTUGAL
CORSICA
SARDINIA
ADRIATIC SEA
BLACK SEA
CASPIAN SEA
ASIA MINOR
EMPIRE
ARCHIPELAGO
GREECE
CANDIA
MEDITERRANEAN SEA
PERSIA
[COPYRIGHT] London, Published by G.W. Bacon & Cº 127 Strand.

‘The Avenger, an Allegorical War Map for 1877’, 1877

George W. Bacon, London

Chromolithograph; border: 447 × 601 mm (17 5/8 × 23 5/8 in)

LIKE THE previous map, this anonymous allegorical map relating to the Russo-Turkish War of 1877 was published in London by George Washington Bacon. For that reason and on stylistic grounds it has sometimes been attributed to Fred W. Rose (see p. 180), but this map presents a totally different political view of the war to that of Rose’s map and the attribution is somewhat unlikely. It is perhaps more probable that Bacon, himself an egalitarian American and a very efficient publicist, wanted something to sell to all of his customers, whatever their views.

In this map, Russia is the ‘Avenger’, a winged, cloaked figure with a shield bearing the Russian Romanov double-headed eagle. Round his neck, he wears a heavy chain with the motto ‘Liberation of the Serfs’, while his sword bears the words ‘Protection to the Oppressed’. Here Russia is the hero coming to the rescue of the Balkan countries, with Rumania on her knees pleading for help, while Turkey recoils in shock, his scimitar still wet with the blood from his Bulgarian atrocities, symbolised by the two skulls and the body, pierced by a sword, on his lap.

The cartoonist commented:

> Russia, represented by an allegorical figure of Progress, is punishing ‘the sick man’ Turkey, for the wanton outrages he has committed. On his breast is a bright jewel, and in his hand a sword, which is likely to prove successful. Hungary, alarmed at the disturbance and pointing out their common danger, is rousing her brother Austria, and asking him whether it would not be advisable for them to interfere in the fray.

In the south, Greece eagerly awaits the outcome, reading the note ‘God helps those who helps themselves’, apparently eager to see what spoils might come his way from the war; ‘Candia’ (Crete) has her hands bound, prisoner of the cruel Turks. France is preoccupied: ‘Power’, the government, is struggling with the people, wearing the cap ‘Liberty Equality Fraternity’, apparently a relic of the civil unrest after the Franco-Prussian War. Germany, heavily armed, seems more preoccupied with France than Russia; the formidable array of artillery is aimed in that direction.

Spain, Portugal, Belgium and Holland are unconcerned by these distant events, while Switzerland is shown as a land of clockmakers. Italy is depicted as the Devil, and a puppet-master, dangling the Pope on strings, while the Sicilians are more interested in the ransoms to be collected from travellers through the island.

Of the countries of the British Isles, it seems that only England is concerned with the outcome of the war. Depicted as St George, England’s patron saint, England is trying to slay the dragon, here the ‘Eastern Question’ – the consequences of the decline of the Ottoman Turkish Empire, commonly known as the ‘sick man of Europe’ – while Scotland is carrying a collected volume of the novels of Sir Walter Scott.

The text is presented in two languages, English and German, so presumably it was intended for export to Germany as well as for sale in Britain.

THE AVENGER
AN ALLEGORICAL WAR MAP.
FOR 1877.
ERKLÄRUNG.
REFERENCE.
BRITISH ISLES
NORWAY
SWEDEN
NORTH SEA
BALTIC SEA
HOLLAND
ENGLISH CHANNEL
PORTUGAL
CORSICA
SARDINIA
SICILY
ITALY
ADRIATIC SEA
GREECE
ARCHIPELAGO
CANDIA
BLACK SEA
CASPIAN SEA
PERSIA
MEDITERRANEAN SEA
AFRICA
EGYPT
BULGARIAN ATROCITIES
London Published by G.W. Bacon & Co 127 Strand
COPYRIGHT.
ALL RIGHTS RESERVED
1035.(302.)

'United States, a Correct Outline', [1880]

Eliza Jane [Lilian] Lancaster, [United States]

Manuscript, ink and colours on card; image: 250 × 375 mm (9 ⅞ × 14 ¾ in); at widest: 261 × 375 mm (10 ¼ × 14 ¾ in)

LILIAN LANCASTER (1852–1939) was a well-known English actress, singer and stage performer (see p. 170), with a notable talent for drawing cartoons and caricatures, often of a cartographic nature. In 1880, such was her success that she embarked on a tour of the United States, and found herself there during the final stages of the Presidential election. That proved an inspiration for two cartoon maps depicting the election.

The two caricatures, part of a group of cartographic cartoon drawings by Lancaster, were acquired relatively recently by the British Library. It is assumed that the caricatures were drawn for publication, as there is a copyright declaration on both images, but no printed versions are known.

On the present map, there is evidence of an erased title at bottom left, but it is only partially legible; the words 'between Hancock and Garfield' are all that can be made out. Imposed on the familiar outline of the United States, this cartoon depicts the course of the Presidential election of 1880, incorporating comic portraits of James A. Garfield (the Republican challenger) and his opponent Winfield Scott Hancock (the Democratic candidate). The two candidates are depicted as squabbling children, both clad in dresses. Hancock is pulling firmly on Garfield's foot, while Garfield appears to be either rubbing tears from his eyes or adopting a defensive boxing stance. Uncle Sam has turned his back on the mayhem; his body forms the eastern seaboard of the United States, with the ruff of his shirt representing the Virginia and Carolina coast, his left leg forming the Florida peninsula, and his right leg extended to form the Texas coast. Behind the fighting figures, the United States flag forms the northern, Midwest and Great Lakes states.

United States.
a correct outline.
BRITISH COLUMBIA
BRITISH AMERICA
CANADA
NEW BRUNSWICK
Pacific City
Lake Superior
Lake Huron
Lake Michigan
Lake Ontario
Falls of Niagara
Lake Erie
Portland
Boston
Newport
New York
Albemarle Sound
Charleston
St. Augustine
C. Canaveral
Tampa B.
Galveston
Mobile Bay
GULF OF MEXICO
MEXICO
ATLANTIC OCEAN
By Lilian Lancaster
Copyright.

'United States: a Correct Outline', 1880

Eliza Jane [Lilian] Lancaster, [United States]

Manuscript, ink and colours on card; image: 243 × 373 mm (9 ⅝ × 14 ¾ in); at widest: 268 mm (10 ½ in)

THE FAMOUS stage performer Lilian Lancaster (1852–1939) was in the United States during the final stages of the Presidential election of 1880, and the visit proved an inspiration for two cartoon maps relating to the election, this one and that on p. 184.

While the first map depicts the campaign, the present map presents the outcome. The image is formed by comic portraits of James A. Garfield (the Republican victor) and his opponent General Winfield Scott Hancock (the vanquished Democratic candidate) superimposed on the familiar outline of the United States. Garfield is kneeling before the female figure of 'Liberty' (as the identification label on her shield has it), presenting a cornucopia to her. She is reclining, clad in a flowing white dress, and holding a laurel leaf, a symbol of victory since ancient times, labelled 'President 1880' over his head.

Hancock, dressed in his military uniform (both men actually held senior rank in the Union Army during the American Civil War), has turned his back on this symbolic scene of the crowning of the victor, his head bowed in defeat.

Hancock forms the eastern seaboard of the United States; the ruff of his shirt represents the Virginia and Carolina coast, while his foot, in some sort of plaster or snow boot, forms the Florida peninsula. Liberty takes the shape of the western part of the United States; the flowing hem of her dress is Texas. Behind Garfield, the United States flag – 'Old Glory' – gives shape to the Midwest and Great Lakes states.

United States
C. Flattery
BRITISH COLUMBIA
BRITISH AMERICA
PRESIDENT
1880
Lake Superior
CANADA
Montreal
New Brunswick
Ottawa
Portland
Boston
Lake Huron
Lake Ontario
Falls of Niagara
Lake Michigan
Lake Erie
Newport
Brooklyn
Philadelphia
Dover
C. Hatteras
C. Lookout
Charleston
Savannah
St Augustine
C. Canaveral
Mouths of the Mississippi R.
Galveston
GULF OF MEXICO
Brownsville
Florida Strait
MEXICO
Santa Barbara Islands
Lilian Lancaster
Nov 4th/80.
Copyright.

'Falmouth Borough Octopus Attempting to Grasp the Parishes of Falmouth and Budock', 1882

Edwin Thomas Olver, London

Chromolithograph; at widest: 449 × 600 mm (17 ⅝ × 23 ⅝ in)

THIS UNUSUAL map is a relatively early reworking of the 'grasping octopus' motif. Here the octopus is no longer a symbol of militaristic expansionism, but of a search for political and financial benefits on a local level from territorial enlargement by democratic means, although the inevitable underlying theme is that of power and empire-building, albeit on a very small scale.

In this case, the octopus is depicted superimposed on an outline map of Falmouth in Cornwall; the map represents Falmouth Town Council (the region coloured in blue), which was responsible for governing the Borough of Falmouth, extending its tentacles over the neighbouring parishes of Falmouth and Budock.

Edwin Thomas Olver (1829–1917) obviously disagreed with the borough's actions. Although Olver spent part of his publishing and printing career in London (he became a freeman of the Stationers' Company by redemption in 1874), he was a Falmouth man born and bred. He later returned to live in Falmouth (or more precisely in Budock) and went on to publish some guidebooks on the area: *Picturesque Cornwall. No. 1. Falmouth and its surroundings* [1894]; *The pictorial guide to Falmouth and its surroundings* (1897); and *Falmouth and its surroundings: historical, descriptive and progressive. A pictorial guide and social souvenir for 1914–1915*. Even the London address given on the map indicates his Cornish origin – Pendennis Castle can be seen at the extreme left.

At the time, the Borough Council governed what was effectively only the town proper: the land extending between Packet Quay and 'Falmouth Quay' (about Quay Street?), with a bit of hinterland; Falmouth Parish, which all but surrounds the borough, contained Falmouth Railway Station and Falmouth Docks (and also modern Custom House Quay). These areas were key to the town's future expansion, both physical and financial, and yet lay outside the council's jurisdiction. The physical case is made clear in the table under the title, which details a variety of pertinent, and interesting, statistics: the borough contained a mere 40 acres (about 16 ha); Falmouth Parish had 651 (265.5 ha); and Budock, labelled in the table 'Penwerris Lighting Company', about 151 acres (61 ha). Added together, they were some twenty times larger than the town itself.

Even more obvious are the financial benefits of a takeover; the revenue of the town was £9,070 per annum and its liabilities £8,639 (nearly a year's income), while the Parish had debts of £8,598 set against an annual income of £19,021 and Budock £2,568 income, with no debt.

The population statistics derive from the 1881 census, so it is likely that the map dates from shortly after that year, probably between 1882 and 1885.

Clearly Olver was responding to perceived attempts by Falmouth Borough to gain control of the outlying parishes of Falmouth and Budock and in this most graphic of forms attempted to campaign against the council. At this remove some aspects of the local politics are not wholly clear, but certainly Falmouth was expanding rapidly in the 1870s, with a lot of residential construction in the hinterland and industrial development around the docks and railway, notably iron-founding and shipbuilding. The enlarged docks brought additional traffic, particularly for fishing vessels, while the military base at Pendennis was greatly expanded. It may be that the council sought greater benefit from the increased prosperity of the region, or simply wished for greater clarity of governance and more effective planning.

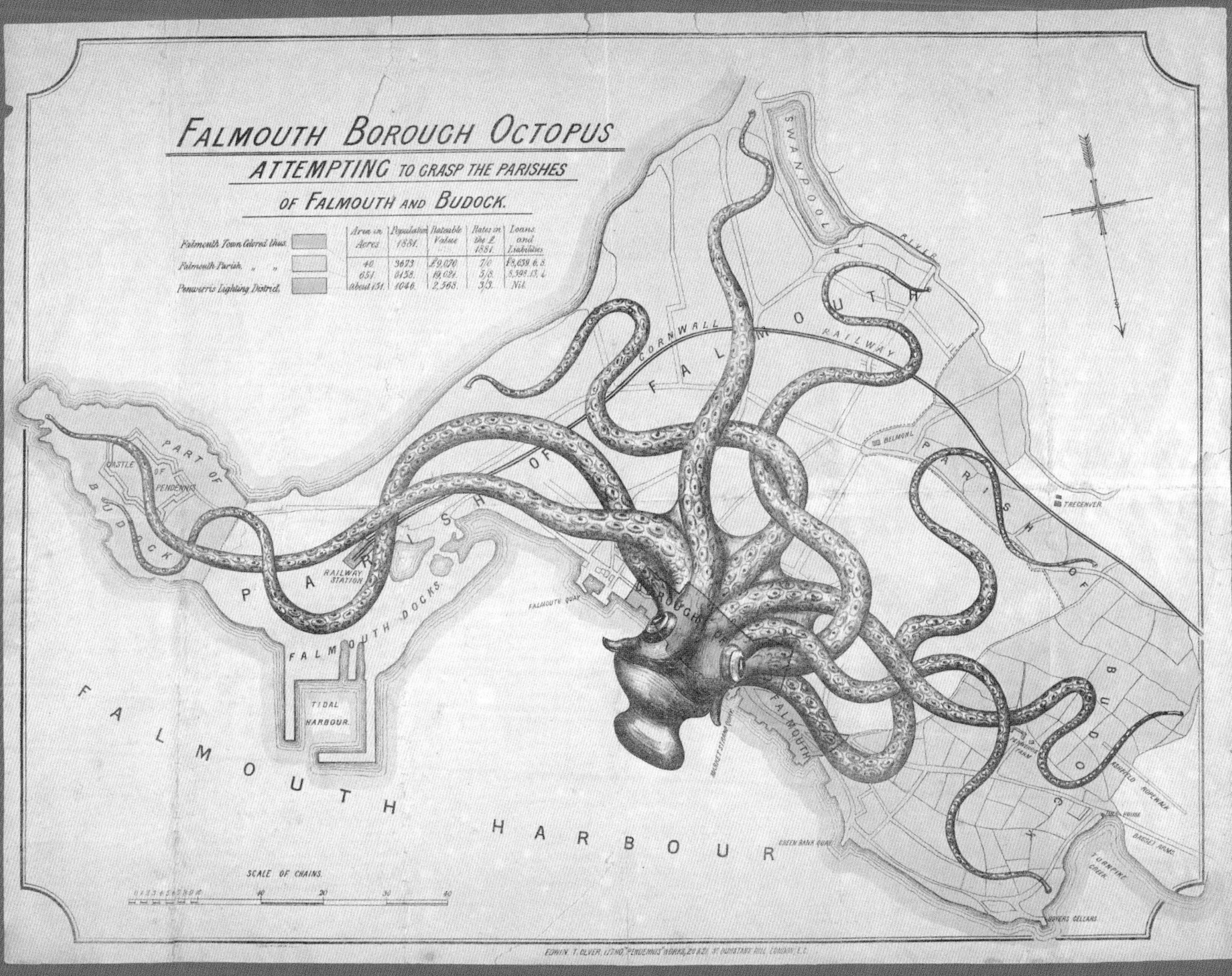
FALMOUTH BOROUGH OCTOPUS
ATTEMPTING TO GRASP THE PARISHES
OF FALMOUTH AND BUDOCK.
Falmouth Town Colored thus.
Falmouth Parish.
Penwerris Lighting District.
SWANPOOL
CORNWALL RAILWAY
PART OF BUDOCK
CASTLE OF PENDENNIS
RAILWAY STATION
FALMOUTH DOCKS
TIDAL HARBOUR.
FALMOUTH HARBOUR
PARISH OF FALMOUTH
PARISH OF BUDOCK
BOROUGH OF FALMOUTH
BELMONT
TREGENVER
SCALE OF CHAINS.

'Map of England. A Modern St George and the Dragon!!!', 1888

William Mecham, London

Chromolithograph; at widest: 502 × 315 mm (19 ¾ × 12 ⅜ in)

ST STEPHEN'S *Review* was a weekly periodical devoted to political and social commentary, published from 1883 to 1892, when it was renamed *Big Ben* for a short period before folding the following year. It was common for the journal to illustrate an article with a colour double-page caricature, commissioned from William Mecham (1853–1902), who went by the pen-name Tom Merry. Mecham was a skilled artist, famed for his cartoons and caricatures; as with Lilian Lancaster, part of his music-hall stage act involved drawing sketches at rapid speed for the audience – 'Lightning Cartoons' – often completed in under a minute.

This geographical cartoon was prepared for issue 274, to a commentary entitled 'The New St George and the Dragon', relating the political struggle between Lord Salisbury, the Prime Minister, and William Ewart Gladstone, his rival and leader of the Liberal Party. Salisbury is depicted as St George. Gladstone is the Dragon; out of his mouth hangs a barbed tongue, on which is written 'Home Rule', referring to the impassioned debate on the governance of Ireland which the domestic politics of the period. It was the split in his own party over Home Rule which led to Gladstone's defeat in a vote in the House of Commons and Salisbury's subsequent victory in the 1886 General Election.

The 'commentary' takes the form of an imagined conversation between the leading characters from Lewis Carroll's *Alice's Adventures in Wonderland*, first published in 1865 and already a classic. Alice, the March Hare, Dormouse and the Mad Hatter rewrite current affairs in the manner of the Mad Hatter's tea party.

> 'No matter,' said the Hatter. 'I was about to add, that great as St. Patrick's task was, it was as nothing compared with St. George's, who had to slay a most truculent dragon.'
>
> 'Well, well,' said the March Hare testily; 'come to the point, man. What does it all amount to?'
>
> The Hatter got up and went stealthily to the door, opened it suddenly to see if anyone was listening, then shut and locked it. He minutely examined every portion of the room which might conceal anyone, and then resumed his seat.
>
> 'He's at it again!' he muttered, in a hoarse whisper.
>
> 'Who? What?' cried the listeners simultaneously.
>
> 'Why, St. George, of course!' snarled the Hatter. 'Who else should it be?
>
> Alice was terribly puzzled. 'Please,' asked she, 'what is he at?'
>
> 'The Dragon,' said the Hatter pouring out another cup of tea.
>
> 'I understand him perfectly,' broke in the March Hare; 'I always do, for I am used to him. It is June now, and I am, therefore, perfectly sane. He refers to Lord Salisbury, as St. George, destroying the old Dragon, Gladstone.'
>
> 'Just so,' said the Dormouse, 'and as it's summer I am awake, so I can tell you the whole thing is contained in the map of England. It is just a puzzle – work it out for yourself; or stay, you might go to sleep, so here I give you the key;' and with that he handed to the astonished Alice a map of England, of which we have faithfully reproduced a copy. (See Cartoon).
>
> 'I don't think much of the Dragon,' said Alice, after gazing at the map.
>
> 'And,' continued the Hatter, who had been ruminating, 'I only hope St. George will prove as strong in the North of England as he is in the map; of course, we know he has killed the Dragon in the South.'

St. Stephen's Review Presentation Cartoon. June 9 th, 1888.
MAP OF ENGLAND.
GERMAN OCEAN.
IRISH SEA.
BRISTOL CHANNEL.
ENGLISH CHANNEL.
Berwick
Holy I.
Cheviot Hills
SOLWAY FIRTH
S. Shields.
Sunderland.
Hartlepool.
Whitehaven.
Whitby.
Douglas.
Scarborough.
Flamborough Head.
Morecambe Bay
LIVERPOOL
Spurn Head.
Humber.
ANGLESEY
Holy I.
The Wash.
Wells
Cardigan Bay.
Yarmouth
Harwich.
Milford Haven.
Caermarthen Bay
Swansea B.
Cardiff
Mouth of R. Thames.
Margate.
Dover.
Hartland Pt.
Dunge Ness.
Hastings
Brighton.
Portsmouth
I. of WIGHT
Tor Bay.
Portland Bill
St. Ives
Plymouth
Lands End.
The Lizard
TOM MERRY, LITH.
A MODERN St. GEORGE AND THE DRAGON!!!

'Angling in Troubled Waters. A Serio-Comic Map of Europe', 1899

Fred W. Rose, London

Chromolithograph; at widest: 500 × 704 mm (19 ¾ × 27 ¾ in)

FRED W. ROSE was the most famous of the Victorian-era artists creating cartographic cartoons. Two maps issued by him in 1899/1900, the present map and that on p. 194, proved hugely popular, and sold in very large numbers, although the fragile paper on which they were printed means that very few examples survive today.

This example of 'Angling in troubled waters' is the earliest printing of the map, most easily identified as such by having the title in English only; later printings had the title also rendered in the other languages – 'Der Fischfang im Trüben', 'La Pêche en eau trouble', 'La Pesca nelle acque turbes'. Demand was substantial: another impression dating from 1899, the first year of publication, refers to 15,000 copies having already being printed.

In this political commentary on the European powers and their overseas possessions and aspirations, Rose has equated burgeoning colonial ambitions with the sport of fishing, and describes the activities of the rival powers in those terms:

> John Bull (who is entirely at one within his own borders), notwithstanding the troubling of the waters by cantankerous neighbours, is satisfied with the fairly good sport he has lately enjoyed, and with his well-equipped bait-can, and the help of the landing net he holds aloft, he may have to land another catch ere long.
>
> In France, the struggle between the civil and military power, not only smirches the clothes of the combatants, but makes it also probable that the Republic is more likely to lose some of the fish it has already taken, than to shine in the angling competition of the day. The shade of Corsica's greatest son is amazed at the spectacle.

Indeed, the Emperor Napoleon is shown standing on Corsica, gazing at the in-fighting in France. The map was drawn at the time of the Spanish–American War, and Spain is shown as a bullfighter brought to his knees by the loss of his possessions in the Americas, and now the Philippines. Rose writes of Germany, 'The German Emperor, not satisfied with his successes in the fields of art, oratory and literature, has taken his pack upon his back, and is looking around to see what advantages he may achieve as an imperial bagman. His fist is no longer mailed.'

British concerns are more fully reflected in the descriptions of Russia and Turkey. John Bull is depicted reeling in Egypt, which was strategically important in the eastern Mediterranean and as a staging point to the Middle East and India, where the British feared Russian intentions. Rose commented:

> [Tsar Nicholas] is offering the olive branch to the world. All honour to him, but if he could discard those toys in his belt, and the store under his right arm, and if we knew exactly what fish he is playing on his line, the world might be more ready to accept his offer.
>
> Turkey, who has lost so much weight as to be scarcely recognizable, is holding his hand to his ear. Would that he might hear the howl of indignation which rises against him for the terrible stain upon his clothes. Russia treads heavily upon him, and he no longer knows the repose of by-gone days. Even the 'present for a good boy', which lies in his pocket, may not bring him much satisfaction.

The skulls in Bulgaria and Armenia – 'the terrible stain' alluded to in the description of Turkey – represent the massacres of the local population by Turkish forces.

ANGLING
IN TROUBLED WATERS
A SERIO-COMIC MAP OF EUROPE
BY
FRED. W. ROSE
AUTHOR OF THE 'OCTOPUS' MAP OF EUROPE
COPYRIGHT - TOUS DROITS RESERVES
KEY MAP OF
EUROPE
RUSSIA
AFRICA
NORTH SEA
BALTIC SEA
BAY OF BISCAY
BLACK SEA
MEDITERRANEAN SEA
PORT ARTHUR
MERV
REFERENCE.
REFERENCE.

'John Bull and His Friends. A Serio-Comic Map of Europe', 1900

Fred W. Rose, London

Chromolithograph; border: 483 × 688 mm (19 × 27 ⅛ in)

THIS IS the second of the two famous political cartoon maps published by Fred W. Rose at the turn of the twentieth century (see p. 180). His focus here is on the politics within Europe, particularly on the ways in which the different European countries responded to the outbreak of the Boer War in South Africa. The Boers were farmers, descendants of Dutch and German settlers, who established two republics in southern Africa and who, in 1899, declared war on Britain. The formal part of the war was won fairly quickly by the British, but the succeeding guerrilla war fought by the Boer commandos was put down only with great effort and no little brutality.

John Bull, in British Army khaki, is shown standing firm against the attack of what the 'Reference' text calls 'two wild cats' – the Boer republics of the Transvaal and the Orange Free State. The store of ammunition behind him reveals the support that he can expect from the countries of the Empire, including Australia, Canada and India. Support is not forthcoming from the Spanish, sulking from their loss of their colonial possessions to America, or from France, still recovering from the Dreyfus Affair and with her own colonial problems. Rose does express his concern at the attitude of the Americans, when he writes: 'The letter at his feet from his friend Uncle Sam, would be more encouraging were it not for the post-script. The Nationalist section in Ireland takes this opportunity to vent her abuse on him, but is restrained by the loyalty of the people' – an allusion to the continuing and inflamed debate in Ireland and Westminster over Irish Home Rule. On the map the postscript reads: '"We wish you success". J. McKinley. P.S. "But hop[e] ye'll get likked." Irish Democrat'.

Germany is preoccupied; Kaiser Wilhelm II is playing with his new fleet. The Kaiserliche Marine (Imperial Germany Navy) was established in 1871 and greatly expanded from 1899 onwards under the leadership of Admiral Alfred von Tirpitz, with the plain intention of challenging Britain's maritime supremacy. This compelled the British to renew their own shipbuilding programme – an arms race that was to lead ineluctably to the First World War, in which two of Rose's sons were to be killed.

In his depiction of Russia Rose has reverted to the octopus delineation, giving a rather more hostile interpretation of Russian activities and intentions than in his 'Angling in troubled waters' map of the previous year (p. 192):

> Russia, in spite of the Tzar's noble effort to impress her with his own peaceful image, is but an Octopus still. Far and wide her tentacles are reaching. Poland and Finland already know the painful process of absorption. China feels the power of her suckers, and two of her tentacles are invidiously creeping towards Persia and Afghanistan, while another is feeling for any vantage where Turkey may be once more attacked.

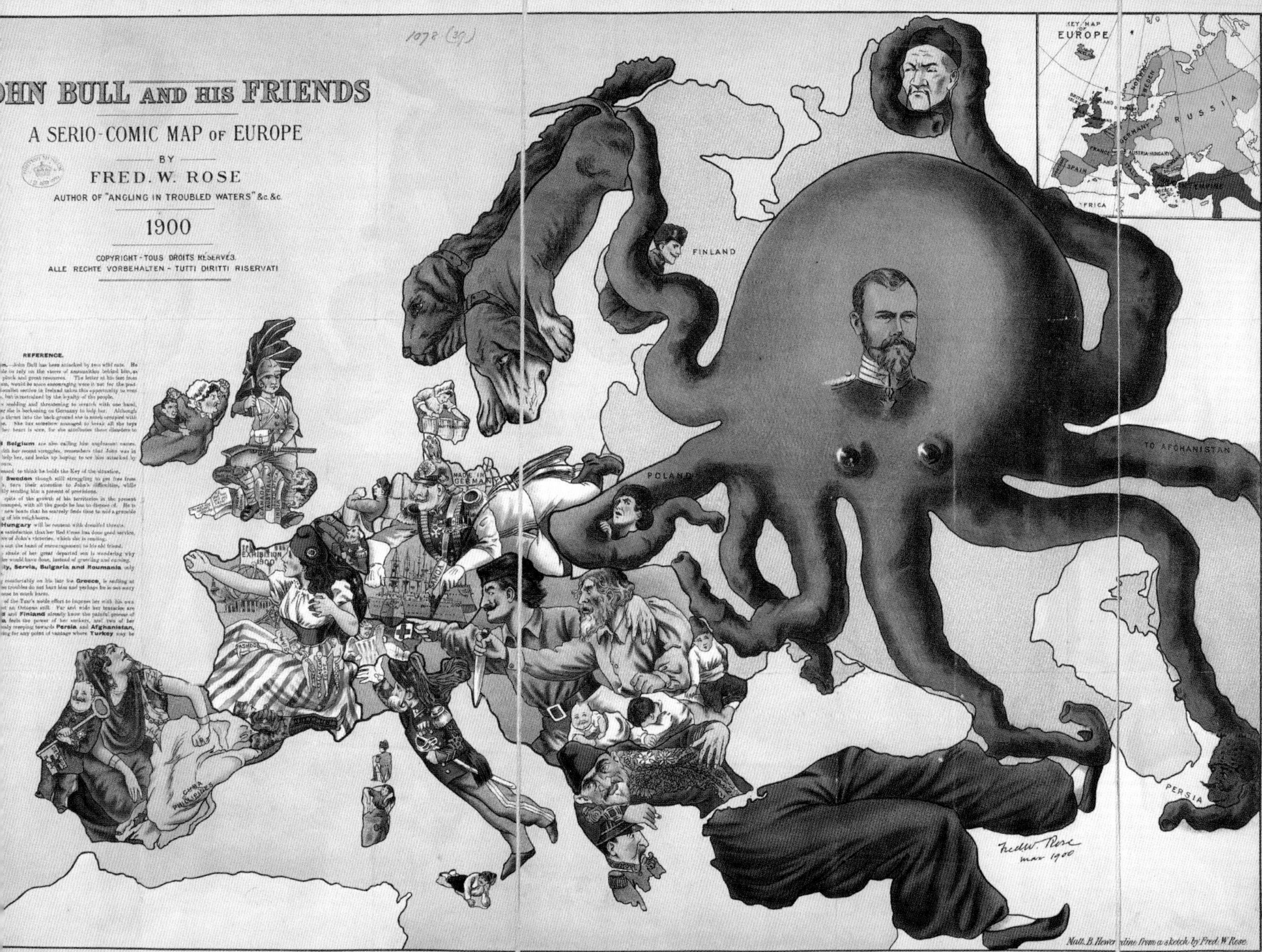
OHN BULL AND HIS FRIENDS
A SERIO-COMIC MAP OF EUROPE
BY
FRED. W. ROSE
AUTHOR OF "ANGLING IN TROUBLED WATERS" &c. &c.
1900
COPYRIGHT - TOUS DROITS RÉSERVÉS.
ALLE RECHTE VORBEHALTEN - TUTTI DIRITTI RISERVATI
REFERENCE.
FINLAND
POLAND
TO AFGHANISTAN
PERSIA
KEY MAP
EUROPE
RUSSIA
Fred W. Rose
mar 1900
Matt. B. Hewerdine from a sketch by Fred. W. Rose
G. W. Bacon & Co., Ltd., 127, Strand, London.

'A Humorous Diplomatic Atlas of Europe and Asia', 1904

Kisaburo Ohara, [Tokyo]

'"Black Octopus" is a name newly given to Russia by a certain prominent Englishman. For the black octopus is so avaricious, that he stretches out his eight arms in all directions, and seizes up every thing that comes within his reach ... But as it sometimes happens he gets wounded seriously even by a small fish, owing to his too much covetousness. Indeed, a Japanese proverb says: "Great avarice is like unselfishness."'

Chromolithograph; printed area 419 × 575 mm (16 ½ × 22 ⅝ in)

KISABURO OHARA was a Japanese student who was studying at Keio University when war broke out between Russia and Japan – the Russo-Japanese War of 1904–1905. Inspired by the octopus depictions of Russia found in the European maps by Fred W. Rose (see pp. 180, 194) and elsewhere, Ohara drew this graphic caricature map of Russia, but broadened the geographical coverage to show the Russian tentacles extending into Asia.

In western Europe, the principal military powers are all shown with their cannons pointing towards Russia, while the German Kaiser fends off a tentacle encircling Finland. Another encircles Poland. Both these countries, as well as the smaller provinces of Montenegro, Serbia, Bulgaria and Rumania are all represented by skulls, to indicate the ills of Russian rule. Turkey in Europe is beset on three fronts – with one tentacle round his waist and another round his ankle, while the Greek crab nips at his elbow.

In Asia one tentacle has encircled the throat of Persia; a second has been fended off by the British in India but instead seeks to drag Tibet from Chinese hands; a third reaches through Manchuria, to 'Port Arther' – Port Arthur on the Yellow Sea, a strategically important natural harbour, a Russian naval base and fortress, constructed at the southern end of the Chinese Eastern Railway (later known as the South Manchuria Railway), to give Russia an open Asian seaport ice-free all the year round. Meanwhile, China looks distinctly unhappy at having a 'friendly' Russian tentacle draped across his shoulders.

As might be expected, Ohara portrays Russia as the aggressor, both generally across the map, and specifically in the war with Japan. Russia had, however, seriously underestimated the military and technological advances that Japan had made in the preceding thirty years. The Russian fleets were defeated on the high seas, and the 'impregnable' fortress of Port Arthur was carried by force of arms. Russia's defeat in the war was the first for a European power at the hands of an Asian one, and heralded an important change in the balance of power in the Far East.

The map was composed in Ohara's native Japanese, but was evidently intended for foreign sale, as key elements are also rendered in English. The title, the panel of text at top left and the place names are in English, specifically to inform a European audience of the perils of underestimating the imperialist tentacles of the Black Octopus.

'Black Octpus' is a name newly given to Russia by a certain prominent Englishman. For the black octpus is so avaricious, that he stretches out his eight arms in all directions, and seizes up every thing that comes within his reach. But as it sometimes happens he gets wounded seriously even by a small fish, owing to his too much covetousness. Indeed, a Japanese proverb says: "great avarice is like unselfishness." We Japanese need not to say much on the cause of the present war. Suffice it to say, that the further existence of the Black Octpus will depend entirely upon how he comes out of this war. The Japanese fleet has already practically annihilated Russia's naval power in the Orient. The Japanese army is about to win a signal victory ove ssia in Corea & Manchuria. And when............ St. Petersburg? Wait & see! The ugly Black Octpus! Hurrah! Hurrah! for Japan.
March, 1904.
Kisaburo Ohara.
法學博士 中村進午先生閲
慶應義塾大學生 小原喜三郎君案
滑稽欧亜外交地圖
A HUMOROUS DIPLOMATIC ATLAS OF EUROPE AND ASIA.
露西亜
RUSSIA
西比利亜
SIBERIA
支那
CHINA
MANTURIA
PEKING
COREA
JAPAN
FORMOSA
PACIFIC OCEAN
TURKESTAN
AFGANISTAN
BERTISTAN
TIBET
INDIA
BURMA
SIAM
ANAM
PERSIA
ARABIA
TURKEY
CASPIAN SEA
BLACK SEA
CREMIA
FINLAND
SWEDEN
NORWAY
DENMARK
POLAND
GERMANY
AUSTRIA
HUNGARY
FRANCE
SPAIN
PORTUGAL
ITALY
SARDINIA
SICILY
GREECE
CRETE
GREAT BRITAIN
ATLANTIC OCEAN
MEDITERRANEAN SEA
大日本帝國萬歳!!!

'How to Get There. An Interesting and Educational Game for 2, 3 or 4 Players', [1908]

J.W.L., London

Chromolithograph; border: 285 × 356 mm (11 ¼ × 14 in)

JUST AS the national railway network formed the basis for race games, so too did the London Underground network. In 1908, the Underground Electric Railways Company of London and four other underground railway companies – the City and South London Railway, the Waterloo and City Railway, the Great Northern and City Railway and Central London Railway – produced a co-ordinated map of the London 'Tube' network using the combined branding 'Underground'– the first such map. It also included the route of London United Trams, other tramways and the routes and termini of the main railway companies. The map can be dated relatively accurately, as it shows the site of the 'Franco-British Exhibition' held at the White City in 1908. The mapmaker, J.W.L. (his initials are found on the map on the box lid but not on the main map), also produced a bronze exhibition medal in 1908.

This map-game also marks the 'Franco-British Exhibition', so can be attributed to precisely the same period; in fact, it is the very map described above, which the printers Johnson, Riddle & Co. converted into a game by the simple expedient of pasting the banner title at the top over the original 'Underground' heading.

The eight lines marked are the Bakerloo, Central London Railway (today's Central Line), City and South London Railway (Northern Line, Bank branch), District, Great Northern & City Railway (now part of the National Rail network), Hampstead Railway (Northern Line, Charing Cross Branch), Metropolitan and Piccadilly lines. A number of stations no longer exist. Some have been renamed, such as Dover Street (Green Park) and Post Office (St Paul's), and some closed: the little-used Down Street was closed in 1932; British Museum was closed when Holborn was expanded in 1933; Mark Lane was replaced by Tower Hill, Marlborough Road was replaced by St John's Wood, and Trafalgar Square is now part of Charing Cross Station.

The game allows up to four players to compete against each other to be the first to reach their destination. The map comes in a box, with the board itself, four carriages (paper cut-outs on metal stands), twenty-five replica Tube tickets (each with a start and destination station and a price), an instruction sheet, pasted to the underside of the lid, and an arrow on a spindle, with a base.

The players would each draw a ticket, and place their carriage on the stop they were instructed to board; then, by spinning the arrow and following the directions indicated, they would make their way towards their final stop, as if using the Tube for real. Although some tickets give a route, the players were free to pick their own route to their final stop. Once there, 'The passenger first arriving at his destination must loudly call the name of his station, adding the words "all change." He is declared the winner and takes the contents of the Booking Office.'

The base of the spindle was divided into sixteen sections, with commands such as 'travel non-stop to Charing Cross', 'travel one station', 'ticket lost take another and start again', 'travel three stations', 'travel non-stop to Bank', 'travel two stations', 'travel non-stop to Oxford Circus', 'travel non-stop to King's Cross', 'signal at danger remain where you are' and 'ticket lost start journey over again'.

While the 'travel non-stop to Charing Cross' option might seem punitive, the passenger could choose not to move if sent in the wrong direction. As with real Tube travel, no overtaking was allowed. If players found a train in front of them, moving in the same direction, they had to remain where they were. Similarly, players had to complete the journey by moving the exact number of stops; if the wheel directed them to move more stops than required, players had to remain at the station until their next turn.

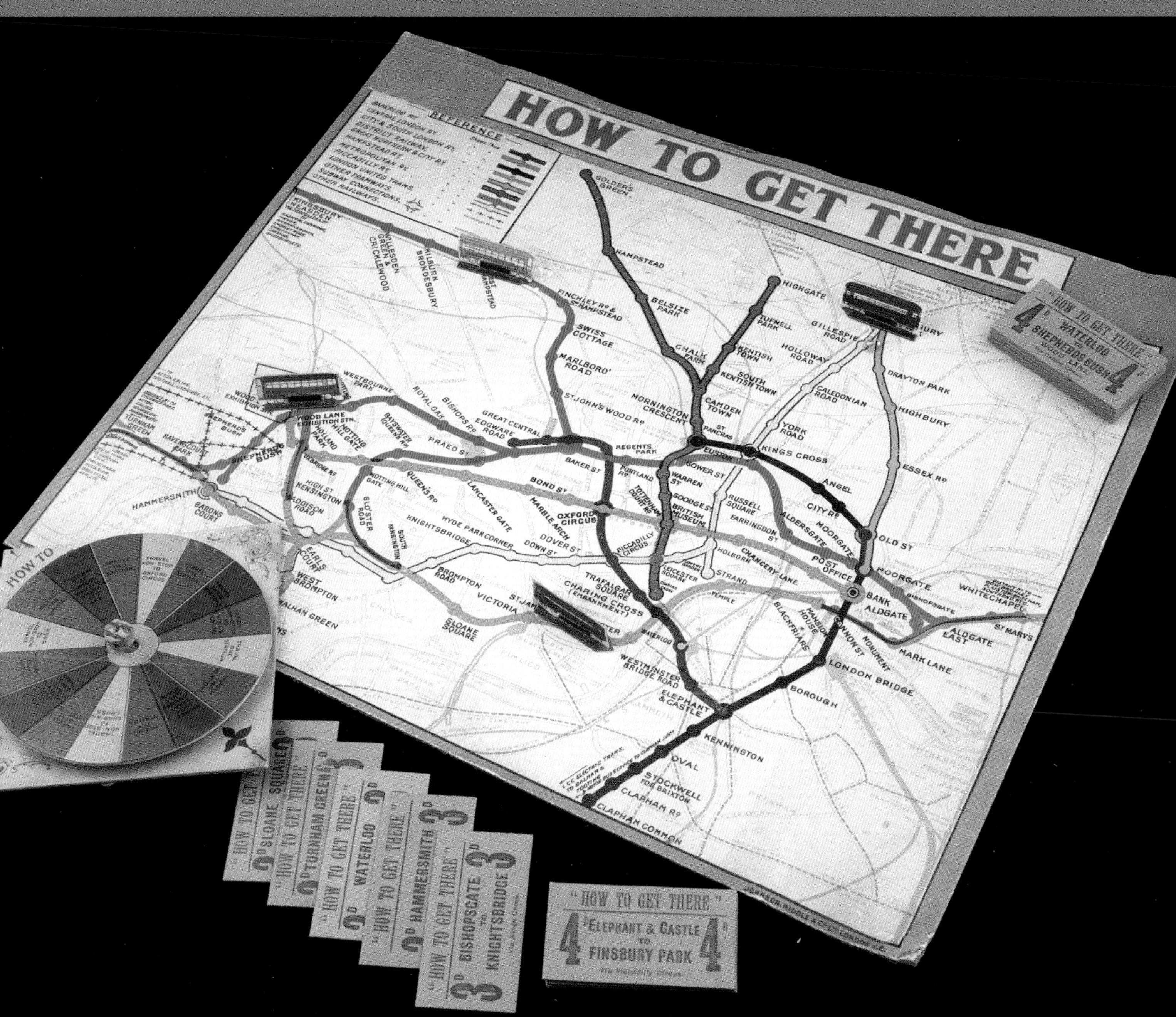
HOW TO GET THERE
REFERENCE
GOLDER'S GREEN
HAMPSTEAD
BELSIZE PARK
HIGHGATE
TUFNELL PARK
KENTISH TOWN
CHALK FARM
SOUTH KENTISH TOWN
CAMDEN TOWN
GILLESPIE ROAD
HOLLOWAY ROAD
DRAYTON PARK
CALEDONIAN ROAD
HIGHBURY
YORK ROAD
KINGS CROSS
EUSTON
ESSEX Rd
ANGEL
CITY Rd
MOORGATE
OLD ST
ALDERSGATE
POST OFFICE
BANK
ALDGATE
WHITECHAPEL
ALDGATE EAST
ST MARY'S
MARK LANE
MONUMENT
CANNON ST
LONDON BRIDGE
BOROUGH
ELEPHANT & CASTLE
KENNINGTON
OVAL
STOCKWELL
CLAPHAM Rd
CLAPHAM COMMON
BLACKFRIARS
WATERLOO
WESTMINSTER BRIDGE ROAD
TRAFALGAR SQUARE
CHARING CROSS (EMBANKMENT)
STRAND
HOLBORN
CHANCERY LANE
FARRINGDON ST
RUSSELL SQUARE
BRITISH MUSEUM
GOODGE ST
WARREN ST
GOWER ST
TOTTENHAM COURT Rd
OXFORD CIRCUS
PICCADILLY CIRCUS
LEICESTER SQUARE
DOVER ST
DOWN ST
HYDE PARK CORNER
KNIGHTSBRIDGE
BROMPTON ROAD
SOUTH KENSINGTON
VICTORIA
SLOANE SQUARE
WEST BROMPTON
EARLS COURT
MARBLE ARCH
BOND St
LANCASTER GATE
QUEEN'S Rd
NOTTING HILL GATE
HIGH St KENSINGTON
ADDISON ROAD
HAMMERSMITH
BARONS COURT
SHEPHERDS BUSH
WOOD LANE EXHIBITION STN.
HOLLAND PARK
WESTBOURNE PARK
ROYAL OAK
BISHOPS Rd
PRAED St
EDGWARE ROAD
GREAT CENTRAL
BAKER St
PORTLAND Rd
REGENTS PARK
St JOHNS WOOD Rd
MORNINGTON CRESCENT
MARLBORO' ROAD
SWISS COTTAGE
FINCHLEY Rd & S. HAMPSTEAD
KILBURN BRONDESBURY
WILLESDEN GREEN & CRICKLEWOOD
TURNHAM GREEN
RAVENSCOURT PARK
"HOW TO GET THERE"
4D WATERLOO TO SHEPHERDS BUSH 4D
WOOD LANE
"HOW TO GET THERE"
SLOANE SQUARE
"HOW TO GET THERE"
TURNHAM GREEN
"HOW TO GET THERE"
WATERLOO
"HOW TO GET THERE"
HAMMERSMITH
"HOW TO GET THERE"
3D BISHOPSGATE TO KNIGHTSBRIDGE 3D
Via Kings Cross.
"HOW TO GET THERE"
4D ELEPHANT & CASTLE TO FINSBURY PARK 4D
Via Piccadilly Circus.
JOHNSON, RIDDLE & Co LTD LONDON S.E.

Map of Belgium, 1912

Eliza Jane [Lilian] Lancaster, London

Chromolithograph; at widest: 148 × 192 mm (5 7/8 × 7 1/2 in)

ELIZABETH HOSKYN'S *Stories of Old* is a collection of twelve tales derived from medieval folklore, legend or history, one each from most of the countries of northern Europe, including England, Scotland, Ireland and Wales, France, Holland, Germany, Russia, Scandinavia and Iceland. Each story is adapted from that country's literary tradition, and recast for the entertainment of the younger reader; to enliven the tale, each is illustrated with a map of the country concerned, with the characters from the story set within the outline of the country's boundaries. The maps are signed, the name hidden within the map area, by Lilian Lancaster (1852–1939; see pp. 170, 172, 184 and 186), working under her married name Lilian Tennant. The embossed cover mistakenly refers to her as 'J. Tennant'.

The seventh tale tells the story of the fabled horse Bayard; Hoskyn's inspiration was a twelfth-century French poem called *Les quatre fils Aymon*, a poem in the Old French oral genre of *chanson de geste* (songs of heroic deeds). While much is uncertain about their origin, these *chansons* celebrated, and commemorated, both actual and legendary deeds from the early history of France, notably from the reign of Charlemagne (i.e. Charles the Great, reigned 768–814), first Holy Roman Emperor, and ruler of large swathes of western Europe.

Les quatre fils Aymon was one of the most popular, and most heavily recycled, of the *chansons* – the story of the four sons of Aymon, Renaud de Montauban and his brothers Alard, Guiscard and Richard. The story is set during the reign of Charlemagne, and resonates with the feudal system of the time – a tiered hierarchy of subjects owing ultimate and personal allegiance to the king.

The story was elaborated over the years; the version that Hoskyn narrates has Aymon as ruler of a region of the Dordogne and therefore a vassal (under-lord) of Charlemagne, but the two men were enemies. Aymon was given a horse by his nephew, Maugis, a magician; the horse was named Bayard and it too had magical powers. None was able to ride it, until Renaud mastered it, and thereafter the horse was his loyal friend. One evening, Renaud was at court, when he got into a quarrel and killed a rival. His brothers came to his aid but, outnumbered, stood little chance against the king's men. Bayard came to the rescue, and carried all four brothers away to safety in the forests of the Ardennes region of Belgium, the scene depicted in the map.

Thereafter the brothers were in open revolt against Charlemagne, but after several years Charlemagne offered peace on the condition that Renaud should surrender Bayard to him, and go on a pilgrimage to the Holy Land; Renaud consented and departed for Jerusalem. Charlemagne ordered that Bayard's four legs be tied together and, weighted down with rocks, the horse was thrown into a river to drown. But Bayard used his teeth to free himself, broke the rocks with his hooves and escaped, returning to the Ardennes, where he roamed free for the rest of his life.

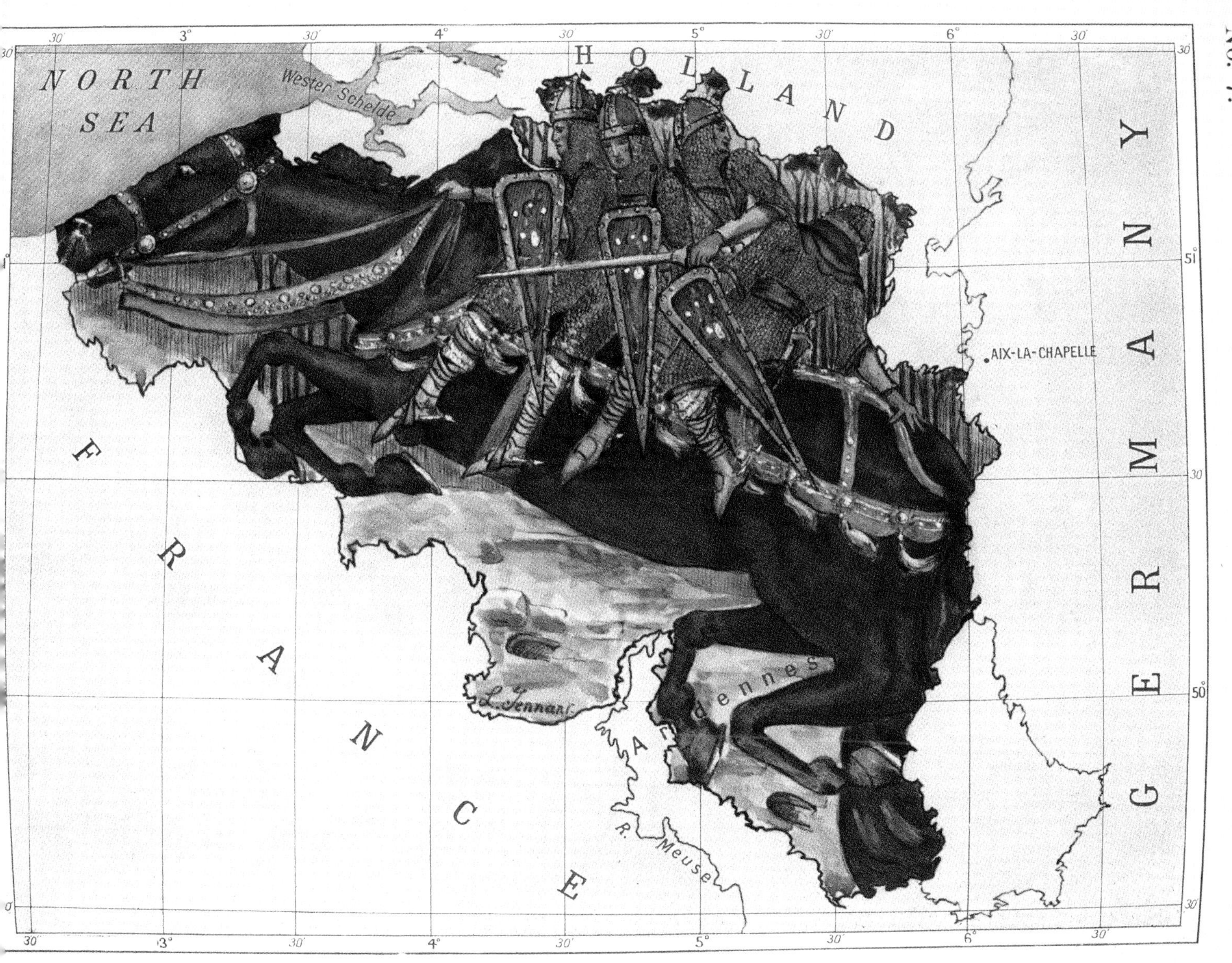
No. 7.
NORTH SEA
Wester Schelde
HOLLAND
GERMANY
AIX-LA-CHAPELLE
FRANCE
Ardennes
R. Meuse
L. Tennant
3°
4°
5°
6°
30'
51°
50°

BELGIUM.

Map of Iceland, 1912

Eliza Jane [Lilian] Lancaster, London

Chromolithograph; at widest: 138 × 185 mm (5 ⅜ × 7 ¼ in)

LIKE THE map on p. 200, this map of Iceland created by Lilian Lancaster (see pp. 184, 186) under her married name of Tennant, comes from Elizabeth Hoskyn's *Stories of Old*. It accompanies a story from Norse mythology, which describes Flóki Vilgerðarson's search for 'Snowland', as recounted in the Landnámabók manuscript, a medieval Icelandic text.

A Viking expedition in about AD 861 discovered an island which their leader, Naddod, called 'Snowland', from the heavy snow that fell on the mountain tops in the distance. Flóki Vilgerðarson decided that he would seek out this new island, and settle there with his family; it was the earliest attempt by the Vikings to colonise the island. He sailed from Norway with his family, food, livestock and supplies for a new home; importantly, so far as his story goes, he also took three ravens with him. Flóki made landfall in the Shetland Islands and then the Faeroe Islands, before sailing on out into the Atlantic. The Vikings were instinctive sailors: they used neither charts nor instruments in their voyages, but instead relied on their experience and learned understanding of the sea to select their course. After a while, Flóki released one of the ravens, knowing that it would head for the nearest land. It did; after circling high above the longboat, it turned back to the Faeroe Islands. When the second bird was released, it circled looking for land, but then flew back to the longboat. Unperturbed, Flóki sailed on. When he released the third bird, it flew up into the sky and then turned to the north-west and flew away. Knowing that land lay in that direction, Flóki altered course and followed the raven, eventually finding 'Snowland'.

When the story of his exploits became famous, people nicknamed him Hrafna-Flóki, or 'Raven-Floki'. Lilian Lancaster's drawing shows him standing in his longboat, with the Norwegian flag, and the third of his ravens about to fly away.

While the summer climate made the island seem hospitable, the hostile winter conditions convinced him that the island was uninhabitable, and so he returned to Norway. During the early spring, while he waited to set out on the return journey, he scaled one of the local peaks and looked down on the coast. In one of the fjords he saw large slabs of drift-ice and, as a parting gift, gave the island a new name that it retains to this day – Iceland.

No. 11.

ICELAND.

‘Changing the Map of Europe: A *Financial Times* Competition’, 1914

Financial Times, London

Lithograph; border: 520 × 602 mm (20 ½ × 23 ¾ in)

THE MAP title explains the competition: ‘the map shows the present demarcation of the countries’, and each player should ‘sketch in the boundaries of states according to your view of probable peace terms’. The rules of the competition are explained on a separate leaf:

> This map is 24 in. × 21 ½ in. in size, and shows clearly the present demarcation of territories, and the principal towns, rivers, mountain ranges, etc ...
>
> Competitors are not restricted in regard to the number of maps they send in, but all maps must reach the Editor of The Financial Times, 72, Coleman Street, London, E.C., on or before 30 November next [the British Library's example is stamped with the acquisition date ‘13 Nov 1914’]. After satisfaction of the prizes, the balance derived from the sale of the maps will be handed to the Prince of Wales' Fund.
>
> The remodelling of the map of Europe has already become a favourite pastime of our French allies.

The First Prize was £25, the second £10 10s. (£10.50), the third £5 5s. (£5.25), with ten additional prizes of a guinea (£1 1s. – £1.05). While an entertaining idea, the competition also has much to say about the attitudes of the organisers; in the rules they write: ‘It should be noted that in making the awards, the management of The Financial Times (whose decision must be final) will not take into consideration estimates of the amount of any *cash* indemnities to be paid, or the disposition of Germany's foreign possessions, as this would unduly complicate the adjudgment of the award.’ In effect, they are stating that, in their opinion, Germany was certainly going to lose the war, have reparations levied against it, and lose its foreign possessions. It was unthinkable, evidently, that the British Isles should be in any way affected by revised boundaries – the north of Scotland and almost all of Ireland are not even included on the map. Whatever happened would be on the European, not the British, side of the Channel.

The First World War began on 28 July 1914 with the Austro-Hungarian invasion of Serbia, and became truly pan-European when Germany invaded Belgium, then France. Britain declared war on 4 August. The war came to an end on 11 November 1918, almost exactly four years after this map was received in the then-British Museum's Map Library.

Although the publishers, and to be fair many others, thought that the war would be fairly short and ‘gentlemanly’, that was not to be the case. It is a matter of sadness to ponder how many light-hearted participants in the *Financial Times*' competition may have lost their lives ‘in Flanders Fields’ or in other theatres in this brutal modern war.

CHANGING THE MAP OF EUROPE
The Financial Times
COMPETITION MAP
EUROPE AFTER THE WAR
THE MAP SHOWS THE PRESENT DEMARCATION OF THE COUNTRIES
INSTRUCTIONS
SKETCH IN THE BOUNDARIES OF STATES
ACCORDING TO YOUR VIEW OF PROBABLE PEACE TERMS
SCALE
50 100 150 200 250 300 MILES
NORWAY
CHRISTIANIA
SWEDEN
STOCKHOLM
PETROGRAD
MOSCOW
RIGA
LIBAU
BALTIC SEA
NORTH SEA
DENMARK
COPENHAGEN
KÖNIGSBERG
EAST P.
DANZIG
WEST P.
RUSSIA
KIEL
STETTIN
HAMBURG
BREMEN
BERLIN
HANOVER
POSEN
WARSAW
POLAND
Vistula
Oder
Elbe
GERMANY
PRUSSIA
SILESIA
BRESLAU
THE HAGUE
HOLLAND
LONDON
DRESDEN
SAXONY
LEMBERG
GALICIA
ANTWERP
BRUSSELS
BELGIUM
COLOGNE
LIEGE
NAMUR
LILLE
COBLENZ
MAINZ
FRANKFURT
NÜRNBERG
PRAGUE
BOHEMIA
MORAVIA
Carpathians
English Channel
REIMS
PARIS
METZ
LORRAINE
ALSACE
STRASSBURG
STUTTGART
BAVARIA
MUNICH
VIENNA
BUDAPEST
AUSTRIA
HUNGARY
TRANSYLVANIA
RUMANIA
BLACK SEA
DIJON
BERNE
SWITZERLAND
TYROL
TRENT
FRANCE
BUCHAREST
CROATIA & SLAVONIA
Danube
Bay of Biscay
LYONS
MILAN
BOSNIA
NISH
SERVIA
BULGARIA
SOFIA
BORDEAUX
DALMATIA
MONT.
CETINJE
CONSTANTINOPLE
TURKEY
MARSEILLES
ITALY
Adriatic Sea
ALBANIA
GREECE
SPAIN
ROME
MEDITERRANEAN SEA

'Hark! Hark! The Dogs Do Bark!' [1914]

Johnson, Riddle & Co., London

Chromolithograph; at widest: 532 × 705 mm (21 × 27 ¾ in)

WHEN THE First World War started in 1914, most commentators thought that the war would be of short duration, and this was reflected in the relatively light-hearted caricature maps issued in the first months of the war. By the second year, when the true scale of the conflict became apparent, such propaganda maps took on an altogether darker tone.

Hark! Hark!, published by G. W. Bacon, depicts the principal protagonists as dogs. The stereotypes are as familiar today as then: the British bulldog, the French poodle and the German dachshund. Serbia, however, is depicted as wasps, stinging the Austrian mongrel. North of the British Isles is a puppet-master in naval garb (sometimes said to be Winston Churchill, at the time First Lord of the Admiralty) who is gradually moving Royal Navy ships on station to blockade Germany by sea.

Walter Lewis Emanuel (1869–1915) contributed the descriptive text outside the lower border; he was a famous English humorist, known for his contributions to the magazine *Punch*, and for a series of anthropomorphic dog books such as *The Dogs of War* (published in 1906 and reprinted in 1913), which was illustrated by the artist Cecil Aldin (1870–1935). Emmanuel's text reads:

> The Dogs of War are loose in Europe, and a nice noise they are making! It was started by a Dachshund that is thought to have gone mad – though there was so much method in his madness that this is doubtful. [Note for the ignorant: The German for Dog is Hund. The English for German in Hun. Dachshunds means badger-dog – and he is sometimes more badgered than he likes.] Mated with the Dachshund, for better or for worse, was an Austrian Mongrel. By the fine unwritten law of Dogdom big dogs never attack little ones. There are, however, scallywags in every community, and, egged on by the Dachshund for private ends, the Mongrel started bullying a little Servian. And then the fat was in the fire, for the little Servian had a great big friend in the form of a Russian Bear, and he stood up for his pal. And that was what the Dachshund wanted. He hoped that a big row would ensue, and in the confusion he intended to steal a bone or two that he had had his eye on. The Dachshund now began to look round for friends, but they seemed strangely scarce. He had relied on an Italian Greyhound, a thoroughbred, named Italia, but Italia dissembled her love in the strangest way, and asserted that War was a luxury which she could not afford just now [...] The Dachshund, to his annoyance, found only one friend, and that was a dog of Constantinople. ...Meanwhile the rest of the European Happy Family looked on, and who shall say how the row will spread? There's the Greek with his knife ready to take a slice of Turkey; there are the Balkans determined not to be baulked of their own little ambitions; there's the Spaniard fond of Bull fighting so long as he is not a John Bull; there's the Portugee just spoiling for a scrap; there's the Swiss suffering from cold feet; there's the Dutchman... All this, and more, may be seen depicted above. Search well and you will find many things. But not Peace. Peace has gone to the Dogs for the present – until a satisfactory muzzle has been found for that Dachshund. Meanwhile the Dachshund's heart bleeds for Belgium – and his nose for Great Britain.

It is interesting to note that a German copy of the map was published in Hamburg early in 1915, presumably in an attempt to highlight and discredit the perceived self-interest of British war aims. It is also interesting to speculate on the market for such a map, originally sold for a shilling (5p); the image is child-oriented, and it may be that the map was aimed at parents and schools as a tool to explain the background to the conflict. When handled in the house or classroom, the maps would have been easily damaged, which explains their relative rarity today.

"HARK ! HARK ! THE DOGS DO BARK !"

WITH NOTE BY WALTER EMANUEL.

and Printed by Johnson, Riddle & Co., Ltd., London, S.E

Published by G. W. Bacon & Co., Ltd., 127, Strand, W.C.

The Dogs of War are loose in Europe, and a nice noise they are making! It was started by a Dachshund that is thought to have gone mad—though there was so much method in his madness that this is doubtful. [NOTE FOR THE IGNORANT: The German for Dog is Hund. The English for German is Hun. Dachshund means badger-dog—and he is sometimes more badgered than he likes.] Mated with the Dachshund, for better or for worse, was an Austrian Mongrel. By the fine unwritten law of Dogdom big dogs never attack little ones. There are, however, scallywags in every community, and, egged on by the Dachshund for private ends, the Mongrel started bullying a little Servian. And then the fat was in the fire, for the little Servian had a great big friend in the form of a Russian Bear, and he stood up for his pal. And that was what the Dachshund wanted. He hoped that a big row would ensue, and in the confusion he intended to steal a bone or two that he had had his eye on for some time. He got what he wanted—and a little more. For the Russian Bear had friends, too. There was a very game little Belgian Griffon, and there was a great big French Poodle, a smart dandified fellow, and there was a Bulldog. Rather a sleepy chap this last one, and the Dachshund despised him because he was not always yapping and snarling. But the Bulldog has a habit of sleeping with one eye open, and, when he is roused, he grips and won't let go.

The Dachshund started by attacking the Belgian Griffon, as being the smallest, and mauled the poor creature cruelly, but was quite unable to kill her. And he was mistaken as to the others. He found that the dandified Poodle could fight, and that the Bulldog had not lost the knack of not letting go, and that Russia, after all, was a Rusher, and soon the Bear idea made the Dachshund tremble. And even the little Servian gave the Austrian Mongrel some nasty bites, and so did a neighbour of his named Monty.

The Dachshund now began to look round for friends, but they seemed strangely scarce. He had relied on an Italian Greyhound, a thoroughbred, named Italia, but Italia dissembled her love in the strangest way, and asserted that War was a luxury which she could not afford just now. All the same Italia loaded her gun, and who knows but what it may go off and whom it may hit—for accidents will happen in the best regulated families. The Dachshund, to his annoyance, found only one friend, and that was a dog of Constantinople. The Dogs of Constantinople are quite well known for being fond of offal.

Meanwhile the rest of the European Happy Family looked on, and who shall say how the row will spread? There's the Greek with his knife ready to take a slice of Turkey; there are the Balkans determined not to be baulked of their own little ambitions; there's the Spaniard fond of Bull fighting so long as he is not a John Bull; there's the Portugee just spoiling for a scrap; there's the Swiss suffering from cold feet; there's the Dutchman, who keeps smiling with difficulty—still some nice meaty bones may come his way, and in any event he may be relied upon to play the game and not to be a Double Dutchman. [ANOTHER NOTE FOR THE IGNORANT:—Holland used to be known as a low lying country, but this title has now been filched by Germany.] And, up North, the Norwegian, the Swede, and the great Dane all have their eyes well skinned.

All this, and more, may be seen depicted above. Search well and you may find many things. But not Peace. Peace has gone to the Dogs for the present—until a satisfactory muzzle has been found for that Dachshund. Meanwhile the Dachshund's heart bleeds for Belgium—and his nose for Great Britain.

‘The Silver Bullet or the Road to Berlin’, Game, 1914

R. Farmer & Son, London

Cardboard sheet, with paper overlay, with design cut in it, pinned to a card base, housed within a wood frame, with glass and a metal ball-bearing; length 240 mm (9 ½ in) × width 162 mm (6 ⅜ in) × height 23 mm (⅞ in)

THIS GAME is not a realistic map; although there is a quasi-geographical structure to it, the labels are schematically rather than accurately placed. Instead, it is a dexterity game, in which the player attempts to manoeuvre the ball along a circuitous, hollowed-out track to reach Berlin, the end of the game. En route, there are a number of hazards – holes in the backing card through which the ball could drop, requiring the player to start again – labelled ‘Bridge Destroyed’, ‘Road Mined’ and ‘Entrenchments’, blocking the easy path from start to finish.

Other hazards are named for German towns, such as Hamburg, Potsdam, Spandau, Dresden, Leipzig and Cologne. No doubt this reflects the risk of an army getting bogged down in street-fighting in built-up areas. Against each of the hazards is marked a number, denoting the casualties that would be incurred: ‘55’ against ‘Entrenchments’ and ‘50’ against Cologne, while ‘bridge destroyed’ is ‘40’. These numbers progressively diminish as the player nears Berlin, perhaps implying that the hardest fighting would be at the start, with resistance weakening as the attacker advanced.

The game could be played both individually and competitively; the individual would simply restart the game every time his ball disappeared into one of the holes, until he was able to reach Berlin (or gave up in disgust). The rules, which are pasted on the underside of the frame, also allow for a group of players to compete against each other:

> Each player starts with an imaginary 500 men, and failure to pass the different fortresses entails a loss of as many men as indicated at the point of failure. The player reaching Berlin with the largest number of men wins the game ... A player losing his full complement of men is thereby put out of action, the next player continuing the attack.

For the game-playing armchair general, the consequences of his actions could be less extreme than for a commander in the field, although the game-maker did caution that: ‘Beginners will be encouraged to know that proficiency generally begets over-confidence, and the player often fails amidst the hearty laughter of the company when he least expects to’.

R. Farmer & Son also produced a number of similar First World War dexterity games, including ‘Trench Football’, ‘Get Rid of the Huns’ (based on a map of Europe), ‘Sky Pirates’ (with zeppelins), and ‘The Submarine Puzzle, or, Outwitting the Pirates’.

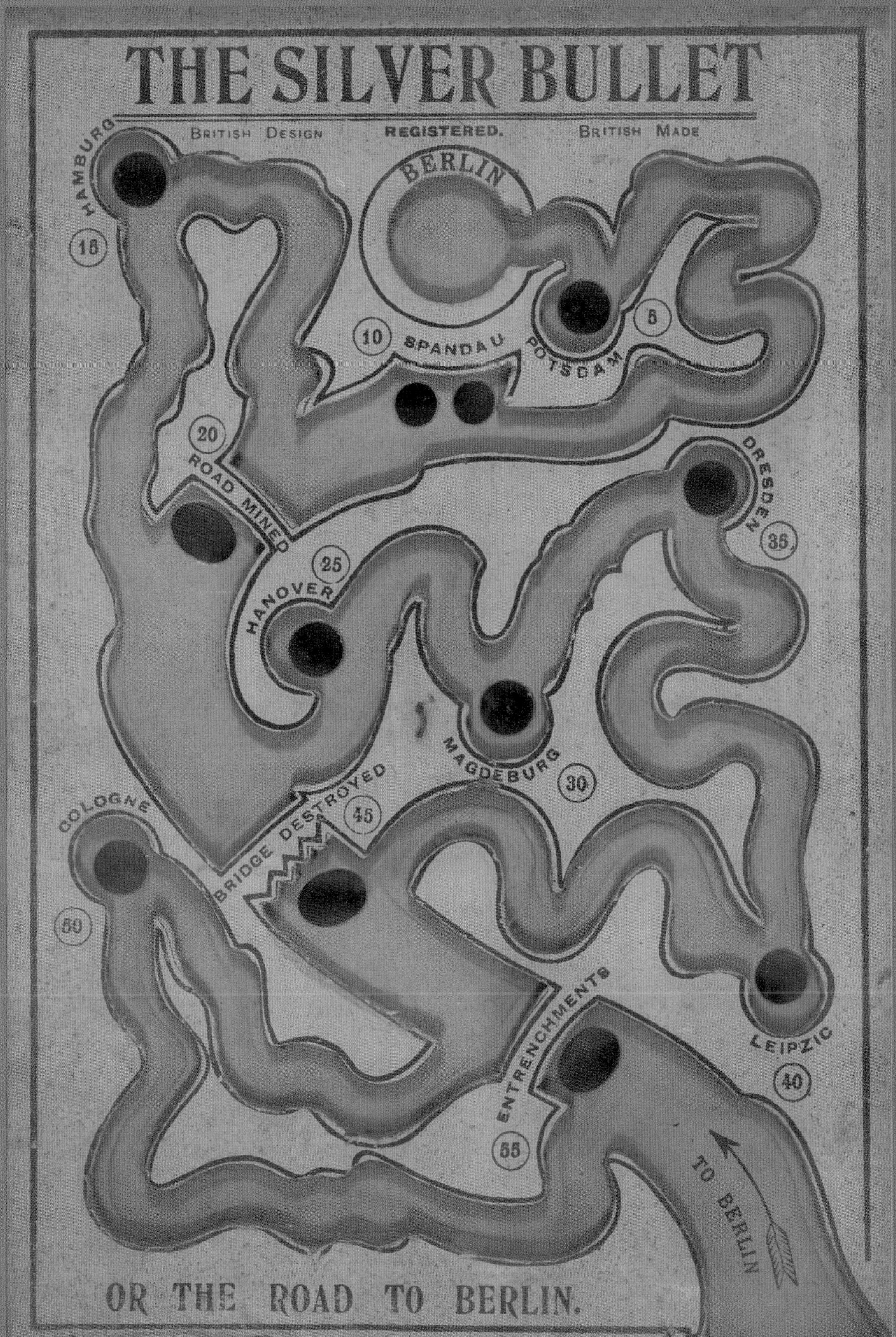
THE SILVER BULLET
BRITISH DESIGN
REGISTERED.
BRITISH MADE
HAMBURG
15
BERLIN
10
SPANDAU
POTSDAM
5
20
ROAD MINED
DRESDEN
35
25
HANOVER
MAGDEBURG
30
BRIDGE DESTROYED
45
COLOGNE
50
ENTRENCHMENTS
LEIPZIC
40
55
TO BERLIN
OR THE ROAD TO BERLIN.

‘Knock Out Germany’, [1914]

Toy Target Company, [London?]

Lithograph; map 365 × 342 mm (14 3/8 × 13 1/2 in); with text: 462 × 346 mm (18 1/4 × 13 5/8 in)

THE TOY Target Company remains wholly obscure. Although a caption states ‘Made in England’, the only business of that name thus far traced had premises in Boston, Massachusetts, some twenty years earlier than the date of this game. The company’s map of Germany is an unusual use of a cartographic theme. There remains some doubt about the map’s purpose, and conclusions are hampered by the rarity of the item, but it appears that the sheet illustrated was intended as an instruction sheet for a much larger map of Germany designed to be mounted on a wood or strawboard backing, then to be used, as the company’s name suggests, as a target in a game – as a cartographical coconut-shy.

The explanation is unclear, but it appears that the extent of Germany, or at least a large part of it including Berlin and Frankfurt, was, in some fashion, cut out, perhaps like a jigsaw, so that a sufficiently accurate strike, or series of strikes, could ‘knock out’ that section of the map, the successful thrower thus winning the game. The text also refers to other ways of scoring points, but this ‘frontispiece’ gives no further clues what they might have been nor, indeed, what sort of missile the game required – darts, small balls, bean bags or the like. The text below the map reads:

> An instructive partial map of Europe with Germany Silhouetted by which means the position of that Country is pronouncedly fixed upon the mind, the sea is indicated by Warships presumably bearing down upon Germany. As predicting the ultimate diminution of the Size of that Country, a section is cut out and partially attached, this embraces Berlin, and Frankfurt, the two principal Towns of Germany, the whole forms a frontispiece of a Target of wood or strawboard and is entitled ‘Knock out Germany’ this wish being ‘parent to the thought’ incites to the sport or pleasure of throwing or shooting a substance at the loose part referred to, and partial, or entire displacement or ‘Knock out’ have their respective scoring points; to the youthful mind this is both instructive and amusing and evokes pleasure in School and Home Circles, the delineation of other countries affording an interesting survey of the positions of Countries engaged in the present European war.

The makers emphasise the educational benefit of the map; while battering the map with their projectiles, the youngsters would learn the positions of the various countries, and the names and locations of their capital cities.

Some of the labels along the lower border of the map have been inked out by an earlier owner; the left-hand label is the imprint of the mapmaker. The other two labels are more likely city names, as can be seen for other countries in the map.

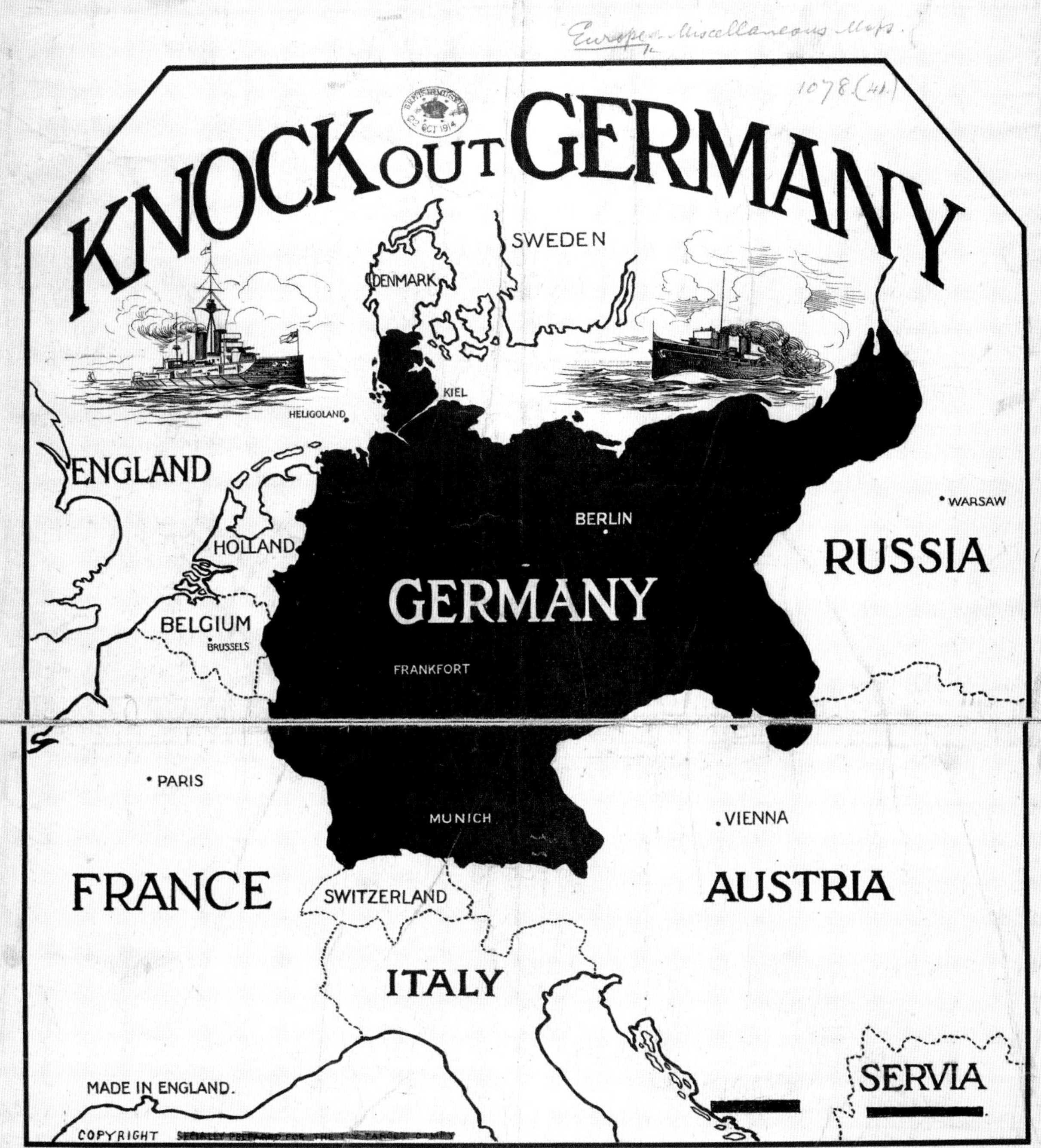

An instructive partial map of Europe with Germany Silhouetted by which means the position of that Country is pronouncedly fixed upon the mind, the Sea is indicated by Warships presumedly bearing down upon Germany. As predicting the ultimate diminution of the size of that Country, a section is cut out and partially attached, this embraces Berlin, and Frankfort, the two principal Towns of Germany, the whole forms a frontispiece of a Target of wood or straw-board and is entitled "Knock out Germany" this wish being "parent to the thought" incites to the sport or pleasure of throwing or shooting a substance at the loose part referred to, and partial, or entire dis-placement or "Knock out" have their respective scoring points; to the youthful mind this is both instructive and amusing and creates pleasure in School and Home circles, the delineation of other countries affording an interesting survey of the positions of Countries engaged in the present european war

'An Anciente Mappe of Fairyland Newly Discovered and Set Forth', 1918

Bernard Sleigh, London

Chromolithograph; border: 475 × 1808 mm (18 ¾ × 71 ⅛ in), on three sheets joined

BERNARD SLEIGH'S (1872–1954) famous map of Fairyland seems to have been inspired by the bedtime tales with which he regaled his two children, to whom the map is dedicated. In the accompanying booklet he writes 'of the Land of Faerie and of the way thereto'. He begins:

> In the Heart of every child, is hidden away a little golden key which unlocks the door of a silent, clean-swept roomful of changing lights and mystic shadows. There, every child enters at times to gaze eagerly upon the one great window, pictured with ancient legends, and glowing with many colours ... and sets trembling fingers to open wide the flashing casements – to stand gazing, awed and silent, upon a sea and sky of gold and crimson, full of winged forms grey against its summer radiance. They see the clear sands piled up against the outer wall, and many boats waiting always to take their willing travellers to the distant gates of Ivory. Some there be who venture forth into a new world, breathing fearlessly its unaccustomed air; enter some dainty, carven shallop and set forth to those rainbow-guarded shores...

He continues:

> Stout is the heart that faces these dangers; daring the soul that braves these troubled seas, to enter at last the hidden entrance to the Harbour of Dreamland. So for the use and guidance of future explorers it has been thought well to make some slight survey of the available paths in this land, where some tracks, though well trodden and seemingly familiar, yet vanish and re-appear – and vanish again in bewildering fashion, – baffling and discouraging even to the most earnest traveller.

The travellers undergo a ritual which will reveal Fairyland to them. They sail to one of the coves, Dreamland Harbour being the largest and busiest, where they would have a magic ointment applied to their eyelids, which allowed them to differentiate between good and bad fairies. Then, having found a guide, the travellers would set out, with this advice: 'You will find the best routes marked out with lines, as red and as fine as the silken clue that guided Theseus through his Cretan labyrinth. If you stray not far from the path, you too will be as safe.' But there were also hazards for the unwary:

> If any other landing is made, these safeguards have to be dispensed with, and great caution is therefore necessary; for you may be frozen into an ice statue, when past the Bridge of Demeter; you may become one of the lost children of the Never Never Land; you may even, if the wind be in the wrong direction, get blown up the Ferlie Firth and swallowed by the Laidly Worm.

The map is laid out in four sections. The left-hand section is best described as a Tolkien-esque dungeons and dragons fantasy land. The second is populated by children's fairy-tale characters: Beauty and the Beast, 'La Belle Dormant' (Sleeping Beauty), Puss in Boots, Humpty Dumpty, Cinderella, Snow White and the Seven Dwarves, Old Mother Hubbard and many others.

The third section features characters from medieval myth, principally the Arthurian legends, with Arthur and Guinevere's tombs, Merlin, the sword 'Excalibore'. The final section is based upon ancient Greek legend: Pegasus the winged horse, the Centaurs, Cerberus, Hercules, the Argonauts, and the Sirens.

‘The New Map Game Motor Chase Across London’, 1925

Geographia Ltd, London

Chromolithograph; at widest: 359 × 479 mm (14 ⅛ × 18 ⅞ in)

WITH THE increasing ubiquity of the motor car in the early twentieth century, the map publishers Geographia Ltd were inspired to introduce a new race game, set on the streets of London. The box label (duplicated on the little box holding the cars and dice) clearly sets the scene in the world of ‘cops and robbers’, showing a getaway car trying to outrun a police car, with one of the officers reaching out to try to grab the criminal. Nowhere else on the game is this made apparent, but on a promotional leaflet found with a later Geographia Ltd race game (see p. 218), it is more clearly stated.

The game is set in the City and West End of London, with the main roads marked, and principal landmarks drawn in silhouette. The players start on the western side of the map and end on the east. It may be that the direction of the game – the criminals all heading for hideouts in the East End – is accidental, but it does bring to mind Charles Booth’s ‘Poverty Map of London’, which refers to the inhabitants of pockets of east London as ‘Lowest class. Vicious, semi-criminal’.

The original rule book is not present, but there are sufficient clues to allow the game to be played. There are indications that the maker conceived the board for two slightly different levels of game: perhaps a longer and shorter version, or versions for the older and younger child. Player ‘1’ starts on the top border of the map on ‘Park Road’, up beyond Lord’s Cricket Ground; but there is a second start for player 1 – the ‘Racer Start’ – coloured blue, and visible north of the ‘Marylebone Road’. However, ‘Home’ for both is along the ‘Old Kent Road’, beyond ‘Albany Road’. Similarly, ‘5’ starts near the junction of the ‘King’s Road’ and ‘Chelsea Embankment’, while the ‘Racer Start’ is on ‘Chelsea Embankment’ near Chelsea Bridge (which is not marked) with ‘Home’ up the ‘Kingsland Road’.

In each case, the ‘Racer Start’ is twelve steps (black dots) closer to ‘Home’. With a single die that represents two throws for the very lucky player, or twelve for one suffering the worst possible luck. So it may have been intended for a shorter game.

Of the others, ‘2’ starts in ‘Maida Vale’ (the ‘Racer Start’ on ‘Edgware Road’) and ends on ‘[Aldgate] High Street’; ‘3’ starts on ‘Bayswater Road’ at the western end of ‘Hyde Park’ (the ‘Racer Start’ near ‘Marble Arch’) and ends on the south side of ‘Tower Bridge’. ‘4’ starts on ‘Kensington Road’, beyond ‘Gloucester Road’ (the ‘Racer Start’ close to the junction of ‘Knightsbridge’ and the ‘Brompton Rd.’), with ‘Home’ on the ‘Hackney Road’.

The routes are all laid out so the players have to go through the City of London; all the roads bear hazard symbols, as listed in the rules, under the title:

Road Signs
TRAFFIC HOLD-UP – Miss one throw
ROAD UNDER REPAIR – Miss two throws
TRAFFIC CONGESTION – Miss two throws
TRAFFIC DIVERTED – Follow direction of arrow except when this would mean ‘going back’

Anyone used to driving in modern London will see that the traffic black-spots have altered little in the ninety years since this map was published: Elephant and Castle, Parliament Square, Trafalgar Square and the Strand, while Victoria, Aldwych, Blackfriars Bridge, Tower Bridge Road and Marylebone Road are highlighted as being particularly notorious.

THE NEW MAP GAME
MOTOR CHASE
ACROSS LONDON
EXCITING ENTERTAINING EDUCATIVE

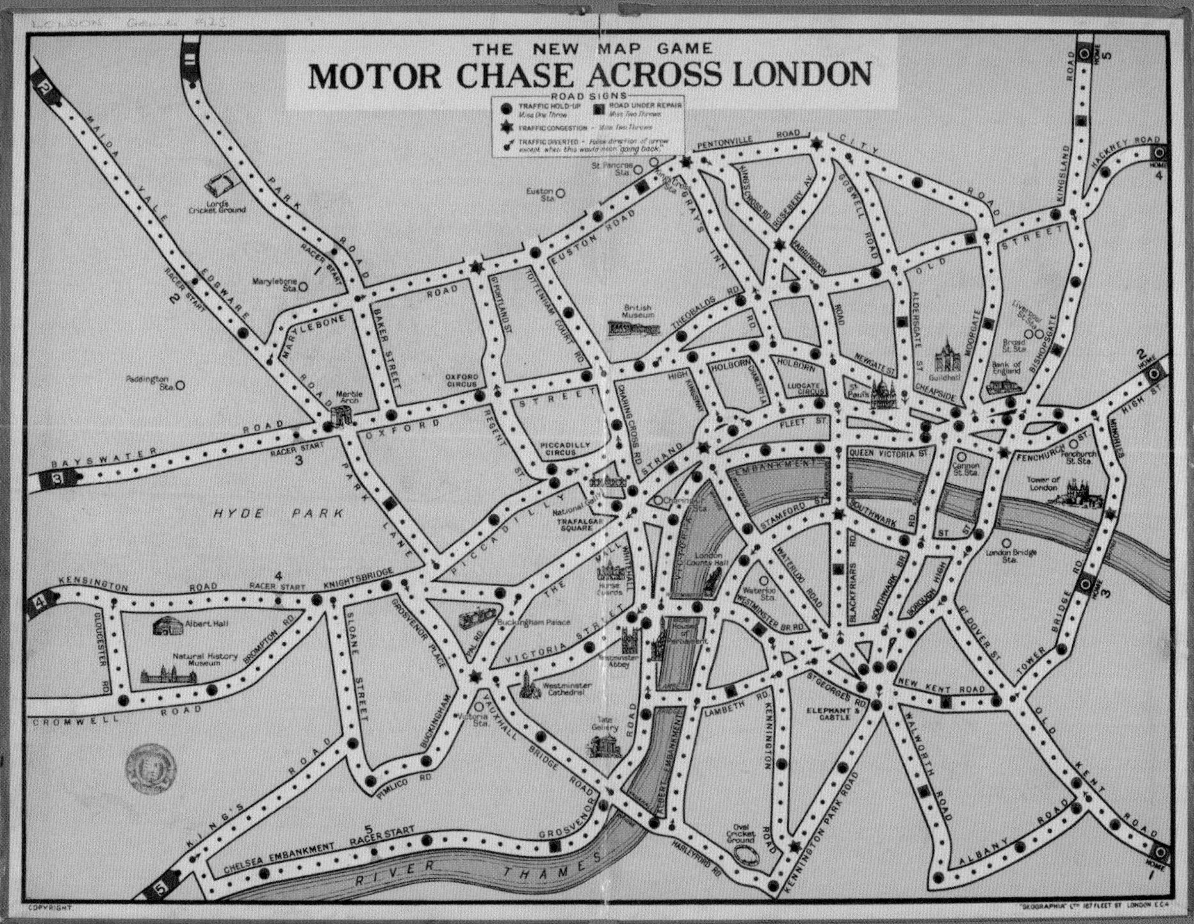
THE NEW MAP GAME
MOTOR CHASE ACROSS LONDON
ROAD SIGNS
HYDE PARK
RIVER THAMES

‘Octopium Landlordicuss (Common London Landlord)’, 1925

William Bacot Northrop, London

‘Octopium Landlordicuss (Common London Landlord) This Fishy Creature lives on Rent Its Tentacles grasp 5 Square Miles of London This Absorbent Parasite sucks £20,000,000 a year from its Victims giving nothing in return. The People must destroy It or be destroyed.’

Lithograph; at widest: 84 × 129 mm (3 ¼ × 5 ⅛ in)

THE OCTOPUS had been a popular symbol for countries seeking to dominate their neighbours from Fred W. Rose onwards (see p. 180); but, in the latter years of the nineteenth century (see Olver p. 188) and the early twentieth century, it began to be used as an allegorical image for financial greed – literally and metaphorically grasping – as can be seen in this unusual political postcard, published in about 1925.

Landlordism – large swathes of property owned by hereditary and aristocratic landlords – was one of the pet hatreds of Lloyd George, British Liberal politician, Chancellor of the Exchequer and wartime Prime Minister. In a famous speech, delivered at Bedford in 1912, he pronounced:

> There is no question more vital ... than the question ... of the land! ... It enters into everything – the food the people eat, the water they drink, the houses they dwell in, the industries upon which their livelihood depends. And to whom does the land belong in Britain? To a handful of rich people! One-third of all the land belongs to members of the House of Lords. Landlordism is the greatest of all monopolies in this land. The power of the landlords is boundless. They may evict their tenants, and devastate the land worse than an enemy would. Now, I am not attacking the landlords either individually or as a class, but can such a state of affairs be allowed to continue?

While there was much merit and conviction in his campaign, it might perhaps also be said that the landlords under attack here were invariably Tories, his political rivals. There were particular issues arising from the consolidation of land and property in London and Westminster in the hands of a tiny number of people, and picking up on Lloyd George’s theme, the American William Bacot Northrop (1898–1965) has created a striking propaganda image. ‘Landlordism’ is depicted as an octopus – termed a parasite – with its tentacles encircling the principal land estates of the capital: the Duke of Westminster’s in Mayfair and Westminster, Earl Cadogan’s in Chelsea and the Duke of Norfolk’s along the Strand, with the Ecclesiastical Commissioners, Lord Portman, Howard de Walden, the Duke of Bedford and Lord Northampton dominating the land north of the Bayswater Road, Oxford Street and Holborn. For each landlord, Northrop gives the acreage of land owned and yearly rental received. The Duke of Westminster received £3,000,000 from the 400 acres (162 ha) he owned, while Howard de Walden generated income of £2,900,000 from his 292 acres (118 ha). The eight landlords ‘named and shamed’ in this card generated £15,140,000 rental income per annum.

There is one very obvious omission: there is no mention of the Royal family. The Crown (now the Crown Estates) and the Prince of Wales (as Duke of Cornwall) also owned (and continue to own) extensive tracts of land in London – although admittedly the Duchy’s estates are mostly south of the Thames, an area largely omitted here. Perhaps that was simply too radical a step.

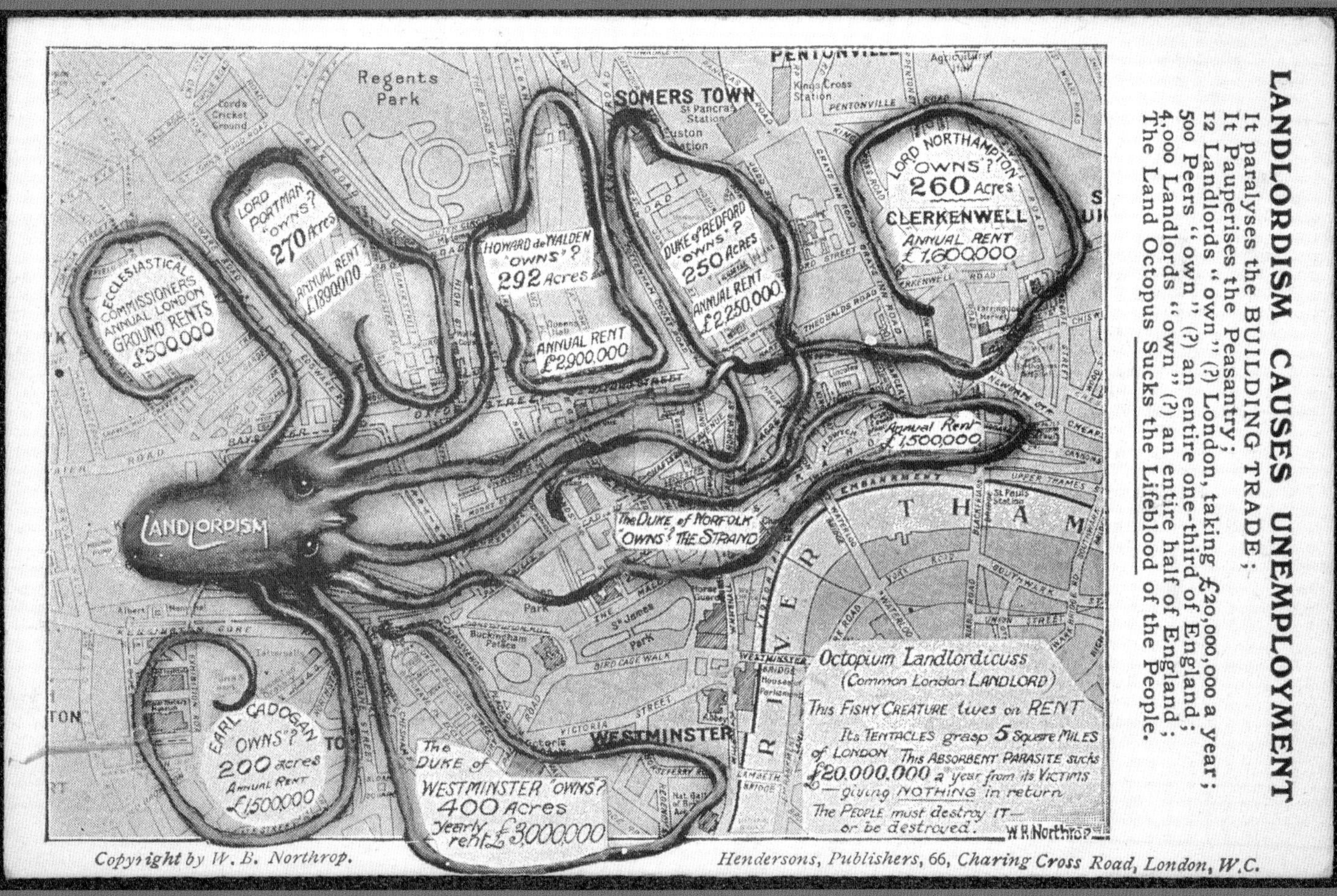
LANDLORDISM CAUSES UNEMPLOYMENT
It paralyses the BUILDING TRADE;
It Pauperises the Peasantry;
12 Landlords "own" (?) London, taking £20,000,000 a year;
500 Peers "own" (?) an entire one-third of England;
4,000 Landlords "own" (?) an entire half of England;
The Land Octopus Sucks the Lifeblood of the People.
Regents Park
SOMERS TOWN
PENTONVILLE
ECCLESIASTICAL COMMISSIONERS ANNUAL LONDON GROUND RENTS £500,000
LORD PORTMAN "OWNS"? 270 Acres ANNUAL RENT £1,890,000
HOWARD de WALDEN "OWNS"? 292 Acres ANNUAL RENT £2,900,000
DUKE of BEDFORD "OWNS"? 250 Acres ANNUAL RENT £2,250,000
LORD NORTHAMPTON "OWNS"? 260 Acres CLERKENWELL ANNUAL RENT £1,600,000
Annual Rent £1,500,000
LANDLORDISM
The DUKE of NORFOLK "OWNS" THE STRAND
T H A M
RIVER
EARL CADOGAN "OWNS"? 200 Acres ANNUAL RENT £1,500,000
The DUKE of WESTMINSTER "OWNS"? 400 Acres yearly rent £3,000,000
WESTMINSTER
Octopium Landlordicuss
(Common London LANDLORD)
This FISHY CREATURE lives on RENT
Its TENTACLES grasp 5 Square MILES of LONDON This ABSORBENT PARASITE sucks £20,000,000 a year from its VICTIMS —giving NOTHING in return
The PEOPLE must destroy IT— or be destroyed.
W.B. Northrop
Copyright by W. B. Northrop.
Hendersons, Publishers, 66, Charing Cross Road, London, W.C.

'Buy British', Game, 1932

Geographia Ltd, London

'A New Map Game. Buy British. An exciting world race and one which will teach the players – Trade Within the Empire'

Chromolithograph; map, at widest: 343 × 478 mm (13 ½ × 18 ⅞ in); paper, at widest: 357 × 478 mm (14 × 18 ⅞ in)

THIS RACE game is set on a Mercator projection map of the world, with the countries of the British Empire coloured in red; these include Canada, British Guiana, Gambia, Sierra Leone, Gold Coast, Nigeria, Union of South Africa, Rhodesia, Nyasa Land, Tanganyika, Kenya, Anglo-Egyptian Sudan, British Somaliland, Aden, India, Federated Malay States, Sarawak, New Guinea, Australia and New Zealand.

Each player, up to a total of five, is given a metal ship as his marker representing one of the nations – England, Canada, South Africa, Australia or India – and that country serves as his home port. The players must navigate around the world, the winner being the first to return home safely. At each port the players pick up an export from that port and land it at the next port they arrive at, then collect a new export to transport onwards.

The format is typical of this type of game. The players follow a circular track around the world. They have to throw an exact number to arrive at a port; en route they can land on points where they can miss a throw, or other points where they have to follow the directions attached to that stop, with hazards and dangers listed along the route, such as 'Proceed to wreck and back', 'Wireless message to call at Auckland' and 'Wireless message to call at Colombo'. Hazards include storms, 'engine trouble' and 'man overboard'.

As well as the five ships, painted red (England), blue (South Africa, but with yellow flags), white (Canada), green (Australia) and brown (India), there are accompanying coloured flags for each country, listing the exports each sends to the next country on the circular track. Thus, the five flags for Great Britain list exports to Canada: Coal, Machinery, Cotton Goods, Iron and Cloth Coal. Canadian exports to South Africa are: Wheat, Rubber Tyres, Paper, Motor Cars and Farm implements. Australia imported Foodstuffs, Asbestos, Tobacco, Ostrich Feathers and Coal from South Africa, and so on.

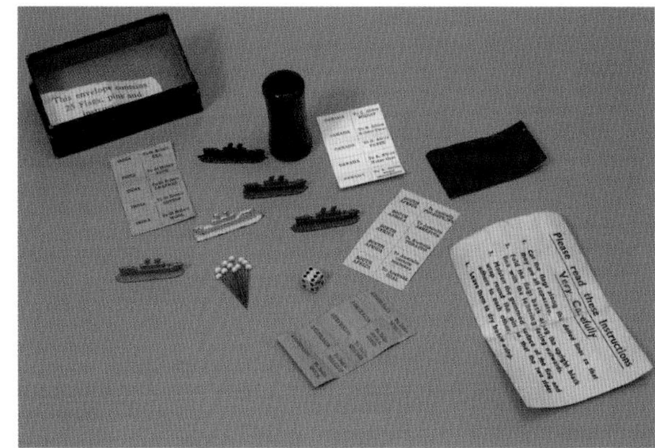

This is an educational game. As the accompanying advertising flyer puts it: 'A Game for young and old "Buy British" – Exciting – Interesting – Educative – A Game, but yet of such a character to teach the players the Trade which goes on within the Empire and the Geography of the World ... Unconsciously you memorise the goods you carry, and the countries and ports you call at and the seas through which you pass.' The flyer also exhorts parents to 'Teach your Children to think British and gain a knowledge of the World'.

The game was sold for 'Price 2/6 [12½p] Complete with dice, ships, flags, etc'.

BUY BRITISH
PENALTIES
THE CORRECT NUMBER MUST BE THROWN TO REACH A POINT MARKED ☐ OR ☐
EACH PLAYER WHOSE THROW PLACES HIM ON A POINT MARKED ◉ Must Miss a Throw
A PLAYER WHOSE THROW PLACES HIM ON A POINT MARKED ✕ Must Proceed According to Instructions Marked Against Same.
GREENLAND
BRITISH ISLES
London
CANADA
LABRADOR
PICK UP PILOT
FOG
NEWFOUNDLD
STORM
DROP PILOT
UNITED STATES
NORTH ATLANTIC OCEAN
RUSSIAN SOVIET REPUBLICS
CHINA
INDIA
Bombay
ARABIA
Aden
DROP PILOT
PICK UP PILOT
EGYPT
TURKEY
PERSIA
PROCEED TO WRECK AND BACK
GAMBIA
SIERRA LEONE
GOLD COAST
NIGERIA
ANGLO EGYPTIAN SUDAN
BRIT. SOMALILAND
COAL AT ADEN
CEYLON
Colombo
WIRELESS MESSAGE TO CALL AT COLOMBO
BRITISH GUIANA
BRAZIL
SOUTH ATLANTIC OCEAN
ARGENTINA
Cape Town
INDIAN OCEAN
Madagascar
ENGINE TROUBLE
AUSTRALIA
Sydney
Auckland
TASMANIA
NEW ZEALAND
NEW GUINEA
FEDERATED MALAY STATES
BRIT. N. BORNEO
PACIFIC OCEAN
STORM
MAN OVERBOARD
WIRELESS MESSAGE TO CALL AT AUCKLAND

Untitled Map of the World, Dissected as a Jigsaw, [1935]

Waddington's, London

'Waddington's Mappa-Mundi "Map of the Earth" The New Travel Game. Exciting, Educational & Fascinating'

Chromolithograph, printed on paper, pasted on card and dissected; 420 × 760 mm (16 ½ × 29 ⅞ in)

IN THEIR heyday, Waddington's were the leading English maker of games and puzzles, active from 1922 to 1994, when the company was taken over by Hasbro. There is much in the company's history of interest; during the Second World War they played a prominent part in helping Allied prisoners of war held by the Germans to escape and make their way back to Britain by concealing items in their games, including fabric maps, compasses and the like.

Waddington's 'Mappa-Mundi' is an entertaining and original variation on the jigsaw puzzle, the game being played in two stages. The game is for up to four players, and the instructions and rules are listed on a sheet of paper, pasted inside the upper lid of the box. The first step was to take the separate cardboard sheets with punched oval shapes, and separate them; these ovals bear the names of the capital cities of the countries of the world, and are colour-coded: green, black, red and blue. They were to be placed in piles by colour, with the names on the underside so they were invisible. Each player would then take a pile or, if there were only two players, they would each take two piles.

Then players had to assemble the jigsaw; at the first time of playing, many of the pieces contained an oval plug, semi-punched out, to make it easier, which needed to be removed, leaving a blank oval space, into which the oval plugs with the names of the cities could be inserted.

Once the jigsaw was assembled, and all the ovals removed from the map pieces, the players would choose one of the piles of plugs, and in rotation – green-black-red-blue – each would pick up one plug and endeavour to place it in the correct position on the jigsaw. Once done, the next would take their go, and so on, until all were placed on the board. At the end of the game, the players would refer to the key map, and tally up how many tiles each player had placed in the right position, and how many were wrong. For every plug placed correctly, the player scored four points; for every one placed incorrectly, the player lost two points. The winner would be the player with the highest cumulative score. There was, however, a second alternative; if players were unsure of the placing of a particular capital and were unwilling to gamble, they could place the plug 'off the map'. For this there was no penalty – the player simply got no points.

In the instructions, Waddington's suggested, after time trials by twenty seasoned players, that one person could make the puzzle in three to four hours, two people in two to three hours and three people in about one-and-a-half to two-and-a-half hours. This sounds a long time, but as Waddington's boasted, 'This puzzle is the largest cardboard puzzle ever produced'.

This game printed and made in England by John Waddington Limited, London and Leeds
GREENLAND
ICELAND
HUDSON BAY
GT. BRITAIN AND IRELAND
NORWAY
SWEDEN
FINLAND
UNION OF SOCIALIST SOVIET REPUBLICS
POLAND
FRANCE
SPAIN
RUMANIA
TURKEY
NEWFOUNDLAND
UNITED STATES
CANADA
ATLANTIC OCEAN
Bermuda Is.
Azores
Canary Is.
Cape Verde Is.
GULF OF MEXICO
WEST INDIES
CUBA
CARIBBEAN SEA
COLOMBIA
BRAZIL
BOLIVIA
PERU
CHILE
ARGENTINA
URUGUAY
St. Helena
MOROCCO
ALGERIA
LIBYA
EGYPT
FRENCH WEST AFRICA
NIGERIA
SUDAN
ABYSSINIA
BELGIAN CONGO
ANGOLA
UNION OF SOUTH AFRICA
MADAGASCAR
PERSIA
ARABIA
ARABIAN SEA
INDIA
BAY OF BENGAL
CEYLON
TIBET
CHINA
MANCHURIA
KOREA
SIAM
JAPANESE EMPIRE
PHILIPPINE ISLANDS
PACIFIC OCEAN
INDIAN OCEAN
EQUATOR
BORNEO
JAVA
AUSTRALIA
TASMANIA
NEW ZEALAND
Oslo
London
Ottawa
St. John's
Rome
Lisbon
Madrid
Copenhagen
Berlin
Vienna
Warsaw
Moscow
Helsinki
Stockholm
Bucharest
Kabul
Teheran
Cairo
Dakar
Accra
Lagos
Nairobi
Cape Town
Colombo
Bangkok
Nanking
Tokyo
Canberra
Wellington
Quito
Bogota
Caracas
La Paz
Asuncion
Monte Video
The World on Mercator's Projection

Map of Poland Jigsaw Puzzle, 1958

Veritas, London

'Polska układanka geograficzna' 'Wydawnictwo K[atolicki] O[środek]. W[ydawniczy]. Veritas w Londynie'

Chromolithograph; at widest: 299 × 306 mm (11 ¾ × 12 in)

THIS MAP of Poland, in Polish, was published in London by a Polish publishing organisation known as Veritas, which survives to this day as the Veritas Foundation, with headquarters in Acton, West London. Veritas was founded in 1947, at the end of the Second World War, by Poles unwilling or unable to return to Poland, then under Soviet Communist domination. Many of those who had fought with the Allied armies, navies and air forces, or had escaped as refugees, remained in Britain. The publishing house was set up to produce all manner of books – from religion to contemporary history and works of literature – to promote Christian (Catholic) education and culture in the Polish community. Many of their books could never have appeared in Poland under the Communist regime. The last page of the booklet that accompanies this jigsaw lists some of their publications from 1958.

This was very much an economical and do-it-yourself jigsaw puzzle. The game is printed on a thin sheet of paper; the purchaser had to paste it to a backing, perhaps cardboard as suggested in the leaflet, and then had to cut it out into a jigsaw – a bit of arts and crafts then a geographical lesson all in one.

The game is aimed at the children of an exiled community – a means of familiarising them with and keeping alive the memory of a lost homeland. The drawing is relatively crude, and the images for the individual cities, towns and regions of Poland are simple, to give the player a quick sense of the nature of each place. The sites include the Cathedral at Poznań, the factories at Katowice and the shipyards and docks of Szczecin and Gdansk. The natural world is also indicated, with a *zubr* (European bison) depicted in the woods at Puszcza Białowieska, farmland in the west and afforested tracts in the east of the country. There are numbers of Poles depicted – women in their bright native dress, a farmer on his tractor.

There is also a drawing of the statue of Bolesław I Chrobry, or Bolesław the Great (992–1025), Duke of Poland from 992 until he briefly became the first King of Poland, at Poznań in 1025 – a nationalist hero to revere at a time of foreign rule. The image of Warsaw is dominated by the statue of King Zygmunt (Sigismund) III Vasa (1566–1632), who relocated Poland's capital from Kraków to Warsaw in 1596. The statue, completed in 1644, was badly damaged by the Germans during the Warsaw Uprising in 1944, but was rebuilt at the end of the war. The statue itself is 2.75 m (9 ft) high; the original column was 8.5 m (28 ft), so it was a most imposing landmark, but the modern column is now even higher: 22 m (75 ft).

MORZE BAŁTYCKIE
GDYNIA
GDAŃSK
SZCZECIN
Kaszubka
TORUŃ
KRÓL BOLESŁAW CHROBRY
GNIEZNO
POZNAŃ
Warta
Odra
Nysa
WROCŁAW
Wisła
Kurpianka
Kormoran
WARSZAWA
WILNO
Wilia
Dźwina
NOWOGRÓDEK
Niemen
ŻUBR
PUSZCZA BIAŁOWIESKA
Prypeć
łoś
Poleszuk
Bug
Wieprz
ŁÓDŹ
Pilica
CZĘSTOCHOWA
KATOWICE
KRAKÓW
LUBLIN
ZAMOŚĆ
LWÓW
BORYSŁAW
szyb
Dniestr
Zbrucz
Góral
POLSKA
układanka geograficzna
Wydawnictwo K. O. W. VERITAS w Londynie
33718 (12)

‘The Afghanistans’, 2008

Anonymous, [Afghanistan]

Rug; total dimensions, 805 × 605 mm (31 ¾ × 23 ⅞ in)

FOR CENTURIES, Afghanistan’s principal export was hand-made carpets. In 2007 this market was worth £128 million per year (about US$ 61 million) but the dislocation caused by the war in Afghanistan, coupled with the worldwide recession, had seen this figure fall to about £22 million by 2011. Part of the problem is that hand-made carpets are considerably more expensive (about four times more) than machine-made carpets produced by Afghanistan’s neighbours.

But, as ever, misfortune can beget an opportunity: Afghan carpet-makers realised that the International Security Assistance Force (ISAF) servicemen drawn from a number of NATO countries, and the accompanying support and diplomatic staff, constituted a new market to be exploited, as quasi-tourists seeking a ‘souvenir’ of their tour of duty in Afghanistan.

A number of designs include maps of Afghanistan as a central part of the decoration and, in 2012, the British Library brought two representative rugs; the first is a general map of Afghanistan, marking the various provinces, with a floral design at bottom right: a typical map that might be on sale in any country of the world. Both have their text in English (or American), which is clearly not a language with which the maker was fully familiar.

The second, shown here, is rather more interesting and presumably created for a specific market – the military. Here the map of Afghanistan is a simple outline of the country; almost every other part of the rug illustrates military materiel – the equipment of an army, including six helicopters, a fighter (or perhaps a reconnaissance drone), four armoured personnel carriers, two Kalashnikov AK-47s, a machine-gun, a grenade and a rocket launcher. There can be little doubt that this was intended for sale to soldiers departing the country, as a real evocation of their tour of duty, for display on their return home.

At the top of the map, there is a grey strip (a road) with what appears to be three low-loader tank transporters, the tanks under bright tarpaulins, and this presumably relates to the Russian withdrawal from Afghanistan at the end of their failed adventure in the country between December 1979 and February 1989.

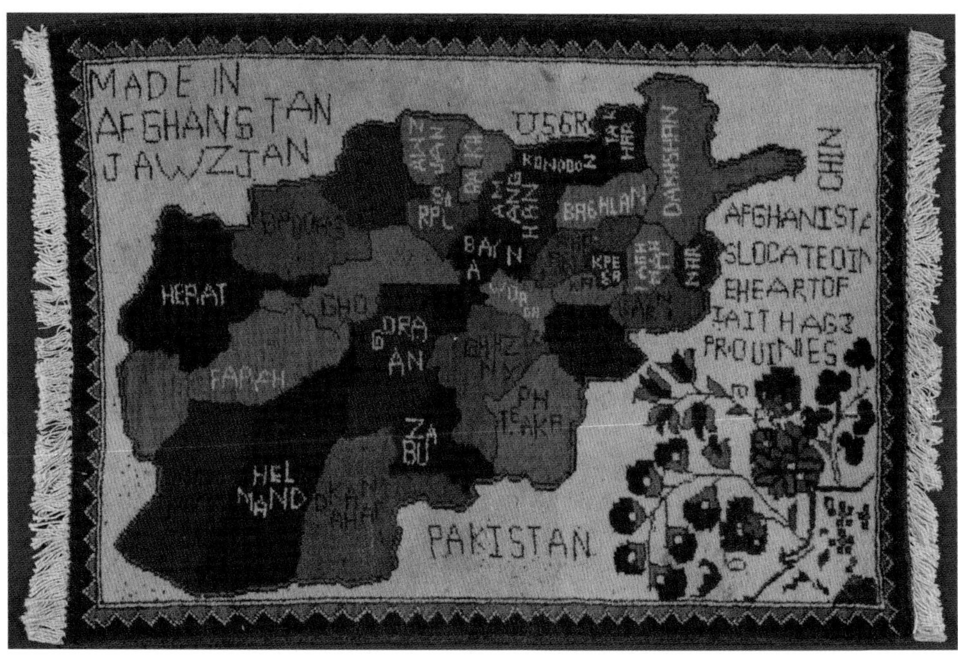

قوای شوروی از افغانستان
خروج است
2008
TANK
AFGHANISTAN
2008
THE AFGHANISTANS
RJOA
ANA
ASHEK
MADE IN AFGANISTANT

Cartobibliographical Details

All maps are from the British Library collections unless otherwise stated.

p. 12

Ptolemaic Map of the World, 1493

Hartmann Schedel, *Liber cronicarum ...* [Nuremberg Chronicle]

Nuremberg: Anton Koberger for Sebald Schreyer & Sebastian Kamermeister, 1493

British Library G.2398

p. 14

Illustration of the Island of Utopia, 1518

Thomas More, *De optimo reip. statu deque noua insula Utopia ...*

Basel, Johannes Froben, 1518

British Library G.2398

p. 16

New Map of the Papist World, 1566

Geneva: Giovanni Battista Trento, 1566

British Library C.160.c.7

p. 20

'The Whole World in a Cloverleaf' [1581]

'Die gantze Welt in ein Kleberblat, Welches ist der Stadt Hannover, meines lieben Vaterlandes Wapen'

Heinrich Bünting, *Itinerarium Sacrae Scripturae ...*

Helmstadt: Jacob Lucius Siebenbürger, 1581

courtesy of Jonathan Potter Ltd.

p. 22

'Europe, the First Part of the World, in Female Form' [1581]

'EUROPA PRIMA PARS TERRÆ FORMA VIRGINIS'

Heinrich Bünting, *Itinerarium et chronicon totius Sacrae Scripturae ...*, Magdeburg, 1598

British Library 3105.a.14

p. 24

'Asia, the Second Part of the World, in the Form of Pegasus' [1581]

'ASIA SECVNDA PARS TERRÆ IN FORMA PEGASI'

Heinrich Bünting, *Itinerarium et chronicon totius Sacrae Scripturae ...*, Magdeburg, 1598

British Library 3105.1.14

p. 26

'The Shape and Position of New Guinea' [1593]

'NOVÆ GVINEÆ Forma, & Situs'

Cornelis de Jode, *Speculum orbis terrararum ...*

Antwerp: Arnold Coninx for Cornelis de Jode, 1593

British Library C.7.c.13.(13b)

p. 30

The Lion of the Low Countries, 1598

'LEO BELGICUS.'

'Iohann van Doetechum fecit.' 'CIVisscher Excudit Anno 1650.'

Text: 'Artificiosa et Geographica tabula sub Leonis figura XVII. inferioris Germaniæ Provincias representans, cui addita sunt singularum insignia, una cum ordinaria Præfectarum distinctione, Longè elimatius quam hactenus unquam expressa. Accesserunt & icones Gubernatorum Generaliū qui utrinque Belgium, Gubernarunt.' ('Artistic and geographical map, in the form of a lion, presenting the XVII Provinces of Germania Inferior (i.e. the Low Countries) ...')

Amsterdam: Claes Jansz. Visscher, 1650

British Library Maps C.9.d.1.(6)

p. 34

Untitled Satirical Map of the World, 1605

'Pag: 18.'

Joseph Hall, William Knight (ed.), *Mundus alter et idem sive Terra Australis ante hac semper incognita longis itineribus peregrine Academici nuperrime lustrate auth: Mercurio Britannico ...*

Frankfurt-am-Main [but London]: [Humphrey Lownes]; sold by the Heirs of Ascanius de Rinialme, 1605

British Library 1079.c.16

p. 36

The Lion of the Low Countries, 1608

'LEO BELGICVS' 'NOVA XVII PROVINCIARVM Germaniæ Inferioris tabula. Leonis effigie, accuratè delineata A NICOLAO IOANNIS PISCATORE.' 'CIV' [monogram of Claes Jansz. Visscher] 'Tot Amstelredam Bij Claes Ianss Visscher in de Kalverstraet A.o 1656.

[text:] 'LEO LOQUITUR. Quanta mihi turget tam vasto corpore membra, Quam densata meo pectore regna vides; Quid foret æterna populus si pace ligatus Alter in alterius commoda feret opem.'

Amsterdam: Claes Jansz Visscher, 1656

British Library Maps C.9.c.1.(7)

p. 38

The Lion of Holland, 1609

'COMITATUS HOLLANDIÆ DENUO FORMÂ LEONIS CURIOSÈ EDITUS A Nicolao Iohannis Visscher Anno 1648' 't'Graefschap Hollandt' 'Edita a Nicolao Iohannis Visscher.' 'Illustrissimo Ornatissimoq Mauritio Dei gratia, Principi Auransiæ, Comiti de Nassau o timo Patriæ nostræ defensori hæc tabula dedicatur, et offertur N.I.V.'

Amsterdam: Claes Jansz. Visscher, 1648

British Library Maps C.9.d.3.(9)

p. 40

Game of the World, 1645

'LE JEU DU MONDE dedié A Monsieur Monsieur le Comte de Vivone Premier Gentilhome de la Chambre du Roy, pour son tres humble et tres obeissant serviteur Du Val.' 'A. Paris chez l'auteur P. Du Val d'Abbeuille Auec Priuillege du Roy 1645 Et se vendent Rue St. Iacques a l'Esperance'

[text:] 'EXPLICATION DE LA FIGVRE. Le premier Cercle marque le Monde Polaire; les 14 suivants les pais d'Amerique; les 15 en suitte depuis 16 jusques a 30 ceux d'Afrique; les 15 autres jusques a 45 ceux d'Asie; et les 18 restants ceux d'Europe. L'assemblage de ces pais se voit aux quatre parties du Monde descrites aux quatre coigns du Ieu.' 'LOIX DV IEV Qui sera recontré d'un autre ira prendre sa place, et luy cedera la sienne; Qui arriuera en France gaignera la partie, qui sil menoit plus des points qu'il ne faut, il retournera en arriere d'autant de points qu'il aura de trop.' 'l'Autheur donnera a ceuz qui le desireront vne grande cognoissance du present ieu, auec les remarques historiques qu'il en a faites, sa demeure est Rue [erasure: 'S.t [illegible].']

British Library Maps*999.(27.)

p. 42

'Map of all the Sea Ports of the World', Game, 1650

'CARTE DE TOUS LES POR[T]S DE MER DU MONDE N. Berey ex.'

[the individual sections:]

[Untitled double-hemisphere map of the World]

Ports de Mer d'Europe

Ports de Mer d'Asie]
Ports de Mer d'Afrique
Ports de Mer d'Amerique.
Ports de Mer de Grece
Ports de Mer de Turquie &c.
Ports de Mer de Danemarq et Süéde.
Ports de Mer d'Allemagne et de Pologne
Ports de Mer d'Italie
Ports de Mer de France
Ports de Mer d'Espagne
Ports de Mer des Isles Britāniques

British Library Maps*974.(3)

p. 44

Draughts Board with Maps of the French Provinces, 1652

'LE JEU DE FRANCE pour les DAMES' Par P. du Val Geographe Ord. du Roy' 'A Paris chez l'Autheur pres le Palais a l'Entree de la Cour St.Eloy 1652 avec privilege du Roy pour 20 ans'

British Library Maps*14425.(1)

p. 46

Untitled Map of the 'Land of Tenderness', 1655

Madeleine de Scudéry, *Clelia, an excellent new romance ...*

London: Henry Herringman; Dorman Newman and others, 1678

British Library Maps 837.m.20

p. 48

The Game of France, 1659

'LE JEU DE FRANCE par P. du Val Geographe Ordinaire du Roy. Explication de la Figure. Elle represente les Provinces de France, auec leurs Villes capitales, Archeveschez, Eveschez, et autres remarques: Les Provinces qui sont vers le Septemtrion, y sont decrites les premieres en suitte, celles qui sont vers le milieu, et enfin celles qui sont vers le Midÿ. ORDRE DU JEU. On y peut joüer auec deux Dez communs, deux, trois, quatre, cinq et six personnes chacun, vne fois et a son rang. selon quil se trouvera placé. LE JEU d'un chacun sera marqué de pieces differentes et avancé dautant de points qu'il en sera decouvert par les Dez:' 'JEU DE FRANCE [the 'N' engraved in reverse] A HAUT et Puissant Seigneur Messire Guillaume de Lamoignon, Chevalier Seigneur de Baville, Conseillier du Roy en tous ses Conseils et Premier President de sa cour de Parlement Par son Tres Hūble et Tres Obeissant serviteur E. Vouillemont 1659.' 'Le tout Gravé et mis au jour par Estienne Voüillemont Graveur Ordinaire du Roy pour les Cartes Geographiques, Plans de Villes et autres tailles doüces. A Paris en L'Isle du palais, au coin de la Rüe du Harlet, a la Fontaine de Iouuence. avec privilege du Roy pour vingt ans 1659.' 'A Paris chez A. de Fer, dans ljsle du Palais ala Sphere Royale 1671. auec privilege du Roy.'

Paris: Antoine de Fer, 1671

British Library Maps 185.n.1.(23.)

p. 50

'Game of the Princes of Europe', 1662

'LES JEU DES PRINCES DE LEUROPE Par P. Du Val Geographe du Roy' 'Chez H Iailot aux deux Globes aur le Quay des Augustins, avec pri du Roy 1670'

[Notes:] 'ORDRE DU JEU. Ce Jeu est appellé le Jeu de l'Europe, par ce que toutes les figures qui le composent representant chacune un païs, un Estat, une Isle de cette partie du monde. On y peut joüer avec deux dex communs, deux, trois, quatre, cinq et Six personnes. Chacun ioüera une fois, et a son rang, selon qu'il se trouvera placé, et marquera de quelques pieces differentes son ieu, qu'il avancera d'autant de points quil y en aura de decouverte apres qu'il aura joüé. Il sera convenu de ce que l'on doit mettre au jeu, d'un liard, d'un sol, s'un Teston, d'une Pistolle, si l'on ueut que chacun mettra sur la Carte de l'Europe qui est au milieu ou l'on mettra aussy les payemens qui seront de la valeur de ceque Chacun aura mis au jeu, et tout cela au profit de Celuy qui gagnera la partie.'

Paris: Alexis-Hubert Jaillot, 1670

British Library Maps*1078.(8.)

p. 52

Geography Reduced to a Game for the instruction of Young Venetian nobles [c.1665]

'GEOGRAFIA RIDOTTA A GIVOCO PER INSTRVTTIONE DELLA GIOVANE NOBILTA VENETIANA. Da Don Casimiro Freschot Dell.'O. D. Sn. B(erasure).' 'Il Cav:re AF: Lucini Fecio.' 'All Illust:mi Signori Padroni miei colend:mi li Signori, Angelo, Costantino, Hieronimo Michiel Figlioli dell Eccell:o Sig: Nicolo Michiel Senatore La premura che ho di corrisponder utilmente all honor che mi fa l'Eccell:mo lor Padre di uolermi in casa sua per assister a lor studi mi ha inspiratorque sto modo di proporli una parte cosi considerabile della lor instruttione, comé é la Geografia, sotto ques ta forma di giuoco, quale dilettandoli, possa seruir a render i loro trattenimenti eruditi Tutte le fatiche mie essendo dedicare a loro, questa uiene authan ricar al mondo l'ambitione che hodie ser conoscuto Della Casa lor Ecell.ma Deuot:mo et Obligat.mo Seruo Don Casimiro Freschot dell' O: D: S: B:' 'In Venetia par Giovanni Pare Libraro All' Fortuna.' 'Si Vende da Giouanni Paré alla Fortuna. / In Venetia Con liceza / de Superiori.'

Venice: Casimir Freschot, [*c*.1665]

Courtesy of Sylvia Ioannou Foundation exemplar

p. 56

Symbolic Rose Map of Bohemia, 1677

'BOHEMIÆ ROSA Omnibus sæculis cruenta in qua plura quàm 80. magna prælia commissa sunt, nunc primum hâc formâ excusa. Chr. Vetter inuen. et delineauit. Wolfgang Kilian sculpsit Augustæ.'

[Text:]

Crevit in Hercijnio Rosa formossima Saltu;
Stat penes armatus pro Statione Leo.
Hæc Rosa non Veneris, Sed crevit Sanguine Martis;
Hic Rhodus, hic saltus, fætáq[ue] terra fuit.
Nil Rosa pulchra time! Hercijnios venit Auster in hortos,
Sub tacitus sileant horrida bella Rosâ.

Bohuslaus Aloysius Balbinus, *Epitome historica rerum Bohemicarum quam ob venerationem christianae antiquitatis, et primae in Bohemia collegialis ecclesiae honorem, Boleslaviensem historiam placvit appellare....*

Prague: Jan Nicolaas Hampel, 1677

Private collection

p. 58

Allegorical Map of the Baltic Sea as Charon, 1701

'part I. pag. 36.'

Olof Rudbeck, *Olavi Rudbeckii Filii Nora Samolad Sive Laponia illustrata et iter per Uplandiam, Gestriciam, Helsingiam, Medelpadiam, Angermanniam, Bothniam, tam occidentalem, & huic annexam Laponiam Lulensem, quam Septentrionalem, cum Laponia Torniensi, & Orientalem, item Finlandiam, Alandiam, &c....*

Uppsala 1701

British Library 432.b.22.

p. 60

Untitled Map of the World, 1718

'Nouvelle Methode de Geographie ou VOIAGE du MONDE par les Villes les plus Considerables de la Terre ou par un JEU On apprend la situation des païs & de Villes, leur dependance & la Religion des peuples avec une Mappemonde ou les routes de ce Voyage sont marquées.' 'A Paris Chez Crêpy ruë S.t Jacques au Lion d'Argent. 1718.'

Paris: Jean Crepy, 1718

British Library CC.5.a.88

p. 64

Map of France Game, 1718

'Nouvelle Methode de Geographie ou VOIAGE Curieux par les Villes les plus Considerables et les principaux Pais de 30. Gouvernements Generaux et les 6. Particuliers du Roiaume de FRANCE Mis en JEU Ou l'on a Marqué les Singularitez des Païs, les Longitudes et les Latitudes des Villes Capitales, les Rivieres qui les Arrosent, ...'

[map entitled:] 'Carte Generale du Royaume de FRANCE Suivant les Nouvelles Observations Se Vend à Paris Chez Crêpy ruë S.t Jacques au Lion d'Argent. 1718.'

Paris: Jean Crepy, 1718

British Library CC.5.a.87

p. 66

'An Accurate Map of Utopia, Which is the Newly Discovered World', 1720

'Accurata UTOPIÆ TABULA Das ist Der Neu entdeckten SCHALCK WELT, oder des so offt benanten, und doch nie erkanten SCHLARRAFFENLANDES Neu erfundene lacherliche Land Tabell Worinnen all undjede laster in besondere Konigreich, Provintzen und Herrschafften ab getheilet Beyneben auch die negst angrentzende Länder Der FROMMEN des Zeitlichen AUFF und UNTERGANGS auch ewigen VERDERBENS Regionen samt einer erklerung anmuthig und nutzlich vorgestellt werden durch Authoren anonymū.'

Nuremberg (?): Anonymous [*c*.1720]

British Library Maps 9 TAB 31.(37)

p. 68

'The Very Famous Island of Mad-head, Lying in the Sea of Shares', 1720

'AFBEELDINGHE van't zeer vermaerde Eiland GEKS-KOP. geligen in de Actie-ze, ontdekt door Mons.r Lau-rens, werende bewoond door een verzameling van alderhande Volkeren. die men dezen generalen Naam (Actionisten) geest.'

[Anonymous] *Het Groote Tafereel der Dwassheid...* (The Great Mirror of Folly)

[L'Honore & Chatelain, Amsterdam, 1720]

British Library Maps CC.5.a.345

p. 70

'An Astronomicall and Chronologicall Clock, Shewing All the Most Usefull Parts of an Almanack', 1725

'An ASTRONOMICALL and CHRONOLOGICALL Clock, shewing all the most usefull parts of an Almanack.' 'IO.s NAYLOR near Namptwich Cheshire.'

[Inset:] [Untitled map of the Northern Hemisphere south to Cuba]

[text:] 'The Explanation March the first 1750/1. The first is a Large Plate 15 Inches square ...'

Nantwich: Joseph Naylor, 1751 [*c*.1751]

Private collection

p. 72

Untitled Map of 'Brobdingnag', North America and 'New Albion', 1726

'Plate, II, Part II Page. 1.'

Jonathan Swift, *Travels into Several Remote Nations of the World. In four parts. By Lemuel Gulliver, first a surgeon, and then a captain of several ships.*

London: Benjamin Motte, 1726

British Library C.59.e.11

p. 74

Allegorical Map of the Siege of the Castle of Love, 1735

'Representation Sÿmbolique et ingenieuse projettée en Siege et en Bombardement comme il faut empecher prudemment les attaques de L'AMOUR. Sÿmbolische Sinnreiche in einer Belagerung u. Bombardirung entworffen Vorstellung wie man den anfällen und Verschungen der LIEBE Klug und tapffer zu begegnen, zur Belustigung u. Sittlicher Belehrung verfertiget von MATTH. SEUTTER S.C. Maj. Geogr. in Augsp.

[key:] Methode pour defendre et conserver son coeur contre les attaques de l'amour. Die Methode sein Hertz wider die Angriffe der Liebe zu bewahren [key in ten columns:] Noms des Bastions et d'autres ouvrages Die Namen derer Boll u: anderer wercke.... [repeated in German]

Augsburg: Georg Matthaüs Seutter Sr

British Library Maps C.26.f.4.(42)

p. 76

The Harz Mountains, Germany, as from the Air, 1749

'PERSPECTIVISCHE VORSTELLVNG des berühmten BLOCKEN ODER BLOKENBERGS mit der jenigen Gegend, so weit solche von dem, der auf der Spitze des Berges stehet, gesehen werden kan. Gezeichnet A.o 1732 von L.S. Bestehorn herausgegeben von Homænn. Erben C.P.S.C.M. 1749....' 'Vue de la montagne de BROKEN située dams le Territoire du Comté de Wernigerode, qui est dans les forêts de Hartz.'

Nuremberg: Homann Heirs, 1749

British Library Maps*30058.(1)

p. 78

Map Screen, 1749

'A MAP OF THE WORLD OR TERRESTRIAL GLOBE IN TWO PLANISPHERES, laid down from the Observations of the ROYAL ACADEMY of SCIENCES, | Wherein as an Introduction to the study of GEOGRAPHY are inserted several things relating to the Doctrine of the EARTHLY GLOBE, as an explanation of the Circles, the Zones & Climates, Longitude & Latitude, the Antipodes Antoeci & Perioeci. An account of the Terms by which the several parts of the EARTH & WATERS are called. Also a Concise Theory of the EARTH describing the Diurnal and | Annual MOTIONS, its Magnitude & the proportions of its parts. Likewise an explanation of the Copernican System, shewing the Magnitudes, Periods & Densities of the PLANETS with their Distances from the SUN and the quantity of the Light & Heat they severally receive, Together with ASTRONOMICAL REMARKS explaining the Causes of Summer and Winter the changes and increase of Day and Night.' 'Engrav'd by Emanuel Bowen Geographer to the KING | LONDON Printed for JOHN BOWLES & SON at the Black Horse in Cornhill 1746.'

London: John Bowles, [1749]

British Library Maps Screen 2

p. 80

'A New & Correct Plan of London including all ye New Buildings &c', Map-Fan, 1760

'A NEW & Correct PLAN, of LONDON,

including all ye New Buildings &c.' 'R. Bennett sculp.'

London: Richard Bennett, [c.1760]

Private collection courtesy Sotheby's

p. 84

Allegorical Map of Spain's Overseas Possessions, in the Form of a Queen, 1761

'ASPECTO SYMBOLICO DEL MUNDO HISPANICO, PUNTUALMENTE ARREGLADO AL GEOGRAFICO, QUE A SU GLORIOSO CATHOLICO REY D. CARLOS TERCERO EL MAGNANIMO DEDICA, Y CONSAGRA D. VICENTE DE MEMIJE, CON IX. THESES, & XC. PROPOSICIONES, QUE A CERCA DE EL DEFIENDE: PRESIDIENDO EL R.P. PASQUAL FERNANDEZ, PUBLICO PROFESSOR DE MATHEMATICAS ENLA UNIVERSIDAD DE MANILA DELA COMPAÑIA DE IESVS AÑO DE 1761.' 'Laur.s Atlas sculp. Man[ill].a

Manilla: Vicente de Memije, 1761

British Library Maps K.Top. 118.19.

p. 86

'North and South America in its Principal Divisions', Jigsaw, 1767

'NORTH and SOUTH AMERICA in its PRINCIPAL DIVISIONS BY J. Spilsbury 1767.'

London: John Spilsbury, 1767

British Library Maps 188.v.15

p. 88

'Asia in its Principal Divisions', Jigsaw, 1767

'ASIA in its Principal Divisions, By J. Spilsbury, 1767.' 'Spilsbury ENGRAVER MAP & PRINT Seller in Russel Court Covent Garden LONDON 1767.'

London: John Spilsbury, 1767

British Library Maps 188.v.13

p. 90

'The Royal Geographical Pastime or the Complete Tour of Europe', Map-Game, 1768

'THE ROYAL GEOGRAPHICAL PASTIME OR THE COMPLETE TOUR OF EUROPE BY Thomas Jefferys GEOGRAPHER to the KING.' 'To His Royal Highness GEORGE PRINCE OF WALES, DUKE OF CORNWALL, &c. &c. &c. and Knight of the Most Noble Order of the Garter. This Plate is BY PERMISSION most humbly Dedicated By his Royal Highnesses most Obedient and Devoted humble Servant T. Jefferys.' 'Publishd according to the Statute of the 7th of George IIId Jan. 1st 1768 by T. Jefferys the corner of St. Martins Lane.'

London: Thomas Jefferys Sr, 1768

Private collection

p. 92

'A Complete Tour Round the World', Map-Game, 1770

'THE ROYAL GEOGRAPHICAL PASTIME Exhibiting A COMPLETE TOUR ROUND THE WORLD in which are delineated the NORTH EAST and NORTH WEST PASSAGES into the SOUTH SEA, and other modern Discoveries, By Thomas Jefferys, GEOGRAPHER to the KING' 'LONDON. 1.st January 1770. Published according to the Statute of the 7th of GEORGE III d by Tho.s Jefferys at the Corner of St. Martin's Lane.' 'Entered in the Hall Book of the Stationers Company, and whoever presumes to Copy it will be prosecuted by the Proprietor, who will reward any Person that shall give Information of it.'

[Welsh Feathers] 'To His Royal Highness GEORGE PRINCE OF WALES, DUKE OF CORNWALL, &c. &c. and Knight of the Most Noble Order of the Garter. This Plate is BY PERMISSION most humbly Dedicated By his Royal Highnesses most Obedient and Devoted humble Servant T. Jefferys.' [printed on a separate label, and pasted over a blank space]

London: Thomas Jefferys Sr, 1770

British Library Maps *950.(22.)

p. 94

'The Royal Geographical Pastime Exhibiting a Complete Tour Thro' England and Wales', 1770

'THE ROYAL GEOGRAPHICAL PASTIME Exhibiting A COMPLETE TOUR THRO' ENGLAND and WALES By Thomas Jefferys GEOGRAPHER to the KING [Welsh Feathers] To His Royal Highness GEORGE PRINCE OF WALES, DUKE OF CORNWALL, &c. &c. and Knight of the Most Noble Order of the Garter. This PLATE is BY PERMISSION most humbly Dedicated By his Royal Highnesses most Obedient and Devoted humble Servant, Thomas Jefferys.' [the dedication printed on a separate sheet, and pasted on] 'LONDON. 1.st January 1770. Published according to the Statute of the 7th of GEORGE III.D by T. Jefferys at the Corner of St. Martin's Lane – Entered in the Hall Book of the Stationers Company and whoever presumes to Copy it will be prosecuted by the proprietor who will reward any person that shall give information of it.'

London: Thomas Jefferys Sr, 1768

British Library Maps*1190.(5.)

p. 96

'Picture of Europe for July 1772', 1772

'Picture of Europe for July 1772.' [the title missing on the BL exemplar]

[London?]: Anonymous, 1772

British Library Maps CC.5.a.569

p. 98

'A Map or Chart of the Road of Love, and Harbour of Marriage' [c.1772]

'A MAP or CHART of the ROAD of LOVE, and HARBOUR of MARRIAGE. Laid down from the latest and best Authorities & regulated by my own Observations; The whole adjusted to the Latitude 51° 30 N. by T.P. Hydrographer, to his Majesty Hymen, and Prince Cupid. NB. The Long:de is reckon'd from ye Meridian of Teens.' 'London. Printed for ROB.T SAYER, No. 53, Fleet Street.'

London: Robert Sayer, [c.1772]

British Library Maps CC.5.a.56

p. 100

Map of the World, Jigsaw, 1787

[title missing] 'By THOMAS KITCHIN Hydrographer to His MAJESTY. Published as the Act directs, Jan.y 1.st 1787 by the Proprietor J. WALLIS, at his Map Warehouse, Ludgate Street, LONDON'

London: John Wallis Sr, 1787

British Library Maps (uncatalogued)

p. 102

'Bowles's Geographical Game of the World', 1790

'BOWLES'S GEOGRAPHICAL GAME OF THE WORLD, IN A NEW COMPLETE AND ELEGANT TOUR through the KNOWN PARTS thereof, LAID DOWN ON MERCATOR'S PROJECTION. [rule] LONDON: Printed for the Proprietor CARINGTON BOWLES, No. 69 St Paul's Church Yard.' 'Published as the Act directs, 12 August, 1790.' 'Entered at Stationers' Hall.'

London: Carington Bowles, 1790

British Library Maps *950.(3.)

p. 104

Untitled Allegorical Map of *The Pilgrim's Progress*, Jigsaw, 1790

'Publ.d May 20th. 1790 by J. Wallis No. 16 Ludgate Street London'

[engraved box label:] 'The Pilgrim's Progress DISSECTED or a Complete View of CHRISTIAN'S TRAVELS from the City of Destruction, to the HOLY LAND. [double rule] Designed as a Rational Amusement, for Youth of both Sexes. [rule] LONDON. Published June 7th 1790 by John Wallis Ludgate Street, Mrs. Newbery St. Paul's Church Yard, Champante and Whitrow Jewry Street, R.V. Brooke Cheapside, and John Binns Leeds.'

London: John Wallis Sr, 1790

British Library C.110.c.20

p. 106

'Geography Bewitched! Or, a Droll Caricature Map of England and Wales', 1793

'Geography Bewitched! or, a droll Caricature MAP of ENGLAND and WALES.' 'Dighton Del.' 'London Printed for Bowles & Carver, No. 69 St Paul's Church Yard.'

London: Henry Carington Bowles & Samuel Carver, [*c*.1795]

British Library Maps C.27.f.15.(1)

p. 108

'Geography Bewitched! Or, a Droll Caricature Map of Scotland', 1793

'Geography Bewitched! or, a droll Caricature MAP of SCOTLAND.' 'Dighton Del.' 'London Printed for Bowles & Carver, No. 69 St Paul's Church Yard.'

London: Henry Carington Bowles & Samuel Carver, [*c*.1795]

British Library Maps C.27.f.15.(2).

p. 110

'Geography Bewitched! Or, a Droll Caricature Map of Ireland', 1793

'Geography Bewitched! or, a droll Caricature MAP of Ireland. This Portrait of LADY HIBERNIA BULL is humbly dedicated to her Husband the great MR JOHN BULL.' 'London Printed for Bowles & Carver, No. 69 St Paul's Church Yard.'

London: Henry Carington Bowles & Samuel Carver, [*c*.1795]

British Library Maps C.29.e.5

p. 112

'A New Map of England & France', 1793

'A new MAP of ENGLAND & FRANCE.' 'The FRENCH INVASION; – or – John Bull, bombarding the Bum-Boats.' 'John Schoebert fecit.' 'Pub.d Nov.r 5th 1793 by H. Humphrey No. 18 Old Bond Street.'

London: Hannah Humphrey, 5 November 1793

British Library Maps 187.L.3.(3.)

p. 114

'Wallis's Complete Voyage Round the World – a New Geographical Pastime', 1796

'WALLIS'S Complete Voyage Round the WORLD. – a New – Geographical Pastime. LONDON, Published Jan.y 20th 1796, by John Wallis, at his Map Warehouse, No. 16, Ludgate Street. [rule] S. Cooke sculp.t 47 Fetter Lane. Of whom may be had on the same Plan. [rule] A Tour through England. 2. A Tour through Europe. 3. A Tour through Scotland. 4. The Genealogy of the Kings of England; — from Egbert 1st King to the present Time. N.B. The above are all 6s. each for the Pocket, on Cloth & Case. or upon a Pasteboard, with Box, Totum & Counters.'

[text imprint:] '...PRINTED FOR JOHN WALLIS, At his Wholesale Juvenile Repository, 13, Warwick-Square, London, By T. Sorrell, 86, Bartholomew-Close, Smithfield.

[engraved slipcase label:] 'WALLIS's New Geographical Game Exhibiting a Voyage round the WORLD.' 'Published Feb. 27. 1802, by John Wallis, at his Map Warehouse, Ludgate Str.t'

London: John Wallis Sr, 1796 [*c*.1805]

British Library Maps C.21.a.20

p. 116

'Allegorical Map of the Track of Youth, to the Land of Knowledge', 1798

'ALLEGORICAL Map of the Track of Youth, to the LAND of KNOWLEDGE' 'V. Woodthorpe sc. 27, Fetter Lane.' 'Published 1st. of Feb.y 1796, by R. Gillet.' 'Enter'd at Stationer's Hall.'

Robert Gillet, *Moral Philosophy and Logic. Adapted to the capacities of youth.*

London: George Sael, 1798

British Library 8463.bbb.12

p. 118

Battle of Trafalgar Commemorative Creamware Jug, 1805

'BATTLE off TRAFALGAR Gained by the British Fleet under Ld. NELSON on the 21 of Oct.r 1805. Against the combined Fleet of France & Spain, in which action the intrepid Nelson fell covered with Glory and renown.'

[Portrait] 'ADMIRAL LORD NELSON Born Sept.r 29th 1758 – died Oct. 21st 1805. Aged 47' // 'England expects every Man to do his Duty'

Worcester?: Anonymous; *c*.1805–1806

Private collection

p. 120

'A Whimsical Sketch of Europe', 1806

'A WHIMSICAL SKETCH OF EUROPE.' 'Publish'd Dec.r 6th 1806, by LAURIE & WHITTLE, 53, Fleet Street, London.'

[text:] 'A POETICAL DESCRIPTION OF THE MAP....'

London: Robert Laurie & James Whittle, 1806

British Library Maps CC.2.f.2

p. 122

'Geographical Recreation, or, a Voyage Round the Habitable Globe', 1809

'GEOGRAPHICAL RECREATION, or, A VOYAGE Round the HABITABLE GLOBE.' 'LONDON, Publish'd Oct.r 1st 1809 BY JOHN HARRIS, at the JUVENILE LIBRARY, Corner of S.t Pauls Church Yard.'

[engraved label:] 'GEOGRAPHICAL RECREATION AN Instructive GAME.' 'Published Oct 20 1809, by J. Harris, Corner St. Paul's Church Yard.'

London: John Harris Sr, 1809

British Library Maps C.43.b.68

p. 124

'Map of Green Bag Land', 1820

'MAP OF GREEN BAG LAND.' 'Printed and Published by J. Onwhyn, || Catherine Street, Strand. – Price 1s.'

London: Joseph Onwhyn, [1820]

British Library Maps CC.2.f.1

p. 126

'Labyrinthus Londoninensis, or The Equestrian Perplexed', [1830]

'LABYRINTHUS LONDINENSIS, or THE EQUESTRIAN PERPLEXED.' 'A PUZZLE Suggested by the Stoppages occasioned by repairing the Streets. The object is to find a way from the Strand to St. Paul's, without crossing any of the Bars in the Streets supposed to be under repair. [double rule] Published by C. Ingrey, 310, Strand, and F. Waller, 49, Fleet Street, London.' 'C. Ingrey lithog.' 'PRICE (with Key) 1s/-'

[text:]

'Mending our Ways, our ways doth oft-times mar,
So thinks the Traveller by Horse or Car,
But he who scans with calm and patient skill
This 'Labyrinthine Chart of London', will
One Track discover, open and unbarred,
That leads at length to famed St. Pauls Church Yard.'

London: Charles Ingrey & Frederick Waller, [*c.*1830]

British Library Crace Port VI, 214

p. 128

'An Illustrative Map of Human Life Deduced from Passages in Sacred Writ', 1833

'An Illustrative Map OF HUMAN LIFE Deduced from Passages in SACRED WRIT. Dedicated to the Rev. Rowland Hill, A.M. [rule] LONDON, Published Jan.y 1st 1833, by JAMES NISBET, 21, Berners St. Oxford St.' 'Drawn & Engraved by John Ping, 8 New St. New Cut Lambeth.' 'Ent. Sta. Hall.' 'PROOF'

London: James Nisbet, 1833

British Library Tab 597a.(69)

p. 130

'Wallis's New Railway Game, or Tour Through England and Wales', 1835

'NEW RAILWAY GAME, or Tour through ENGLAND AND WALES. [rule] LONDON, Published by E. Wallis, 42, Skinner Street.'

Text imprint: '... PASSMORE, PRINTER, GREAT GUILDFORD STREET, SOUTHWARK.'

Embossed cover title: 'WALLIS'S RAILWAY GAME OR TOUR THROUGH ENGLAND & WALES. [image of a train, labelled 'VICTORIA']'

London: John Passmore, [*c.*1855]

British Library Maps 6.aa.42

p. 132

'A Voyage of Discovery; or, the Five Navigators. An Entirely New Game', 1836

'A VOYAGE OF DISCOVERY; OR, THE FIVE NAVIGATORS. AN ENTIRELY NEW GAME.' 'Printed by Lefevre & Kohler.' 'LONDON: PUBLISHED BY WILLIAM SPOONER, 259 REGENT STREET, OXFORD STREET, 1836.'

[engraved label:] 'A VOYAGE OF DISCOVERY; OR, THE FIVE NAVIGATORS. // William Spooner, Regent Street, London.'

London: Louis Maria Lefevre & William Kohler for William Matthias Spooner

British Library Maps C.43.b.74

p. 134

'The Journey or, Cross Roads to Conqueror's Castle. A New and Interesting Game', [1837]

'THE JOURNEY or, Cross Roads to Conqueror's Castle.' 'A NEW AND INTERESTING GAME.' 'ENTERED AT STATIONERS HALL.' 'LONDON BY W. SPOONER 337 STRAND'.

London: William Matthias Spooner, [*c.*1837]

British Library Maps C.43.b.72

p. 136

'The Travellers, or, a Tour Through Europe', 1842

'THE TRAVELLERS, OR, A TOUR THROUGH EUROPE.' 'W. Clerk, Lithog.r 202, High Holborn.' 'London: Published by William Spooner, 377 Strand. Dec.r 1st 1842.'

[engraved title label:] 'THE TRAVELLERS OF EUROPE. LONDON: PUBLISHED BY WILLIAM SPOONER, 377, STRAND.'

London: William Matthias Spooner, 1842

British Library Maps 197.b.35

p. 138

'Game of Star-Spangled Banner, or Emigrants to the United States', [1842]

Game of Star-Spangled Banner, – OR – EMIGRANTS To the UNITED STATES. E. Wallis, Skinner St. London. Ent.d at Stationers Hall.

[embossed slipcase title:] [United States flag with twenty-six stars] 'THE STAR-SPANGLED BANNER'

London: Francis William Passmore for Edward Wallis, [1842]

British Library Maps C.29.b.10

p. 140

'Wallis's New Game of Wanderers in the Wilderness', [1844]

'Wallis's New GAME of WANDERERS in the WILDERNESS.' 'J. H. Banks' 'London Edward Wallis, 42, Skinner Street.'

[embossed slipcase title:] 'WANDERERS IN THE WILDERNESS'

London: Francis William Passmore for Edward Wallis, [1842]

British Library Maps C.29.b.9

p. 142

'Picturesque Round Game of the Produce and Manufactures of the Counties of England and Wales', [1844]

'WALLIS'S PICTURESQUE ROUND GAME OF THE PRODUCE & MANUFACTURES, OF THE COUNTIES OF ENGLAND & WALES.' 'London. Published by EDWARD WALLIS, 42, Skinner Street. Entered at Stationers Hall.'

[cover title:] 'PICTURESQUE ROUND GAME OF THE PRODUCE AND MANUFACTURES OF THE COUNTIES OF ENGLAND AND WALES'

[index booklet:] 'EXPLANATION TO THE PICTURESQUE ROUND GAME OF THE PRODUCE AND MANUFACTURES OF THE COUNTIES OF ENGLAND AND WALES. LONDON: EDWARD WALLIS, 42, SKINNER STREET.'

London: Edward Wallis, [*c.*1844] ; the rule book printed by Francis William Passmore

British Library Maps, C.29.b.16

p. 144

'Spooner's Pictorial Map of England & Wales Arranged as an Amusing and Instructive Game for Youth', 1844

'SPOONER'S PICTORIAL MAP OF ENGLAND & WALES Arranged as AN AMUSING AND INSTRUCTIVE GAME FOR YOUTH. ILLUSTRATED WITH UPWARDS OF One Hundred & Twenty Views. LONDON: Published by William Spooner, 377, Strand, – Nov.r 5.th 1844 – [rule]'

[engraved title label:] 'THE TRAVELLERS OF ENGLAND AND WALES. LONDON: PUBLISHED BY WILLIAM SPOONER, 377, STRAND.'

London: William Matthias Spooner, 1844

British Library Maps C.44.b.60

p. 146

'The Cottage of Content or Right Roads and Wrong Ways. A Humorous Game', 1848

'THE COTTAGE OF CONTENT OR RIGHT ROADS AND WRONG WAYS.' 'A HUMOROUS GAME.' 'LONDON, PUBLISHED BY Wm. SPOONER, 369, STRAND, NOVr. 1st 1848.'

[lithographed cover title:] 'THE COTTAGE OF CONTENT OR RIGHT ROADS AND WRONG WAYS. A GAME. LONDON: PUBLISHED BY WILLIAM SPOONER, 379, STRAND.'

London: William Matthias Spooner, 1848 .

British Library Maps C.43.b.73

p. 150

'The New Game of The Royal Mail or London to Edinburgh by L. & N.W. Railway', 1850

'THE NEW GAME OF The Royal Mail or London to Edinburgh BY L. & N.W. RAILWAY. [double rule] PUBLISHED BY JOHN JAQUES & SON, LONDON.'

London: John Jacques, *c.*1850

British Library Maps C.44.d.82

p. 152

'Dame Venodotia, Alias Modryb Gwen; A Map of North Wales', 1851

'DAME VENODOTIA, ALIAS MODRYB GWEN; A Map of North Wales. [key]. PUBLISHED BY H. HUMPHREYS, CASTLE SQUARE, CARNARVON.' 'Designed by H. Hughes, and Drawn on Stone by J.J. Dodd.'

Caernarfon: Hugh Humphreys, [*c.*1851]

British Library Maps 6096.(12.)

p. 154

'The Crystal Palace Game, Voyage Round the World, an Entertaining Excursion in Search of Knowledge, Whereby Geography Is Made Easy', *c.*1854

'THE CRYSTAL PALACE GAME, VOYAGE ROUND THE WORLD, an entertaining excursion in search of knowledge, whereby GEOGRAPHY IS MADE EASY. [triple rule] By Smith Evans, F.R.G.S.' 'L'ENFANT LITH.' 'ALFRED DAVIS & Co. 58, 59, & 60, HOUNDSDITCH, LONDON.'

[cover title:] 'THE CRYSTAL PALACE GAME A VOYAGE ROUND THE WORLD' [vignette of a ship] 'AN ENTERTAINING EXCURSION IN SEARCH OF KNOWLEDGE WHEREBY GEOGRAPHY IS MADE EASY'

London: Alfred Davis, [*c.*1854]

British Library Maps 28.bb.7

p. 156

'Betts's New Portable Terrestrial Globe', 1852

'By the Queen's | Royal Letters Patent, | BETTS'S NEW PORTABLE TERRESTRIAL GLOBE | Compiled from | THE LATEST AND BEST AUTHORITIES. [rule] | London, John Betts, 115 Strand.'

[ox title:] 'BETTS'S PATENT PORTABLE GLOBE. 115 Strand, London W.C.'

London: John Betts, [*c.*1852]

British Library Maps G.46

p. 158

'Comic Map of the Seat of War with Entirely New Features', 1854

'COMIC MAP OF THE SEAT OF WAR WITH ENTIRELY NEW FEATURES.' 'DONE BY T[homas]. O[nwhyn].' 'May 30.th 1854 PUB.d BY ROCK BROTHERS & PAYNE LONDON.'

London: [William Frederick] Rock, [Henry] Rock [Jr.] and [John] Payne, 1854

British Library Maps X.6168

p. 160

Game Based on the Geography of France, [*c.*1855]

The title is set within an elaborate frame centred on an allegorical depiction of 'la belle France' seated on a throne with the paraphernalia of war around her; there are four other insets: apparently geography (bottom right), art (top right), music (top left) and trade (bottom left). One of the sacks at bottom left bears the name 'Baetien aine', while the delineation of France is signed 'J. Gaildrau' and 'Im-Lemercier Band & C.' (?) although both are hard to read.

British Library Maps (uncatalogued)

p. 162

'Map of Europe, 1859. Illustrative War Scene', 1859

'KAART VAN | EUROPA 1859.' 'AANSCHOUWELIJK OORLOGSTOONEEL, zamengesteld naar de beste telegraphische berigten, waarop met een oogopslag de politieke toestand en ligging der grootere en kleinere Mogendheden van Europa kunnen erkend en beoordeeld worden.' 'Lith. v. Emrik & Binger, Haarlem.' 'UITGAVE van J.J. VAN BREDERODE.' 'PRIJS 15 CENTS.' (MAP OF | EUROPE 1859. ILLUSTRATIVE WAR SCENE, based on the best telegraphic messages, in which one can instantly recognise and judge the political situation and locations of the greater and smaller powers of Europe.)

Haarlem: Emrik & Binger for Jacobus Johannes van Brederode

British Library Maps CC.5.a.476

p. 164

'The Evil Genius of Europe', 1859

'THE EVIL GENIUS OF EUROPE On a careful examination of the Panorama the Genius will be discovered struggling hard to pull on his Boot. It will be noticed, he has just put his foot in it. Will he be able to wear it?' 'W. Nicholson Lith. 3. Bell Yard, Gracechurch St.' 'LONDON: W. CONEY, 61, WARDOUR St. OXFORD St.'

London: William Coney, 1859

British Library Maps 1078.(24.)

p. 166

'New Map of France', 1862

'NOUVELLE CARTE DE FRANCE, A l'Usage de la Géographie versifiée de la France PAR M. V[ICT].OR GUILLON.' 'Imp[rimée] Lemercier, Paris.' 'Paris, A. Logerot, Editeur, Quai des Augustins 55.'

[cover title:] 'CARTE ET JEU MÉTHODIQUE A L'APPUI DE LA GÉOGRAPHIE VERSIFIÉE DE LA FRANCE [diagram] MÉTHODE DE CADRAN.'

Paris: Augustine Logerot, [1862]

British Library Maps C.27.f.21

p. 168

'The Dissected Globe', [1866]

London: Abraham Nathan Myers, [1866]

British Library Maps G.3. (Globe 3)

p. 170

Map of Prussia, [1868]

'PRUSSIA.' 'Vincent Brooks, Day & Son, Lith. London, W.C.'

His Majesty of Prussia – grim and old –
Sadowa's King – by needle guns made bold;
With Bismarck of the royal conscience, keeper,
In dreams political none wiser – deeper.

Aleph, *Geographical Fun: being humourous*

outlines of various countries with an introduction and descriptive lines ...

London: Vincent Brooks, Day & Son, [1868]

British Library Maps 12.d.1

p. 172

Map of Spain and Portugal, [1868]

'SPAIN & PORTUGAL.' 'Vincent Brooks, Day & Son, Lith. London, W.C. 'Vincent Brooks, Day & Son, Lith. London, W.C.'

'These long divided nations soon may be, By Prims' grace, joined in lasting amity. And ladies fair – if King Ferdinando rules, Grow grapes in peace, and fatten their pet mules.'

Aleph, *Geographical Fun: being humorous outlines of various countries with an introduction and descriptive lines...*

London: Vincent Brooks, Day & Son, [1868]

British Library Maps 12.d.1

p. 174

'Novel Carte of Europe Designed for 1870', 1870

'THE COMIC CARTE OF EUROPE FOR 1870.' 'NOVEL CARTE OF EUROPE, DESIGNED FOR 1870.' 'J.G. Lith. DUBLIN.' '(ENTERED AT STATIONERS HALL).'

[text:] 'England, isolated, filled with rage, and almost forgetting Ireland whom she holds in leash. Spain smokes, resting upon Portugal. France repulses the invasion of Prussia, who advances with one hand on Holland, the other on Austria. Italy, also, says to Bismarck, "Take thy feet from hence." Corsica and Sardinia, being a true Gavroche, laughs at everything. Denmark, who has lost his legs in Holstein, hopes to regain them. Turkey-in-Europe yawns and awakens. Turkey-in-Asia smokes her opium. Sweden, bounding as a panther. Russia resembles an old bogy who would wish to fill his basket.'

Dublin: Joseph Goggins, [1870]

British Library Maps 1078.(27.)

p. 176

Satirical Map of Europe, 1871

'L'EUROPA GEOGRAFICO-POLITICA VEDUTA A VOLA D'OCA.' 'Bologna: presso Manfredi Manfredo editore Via Venezia N. 1749

Bologna: Manfredo Manfredi, [c.1871]

British Library Maps CC.5.a.529

p. 178

Puzzle Game Blocks [1875]

[box lid with the title:] 'ATLAS'

'CARTE DE L'AMÉRIQUE SEPTENTRIONALE dressée ET dessinée SOUS LA DIRECTION de Mr J.G. Barbié du Bocage.' 'Ch. Smith Sculpt.'

'CARTE DE L'AMÉRIQUE MÉRIDIONALE...'

'CARTE DE L'AFRIQUE...'

'CARTE DE L'ASIE...'

'CARTE DE L'EUROPE...'

'FRANCE Divisée EN 89 DÉPARTEMENTS avec Sièges

Paris, c.1875

British Library Maps C.21.f.1

p. 180

'Serio-Comic War Map for the Year 1877', 1877

'SERIO-COMIC WAR MAP FOR THE YEAR 1877. BY F.W.R.' 'London, Published by, G.W. Bacon & Co. 127, Strand.' 'COPYRIGHT.'

[text:] 'REFERENCE. [rule] THE OCTOPUS – Russia – forgetful of the wound it received in the Crimea, is stretching forth its arms in all directions. Having seized hold of the Turk, it is eagerly pushing forward in the hope that it may overwhelm him, as it has already done Poland. At the same time, Greece seems likely to annoy the Turk in another quarter. Hungary is only prevented from attacking his neighbour, Russia, through being held back by his sister Austria. The Frenchman, remembering his late defeat, is carefully examining his weapons; and Germany is naturally interested in his movements, and holds himself in readiness for any emergency. Great Britain and Ireland are eagerly watching the fray – ready at any moment, at least, to prevent Russia from seizing the Turk's watch, or interference with Suez. Spain is taking his much required rest. Italy is ruthlessly making a toy of the Pope; and the wealthy King of Belgium is taking care of his treasure. Denmark's flag is small, but she has reason to be proud of it.'

London: George Washington Bacon, 1877

British Library Maps*1078.(35.)

p. 182

'The Avenger, an Allegorical War Map for 1877', 1877

'THE AVENGER AN ALLEGORICAL WAR MAP FOR 1877.' 'London. Published by G.W. Bacon & Co. 127, Strand. COPYRIGHT.' 'ALL RIGHTS RESERVED.'

London: George Washington Bacon, [1877]

British Library Maps 1035.(302.)

p. 184

'United States, a Correct Outline', [1880]

'United States a correct outline.' | 'By Lilian Lancaster Copyright.' | [5e]

British Library Maps cc.5a.230

p. 186

'United States a Correct Outline', 1880

'United States a correct outline.' | 'Lilian Lancaster Novr. 4th 1880.' 'Copyright.'

British Library Maps cc.5a.229

p. 188

'Falmouth Borough Octopus Attempting to Grasp the Parishes of Falmouth and Budock', 1882

'FALMOUTH BOROUGH OCTOPUS [double rule] ATTEMPTING TO GRASP THE PARISHES [rule] OF FALMOUTH AND BUDOCK. [rule] [colour key to the areas described, with table of areas, etc.]' 'EDWIN T. OLVER, LITHO, "PENDENNIS" WORKS, 20 & 21 ST. DUNSTANS HILL, LONDON E.C.'

London: Edwin T. Olver, 1882

British Library Maps CC.5.a.600

p. 190

'Map of England. A Modern St George and the Dragon!!!', 1888

'MAP OF ENGLAND. [double rule]' 'A MODERN S.T GEORGE AND THE DRAGON!!!' 'TOM MERRY. LITH.' 'WITH ACKNOWLEDGEMENTS TO LILLIE TENNANT.' 'St. Stephen's Review Presentation Cartoon, June 9th, 1888.'

St. Stephen's Review..., issue 274, for 9 June.

London: Constitutional News Association, 1888

British Library Maps cc.5.a.578

p. 192

'Angling in Troubled Waters. A Serio-Comic Map of Europe', 1899

'ANGLING IN TROUBLED WATERS [double rule] A SERIO-COMIC MAP OF EUROPE – BY – FRED. W. ROSE AUTHOR OF THE "OCTOPUS" MAP

OF EUROPE COPYRIGHT – TOUS DROITS RÉSERVÉS. [rule] 'MATT. HEWERDINE FROM DESIGN BY Fred.W. Rose.' 'G.W. Bacon & C.o, Ltd., 127, Strand, London.'

[printed label:] 'A Serio-Comic Map of EUROPE [double rule] ANGLING IN TROUBLED WATERS BY FRED. W. ROSE [rule] PRICE ONE SHILLING [rule] LONDON: G.W. BACON & CO., Ltd., 127, STRAND. 1899.'

London: George Washington Bacon & Co., [1899]

British Library Maps *1078.(35.)

p. 194

'John Bull and His Friends. A Serio-Comic Map of Europe', 1900

[rule] JOHN BULL AND HIS FRIENDS [rule] A SERIO-COMIC MAP OF EUROPE – BY – FRED. W. ROSE AUTHOR OF "ANGLING IN TROUBLED WATERS" &c. &c. [rule] 1900 COPYRIGHT – TOUS DROITS RÉSERVÉS. ALLE RECHTE VORBEHALTEN – TUTTI DRITTI RISERVIATI [double rule]' 'Fred. W. Rose Mar 1900 [facsimile signature]' 'Matt. B. Hewerdine from a sketch by Fred. W. Rose.' 'G.W. Bacon & Co., Ltd., 127, Strand, London.'

London: George Washington Bacon, 1900

British Library Maps *1078.(39.)

p. 196

'A Humorous Diplomatic Atlas of Europe and Asia', 1904

'A HUMOROUS DIPLOMATIC ATLAS OF EUROPE AND ASIA.'

[text:] '"Black Octopus" is a name newly given to Russia by a certain prominent Englishman. For the black octopus is so avaricious, that he stretches out his eight arms in all directions, and seizes up every thing that comes within his reach. But as it sometimes happens he gets wounded seriously even by a small fish, owing to his too much covetousness. Indeed, a Japanese proverb says: "Great avarice is like unselfishness." We Japanese need not to say much on the cause of the present war. Suffice it to say, that the further existence of the Black Octopus will depend entirely upon how he comes out of this war. The Japanese fleet has already practically annihilated Russia's naval power in the Orient. The Japanese army is about to win a signal victory over Russia in Corea & Manchuria. And when... St. Petersburg? Wait & see. The ugly Black Octopus! Hurrah! Hurrah! for Japan. Kisaburo Ohara. March, 1904.'

[Tokyo]: [Kisaburo Ohara], March 1904

British Library Maps 1035.(107.)

p. 198

'How to Get There. An Interesting and Educational Game for 2, 3 or 4 Players', [1908]

'HOW TO GET THERE' 'JOHNSON, RIDDLE & Co. Ltd. LONDON, S.E.' Manufactured entirely in England. [J.W.L.]' '[London. ENTERED AT STATIONERS' HALL.'

[box title:] 'HOW TO GET THERE [Tube map of central London] An Interesting and Educational Game for 2, 3 or 4 Players.' 'J.W.L.' 'LONDON. JOHNSON, RIDDLE & Co. L.TD, LONDON, S.E.' 'PROV. PROT. 235763' 'SELECT A TICKET [rule] PAY THE FARE [rule] GET THERE FIRST & WIN THE CONTENTS OF THE TILL LEARN THE QUICKEST WAY TO GET ABOUT LONDON'

London: Johnson, Riddle & Co., [1908]

British Library Maps 188.v.32

p. 200

Map of Belgium, 1912

'BELGIUM.' 'L. Tennant.' 'No. 7.'

Elizabeth Louisa Hoskyn, *Stories of Old*

London: Adam & Charles Black, 1912

British Library 09008.bb.10

p.202

Map of Iceland, 1912

'ICELAND.' 'L. Tennant.' 'No. 11.'

Elizabeth Louisa Hoskyn, *Stories of Old*

London: Adam & Charles Black, 1912

British Library 09008.bb.10

p. 204

'Changing the Map of Europe: A *Financial Times* Competition', 1914

'CHANGING THE MAP OF EUROPE The Financial Times COMPETITION MAP EUROPE AFTER THE WAR THE MAP SHOWS THE PRESENT DEMARCATION OF THE COUNTRIES. – INSTRUCTIONS – SKETCH IN THE BOUNDARIES OF STATES ACCORDING TO YOUR VIEW OF PROBABLY PEACE TERMS [scale bar].'

text leaf: 'CHANGING THE MAP OF EUROPE. [rule] A Financial Times COMPETITION. [rule] IN aid of the PRINCE OF WALES' FUND, THE FINANCIAL TIMES offers the following Prizes for the most accurate forecast of the map of Europe as it will appear after the War, in virtue of the first definitive peace agreement between the European Powers:–...'

London: *The Financial Times*, [1914]

British Library Maps 1078.(40)

p. 206

'Hark! Hark! The Dogs Do Bark!' [1914]

'HARK! HARK! THE DOGS DO BARK! WITH NOTES BY WALTER EMANUEL.' 'Designed and Printed by Johnson, Riddle & Co., Ltd., London, S.E.' 'Published by G.W. Bacon & Co., Ltd., 127, Strand, W.C.'

Johnson, Riddle & Co., London, [1914]

British Library Maps 1078.(42)

p. 208

'The Silver Bullet or the Road to Berlin', Game, 1914

'THE SILVER BULLET OR THE ROAD TO BERLIN.' 'BRITISH DESIGN' 'BRITISH MADE' 'REGISTERED'

[rule label:] 'THE NEW WAR GAME. The Silver Bullet Or THE ROAD TO BERLIN. REGISTERED. British Design. British Manufacture. RULES OF THE GAME ...'

London: F.R. & S., [*c*.1914]

British Library Maps (uncatalogued)

p. 210

'Knock Out Germany', [1914]

'KNOCK OUT GERMANY' 'S[P] ECIALLY PREPARED FOR THE TOY TARGET COMPANY.' 'MADE IN ENGLAND' 'COPYRIGHT'

[London?]: The Toy Target Company, [1914]

British Library 1078.(41.)

p. 212

'An Anciente Mappe of Fairyland Newly Discovered and Set Forth', 1918

'AN ANCIENTE MAPPE of FAIRYLAND newly discovered and set forth.' 'DESIGNED BY BERNARD SLEIGH.' 'W.H.D. Writer.' PUBLISHED BY SIDGWICK & JACKSON, LTD., 3, ADAM STREET, ADELPHI, LONDON, W.C.'

[booklet entitled:] 'A GUIDE TO THE MAP OF FAIRYLAND [wood engraving of a fairy, monogrammed 'S'[leigh]] DESIGNED & WRITTEN BY BERNARD SLEIGH. LONDON: SIDGWICK & JACKSON.'

London: Sidgwick & Jackson, [1918]

British Library L.R.270.a.46

p. 214

'The New Map Game Motor Chase Across London', 1925

'THE NEW MAP GAME MOTOR CHASE ACROSS LONDON' '"GEOGRAPHIA" LTD. 167 FLEET ST LONDON E C 4' 'COPYRIGHT'

[box label:] 'THE NEW MOTOR GAMES MOTOR CHASE ACROSS LONDON EXCITING – ENTERTAINING – EDUCATIVE COPYRIGHT "GEOGRAPHIA" LTD. 167 FLEET ST., LONDON. E.C.4.'

London: Geographia Ltd., [1925]

British Library Maps 162.p.9

p. 216

'Octopium Landlordicuss (Common London Landlord)', 1925

'Octopium Landlordicuss (Common London LANDLORD) This FISHY CREATURE lives on RENT Its Tentacles grasp 5 Square MILES of LONDON This ABSORBENT PARASITE sucks £20,000,000 a year from its VICTIMS giving nothing in return. [rule] The PEOPLE must destroy IT or be destroyed.' 'W.B. Northrop.' 'Hendersons, Publishers, 66, Charing Cross Road, London, W.C.' 'Copyright by W.B. Northrop.'

[text:] 'LANDLORDISM CAUSES UNEMPLOYMENT It paralyses the BUILDING TRADE; It pauperises the Peasantry; 12 landlords "own" (?) London, taking £20,000 a year; 500 peers "own" (?) an entire one-third of England; 4,000 Landlords "own" (?) an entire half of England; The Land Octopus Sucks the Lifeblood of the People'

London: Hendersons, [1925]

British Library Maps C.12.c.1.(1607.)

p. 218

'Buy British', Game, 1932

'BUY BRITISH' 'Published by "GEOGRAPHIA" LTD. 55 Fleet Street, London, E.C.4.'

[box title:]'A NEW MAP GAME BUY BRITISH An exciting world race and one which will teach the players – TRADE WITHIN THE EMPIRE' 'Exciting Interesting Educative' 'A Game for any number of Players up to Five' 'COPYRIGHT, GEOGRAPHIA LTD. 55 FLEET ST. E.C. 4.' 'MADE IN ENGLAND.'

[rules pasted inside the lid:] 'A New Map Game [rule] "BUY BRITISH" Trade within the Empire Exports and Imports. A game for any number of players up to five. The object of this Game is to complete a voyage to the largest of the British Dominions and land imports and take away exports, and the winner is the one who first gets home to his starting point.'

London: Geographia Ltd, [1932]

British Library Maps 162.p.6

p. 220

Untitled Map of the World, Dissected As a Jigsaw, [1935]

[index map:] 'MAPPA-MUNDI KEY SHEET SHOWING CAPITALS IN THEIR CORRECT POSITIONS.' 'ACTUAL SIZE OF JIG-SAW PUZZLE' 'This game printed and made in England by John Waddington Limited, London and Leeds' 'This map compiled and reproduced from copyright maps by the kind permission of Messrs. W. & A.K. Johnston Ltd., Edinburgh and London.' 'Patent No. 35347'

[box title:] 'WADDINGTON'S MAPPA-MUNDI "MAP OF THE EARTH" THE NEW TRAVEL GAME' 'EXCITING, EDUCATION & FASCINATING.'

London: Waddington's, [c.1935]

British Library Maps C.29.e.8.

p. 222

Map of Poland Jigsaw Puzzle, 1958

'POLSKA układanka geograficzna' 'Wydawnictwo K[atolicki] O[środek]. W[ydawniczy]. VERITAS w Londynie' ['Poland geographical jigsaw puzzle' 'Veritas, the Catholic Publishing House Centre, in London']

[leaflet:] 'NASZ KRAJ – POLSKA UKŁANDANKA GEOGRAFICZNA WYDAŁ KATOLICKI OŚRODEK WYDAWNICZY "VERITAS" W LONDYNIE [rule]...' [Our Country – Poland Geographical jigsaw puzzle; Centre of The Roman Catholic Church "Veritas" In London ...]

London: Veritas, [1958]

British Library Maps 33718.(12.)

p. 224

'The Afghanistans', 2008

'THE AFGHANISTANS' 'MADE IN AFGHANISTANI' '2008'

British Library Maps (uncatalogued)

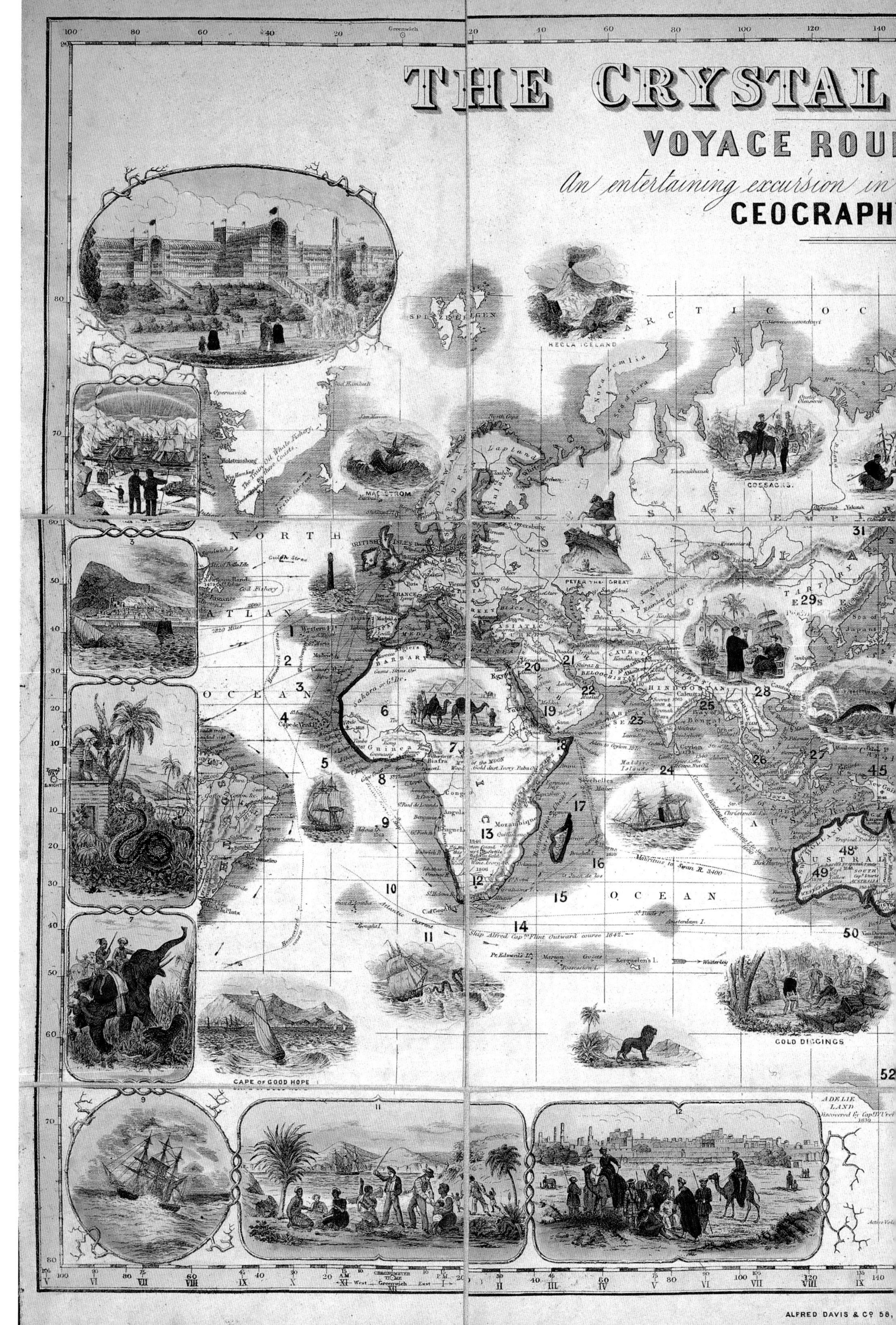

THE CRYSTAL
VOYAGE ROU
An entertaining excursion in
GEOGRAPH
HECLA ICELAND
SPITZBERGEN
MAELSTROM
COSSACKS.
PETER THE GREAT
CAPE of GOOD HOPE
GOLD DIGGINGS
ADELIE LAND
NORTH
ATLAN
OCEAN
BRITISH ISLES
FRANCE
Lapland
Nova Zemlia
Sea of Kara
Cod Fishery
BARBARY
Sahara
Guinea
Congo
Angola
Benguela
Mozambique
Seychelles
Maldive Islands
Ceylon
Calcutta
HINDOOSTAN
Christmas I.
Amsterdam I.
Kerguelen's I.
Pr. Edward's I.
St. Helena
Rio Janeiro
La Plata
Ship Alfred Capt. Flint Outward course 1842.
Mauritius to Swan R. 3400
CHRONOMETER
ALFRED DAVIS & Co. 58,

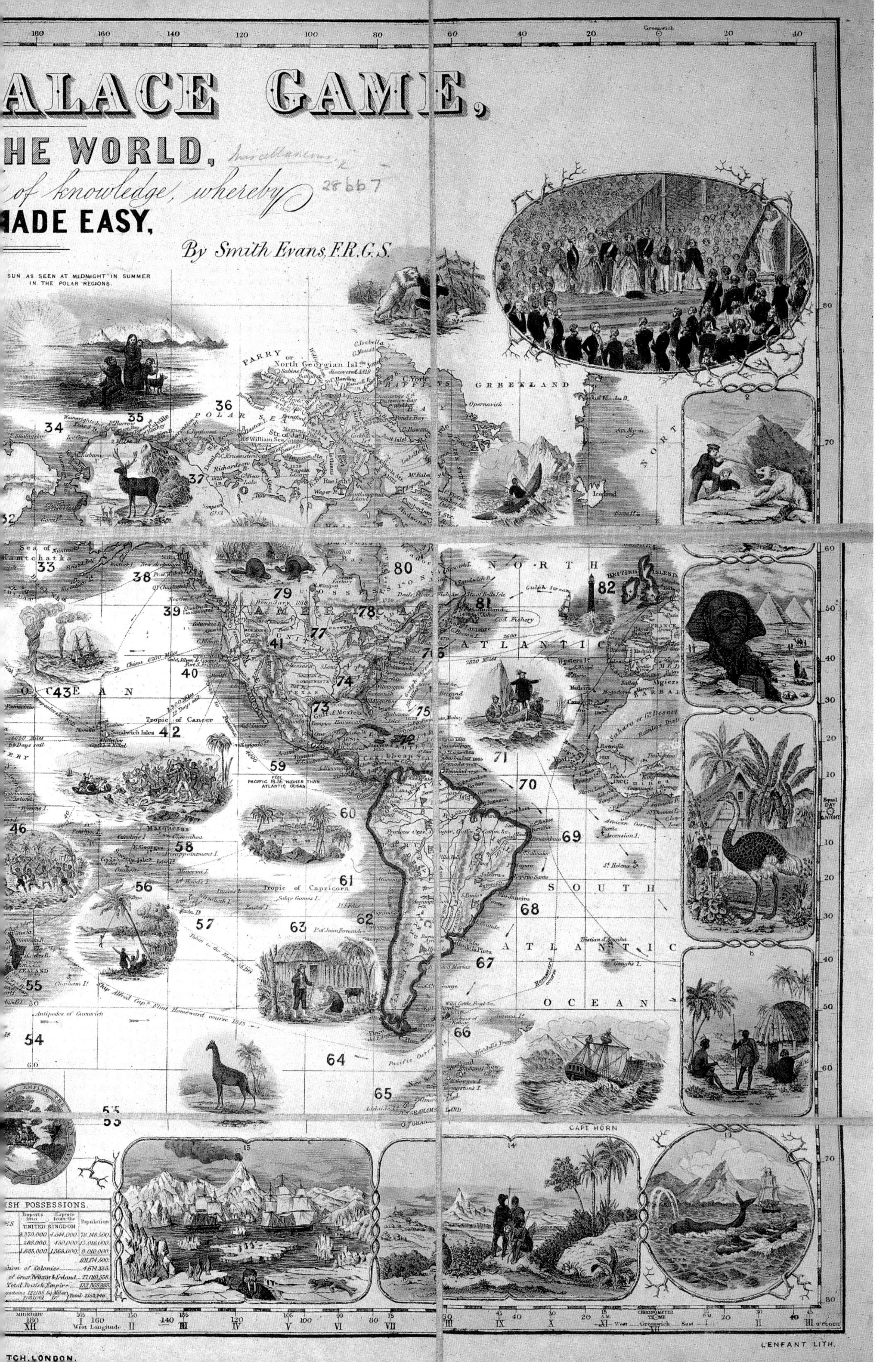

ALACE GAME,
HE WORLD,
of knowledge, whereby
MADE EASY,
By Smith Evans, F.R.G.S.
SUN AS SEEN AT MIDNIGHT IN SUMMER IN THE POLAR REGIONS.
Greenwich
PARRY or North Georgian Isl.ds
POLAR SEA
GREENLAND
Iceland
Sea of Kamtchatka
NORTH ATLANTIC
OCEAN
Tropic of Cancer
Sandwich Isles
Gulf of Mexico
Caribbean Sea
PACIFIC 18.35 FEET HIGHER THAN ATLANTIC OCEAN
Marquesas
Tropic of Capricorn
Ascension I.
St. Helena
SOUTH ATLANTIC OCEAN
Tristan d'Acunha
Antipodes of Greenwich
CAPE HORN
ISH POSSESSIONS.
West Longitude
CHRONOMETER TIME
L'ENFANT LITH.
TCH. LONDON.

Index